Fundamentals of Technical Mathematics

Fundamentals of Technical Mathematics

Sarhan M. Musa
Prairie View A&M University

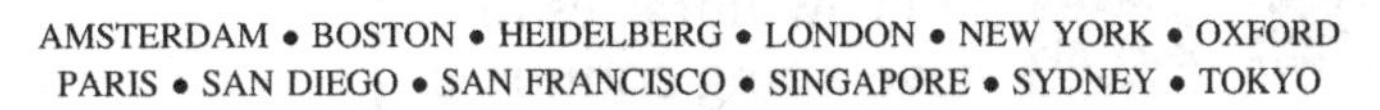
AMSTERDAM • BOSTON • HEIDELBERG • LONDON • NEW YORK • OXFORD
PARIS • SAN DIEGO • SAN FRANCISCO • SINGAPORE • SYDNEY • TOKYO

Academic Press is an imprint of Elsevier

Academic Press is an imprint of Elsevier
125 London Wall, London EC2Y 5AS, UK
525 B Street, Suite 1800, San Diego, CA 92101-4495, USA
225 Wyman Street, Waltham, MA 02451, USA
The Boulevard, Langford Lane, Kidlington, Oxford OX5 1GB, UK

ISBN: 978-0-12-801987-0

British Library Cataloguing in Publication Data
A catalogue record for this book is available from the British Library

Library of Congress Cataloging-in-Publication Data
A catalog record for this book is available from the Library of Congress

For information on all Academic Press publications
visit our website at http://store.elsevier.com/

To my late father, Mahmoud; my mother, Fatmeh; my wife, Lama; and my children, Mahmoud, Ibrahim, and Khalid

Contents

Preface

Fundamentals of Technical Mathematics introduces applied mathematics for engineering technologists and technicians. Through a simple, engaging approach, the book reviews basic mathematics including whole numbers, fractions, mixed numbers, decimals, percentages, ratios, and proportions. The text covers conversions to different units of measure (standard and/or metric) and other topics as required by specific businesses and industries. Building on these foundations, it then explores concepts in arithmetic; introductory algebra; equations, inequalities, and modeling; graphs and functions; measurement; geometry; trigonometry; and matrices, determinants, and vectors. It supports these concepts with practical applications in a variety of technical and career vocations, including automotive, allied health, welding, plumbing, machine tool, carpentry, auto mechanics, HVAC, and many other fields. In addition, the book provides practical examples from a vast number of technologies and uses two common software programs, Maple and Matlab.

This book has eight chapters. Chapter 1 provides basic concepts in arithmetic. Chapter 2 introduces algebra. Chapter 3 presents equations, inequalities, and modeling. Chapter 4 presents graphs and functions. Chapter 5 presents measurement. Chapter 6 introduces geometry. Chapter 7 reviews trigonometry. Finally, Chapter 8 introduces matrices, determinants, and vectors.

Sarhan M. Musa

Preface

Acknowledgments

It is my pleasure to acknowledge the outstanding help and support of the team at Elsevier in preparing this book, especially from Cathleen Sether, Steven Mathews, Katey Birtcher, Sarah J. Watson, Amy Clark, and Anitha Sivaraj.

Thanks for professors John Burghduff and Mary Jane Ferguson for their support, understanding, and being great friends.

I would also like to thank Dr Kendall T. Harris, his college dean, for his constant support. Finally, this book would never have seen the light of day if not for the constant support, love, and patience of our family.

Chapter 1

Basic Concepts in Arithmetic

You must do the thing you think you cannot do.

Eleanor Roosevelt

■ Alessandro Volta (1745—1827), an Italian physicist, chemist and a pioneer of electrical science. He is most famous for his invention of the electric battery. Alessandro has a special talent for languages. Before he left school, he had learned Latin, French, English and German. His language talents helped him in later life, when he traveled around Europe, discussing his work with scientists in Europe's centers of science.

INTRODUCTION

In this chapter, we will discuss the most useful and important basic topics in arithmetic (science of numbers), which are essential refreshment for people who have forgotten or do not like math. Mastering the fundamental concepts in mathematics is vital in strengthening the foundation for learning advanced material and forming a like for mathematics.

1.1 BASIC ARITHMETIC

In *arithmetic* numbers used are known. The purpose of this section is to introduce the basic principles/laws of arithmetic such as properties for the

Fundamentals of Technical Mathematics. http://dx.doi.org/10.1016/B978-0-12-801987-0.00001-0

operations of addition (+), subtraction (−), multiplication (×), and division (÷). Next, we introduce fractions, decimals, and percents.

Numbers are categorized into six categories. First is *real* numbers, which includes all numbers. *Whole* numbers are numbers 0, 1, 2, 3, ..., while *natural* numbers do not include zero, i.e., 1, 2, 3,.... *Integer* numbers are whole numbers including their negative counterparts, i.e., ..., −3, −2, −1, 0, 1, 2, 3, ... *Rational* numbers, i.e., $\frac{1}{2}, \frac{-4}{3}, 0.75, 0.111$. *Irrational* numbers are decimals that are not negatives and have no end such as π (pi) and $\sqrt{2}$.

1.1.1 Arithmetic operations

1.1.1.1 Addition (+)

Addition is the process of finding the sum of two or more numbers. The symbol (+) represents the addition operation.

Example 1

(a) How much is 3 + 5?
(b) How much is 3 + 5 + 13?
(c) How much is 12 + 15 + 28 + 13?

Solution

(a) $3 + 5 = 8$
(b) $3 + 5 + 13 = 21$
(c) We can use columns of numbers to get the total or sum many numbers:

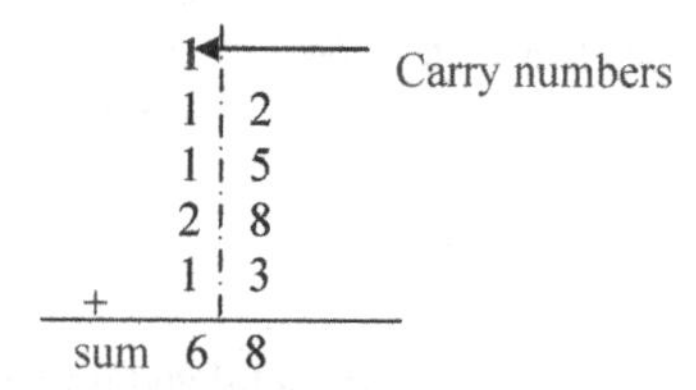

Therefore, 12 + 15 + 28 + 13 = 68. We write the numbers in columns, and then we add the numbers in the columns from right to left. If the sum of the numbers in any column is 10 or more, we write the ones number under the sum line, carry the tens number to the next column, and add it to the numbers in that column.

1.1.1.2 Subtraction (−)

Subtraction is taking away a number from another number. The symbol (−) represents the subtraction operation.

Example 2

(a) How much is 3 subtracted from 7?
(b) How much is 7 – 2?
(c) How much is 21 – 72?
(d) How much is 23 – 82?

Solution

(a) $7 - 3 = 4$.
When we subtract a small number from a bigger number, we get a positive number (+).
Note that, positive numbers do not require the "+" sign in front of the number.
(b) $2 - 7 = -5$.
When we subtract a big number from a smaller number, we get a negative number (–).
(c) We can use columns of numbers to get the subtraction:

$$\begin{array}{r} 72 \\ -\;21 \\ \hline \text{difference}\;\; 51 \end{array}$$

We write the numbers in a vertical column and then subtract the bottom numbers from the top numbers. We start from right to left.
(d)

$$\begin{array}{r} {\scriptstyle 7\;\;12} \\ \not{8}\,\not{2} \\ -\;23 \\ \hline \text{remainder}\;\; 59 \end{array}$$

When the bottom number is larger than the top number, we borrow from the number in the top of the next column and add ten to the top number before subtracting, at the same time we reduce the top number in the next column by 1.
Since 2 is smaller than 3, 2 borrows 1 from 8 to become 12, and 8 becomes 7.

1.1.1.3 Multiplication (×)

Multiplication is a quick way to add many similar numbers. The symbol (×) represents the multiplication operation.

For example, 7×5 is the same as adding 7 for 5 times, $7 + 7 + 7 + 7 + 7 = 35$. This is the same as $7 \times 5 = 35$ by multiplication table.

Example 3

(a) Multiply 23 with 12.
(b) Multiply 57 with 18.

Solution

(a)

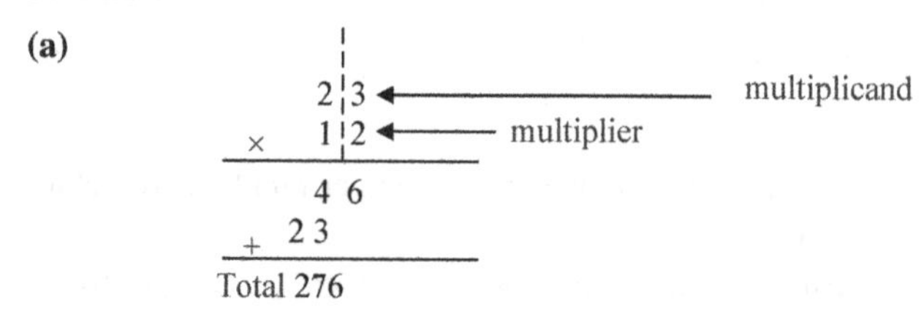

We write the numbers in a vertical column and then we multiply each number in the multiplicand from right to left by the rightmost multiplier. If the product is greater than 9, add the tens number to the product of numbers in the next column. Next, we multiply each number in the multiplicand by the next number in the multiplier and we place the second partial product under the first partial product, by moving one space to the left. We continue the process for each number in the multiplier, and then we add the partial products to get the final answer.

(b)

5 ← Carry number
5 7
× 1 8
45 6
+ 57
Total 1026

Signs play very important roles in arithmetic and algebra; we have summarized the sign rules for multiplication in the table below.

Properties of signs for multiplication

Multiplication sign	Result
$(+b) \times (+a)$	$+ab = ab$
$(-b) \times (+a)$	$-ab$
$(+b) \times (-a)$	$-ab$
$(-b) \times (-a)$	$+ab = ab$

We see from the table that, like signs give (+) and unlike signs give (−).

Any number written without a sign except 0 is considered to be positive, for example, 3 = +3.0 is a neural number, neither positive nor negative.

Example 4

(a) $(3) \times (4) = 12$
(b) $(-2) \times (5) = -10$
(c) $(3) \times (-8) = -24$
(d) $(-3) \times (-6) = 18$

1.1.1.4 Division (÷)

Division is used for separating a number to several equal groups of numbers. The symbol (÷) represents the division operation.

Example 5

(a) Divide 22 by 2.
(b) $634 \div 28$.

Solution

(a) $22 \div 2 = 11$
By dividing 22 by 2 we get 11 without remainder.
When we have very large numbers to divide, long division is performed with the method as shown below:

(b)

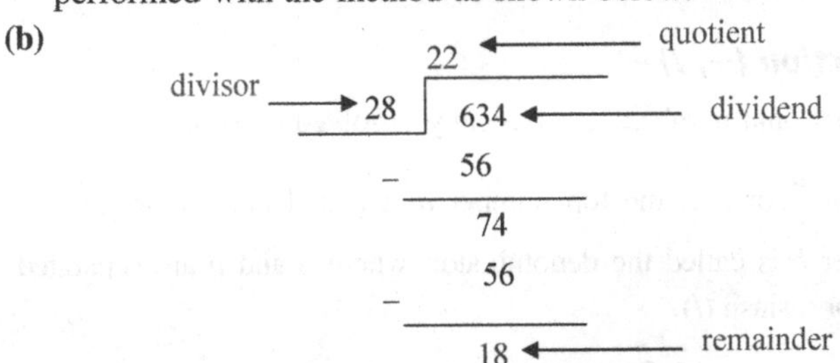

Long division allows us to use two numbers of the divided at a time, making the division process easier. We divide 63 by 28, we get 2 of 28 in the 63, we put the 2 as the quotient. We multiply 2 by 28 to give us 56. Then we subtract 56 from 63 to give us 7. The 4 is brought down so we have 2 numbers to divide. So, $634 \div 28 = 22 + (18/28)$. This means that 634 can be divided into 22 groups of 28 and there will be 18 left over, which is not enough to make a 23rd group of 28.

Signs are important in division as in multiplication. The table below summarizes the rules of signs in division.

Properties of signs for division

Division sign	Result
$\frac{+b}{+a}$	$+\frac{b}{a} = \frac{b}{a}$
$\frac{-b}{+a}$	$-\frac{b}{a} = \frac{-b}{a}$
$\frac{+b}{-a}$	$-\frac{b}{a} = \frac{-b}{a}$
$\frac{-b}{-a}$	$+\frac{b}{a} = \frac{b}{a}$

Example 6

(a) $\frac{8}{4} = 2$

(b) $\frac{-21}{7} = -3$

(c) $\frac{32}{-2} = -16$

(d) $\frac{-15}{-3} = 5$

1.1.1.5 Fraction (—, /)

Fractions are rational numbers and basically unsolved division.

For fraction of $\frac{a}{b}$ or a/b, the top number a is called numerator and the bottom number b is called the denominator, where a and b are separated by a bar (—) or a slash (/).

$\frac{a}{b}$ — a ← Numerator; b ← Denominator

For example, 3/16 is a fraction. Also, the denominator of any fraction cannot be zero, that is, $b \neq 0$. But, when the numerator of a fraction is zero, then the value of the fraction is zero. For example, $0/5 = 0$. Any whole number has a denominator equal to 1, that is, $a/1 = a$.

Example 7

Write the fraction for each shape in following spectrum.

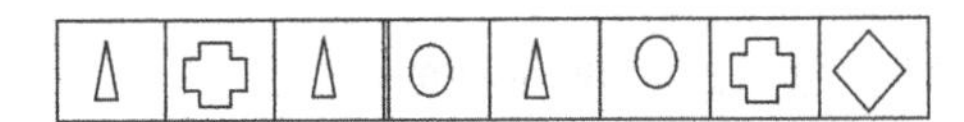

Solution

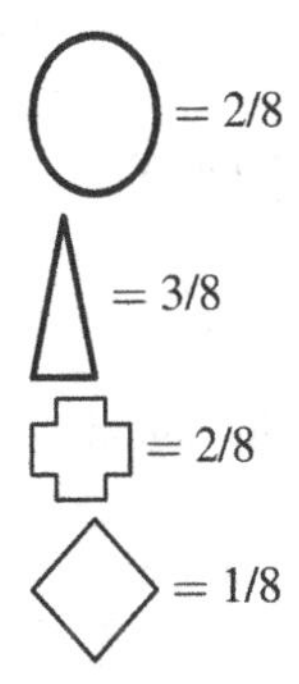

There are two types of fractions: proper and improper.

Definition of proper fraction

A proper fraction is a fraction with a numerator less than its denominator.

For example, $\frac{1}{2}, \frac{2}{7}$, and $\frac{5}{9}$ are proper fractions.

Definition of improper fraction

An improper fraction is a fraction with a numerator greater than or equal to its denominator.

For example, $\frac{9}{7}, \frac{5}{2}$, and $\frac{12}{7}$ are improper fractions.

If the numerator $a =$ denominator b, then the fraction is equal to 1, such as $a/a = 1$, where a is a *whole* number.

Fractions can be at *lowest term* when both the numerator and denominator can only be divided by 1. Therefore, to reduce a fraction to *lowest terms*, we need to divide the numerator and the denominator by the largest number in which both can be divided equally. This process is reducing or simplifying.

Example 8

(a) Reduce the fraction $\frac{15}{35}$ to lowest terms.

(b) Simplify the fraction $\frac{12}{30}$ to lowest terms.

Solution

(a) We divide both the numerator and the denominator by 5, as
$\frac{15}{35} = \frac{15/5}{35/5} = \frac{3}{7}$. $\frac{3}{7}$ is in simplest form because it cannot be divided by a number other than 1.

(b) We divide both the numerator and the denominator by 6, as
$\frac{12}{30} = \frac{12/2}{30/2} = \frac{6}{15} \Rightarrow$ is not the simplest form, so divide again.
$\frac{6/3}{15/3} = \frac{2}{5}$. Now, $\frac{2}{5}$ is the simplest form because it cannot be divided equally further.
Or
$\frac{12}{30} = \frac{12/6}{30/6} = \frac{2}{5}$.

Definition of mixed number
A mixed number is a number that consists of a whole number and a fraction.

For example, $3\frac{1}{2}$ and $5\frac{3}{4}$ are mixed numbers.

To convert an improper fraction into an equivalent mixed number, we need to divide the denominator into the numerator and write the answer as a whole number, and any remainder (which is smaller than the numerator) becomes the numerator of the fraction part of the mixed number.

Example 9

(a) Convert $\frac{5}{3}$ into a mixed number.

(b) Convert $\frac{9}{2}$ into a mixed number.

Solution

(a) We divide 5 by 3, as

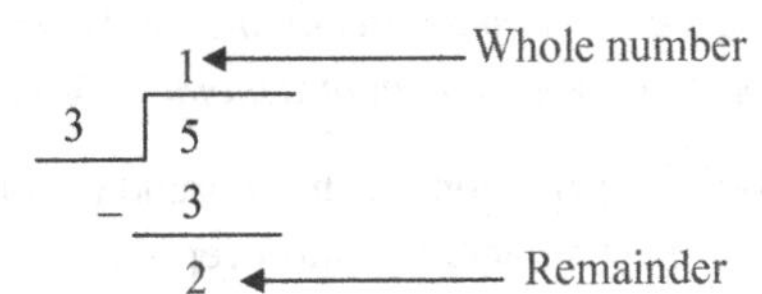

Therefore, the final solution is:

$$\frac{5}{3} = 1\frac{2}{3}$$

(b)

$$\begin{array}{r|l} & 4 \quad \leftarrow \text{Whole number} \\ \hline 2 & 9 \\ - & 8 \\ \hline & 1 \quad \leftarrow \text{Remainder} \end{array}$$

That is, $\frac{9}{2} = 4\frac{1}{2}$.

To convert mixed numbers into improper fractions, you need to multiply the whole number by the denominator of the fraction and add the answer to the numerator of the fraction, then use the answer as the new numerator and keep the same denominator as it is from before.

Example 10

(a) Convert $5\frac{2}{3}$ into an improper fraction.

(b) Convert $7\frac{1}{5}$ into an improper fraction.

Solution

(a) Multiply the whole number by the denominator of the fraction and add the answer of it to the numerator of the fraction, then use it on the numerator and keep denominator as it is from before:

$$5\frac{2}{3} = \frac{3 \times 5 + 2}{3} = \frac{17}{3}$$

(b) Multiply the whole number by the denominator of the fraction and add the answer of it to the numerator of the fraction, then use it on the numerator and keep the denominator as it is from before:

$$7\frac{1}{5} = \frac{5 \times 7 + 1}{5} = \frac{36}{5}$$

1.1.1.6 Adding or subtracting of fractions

When we want to add or subtract fractions, we need to have the same denominator, which is called a *common denominator*. When we add and subtract fractions, we use the *lowest common denominator* (LCD).

When adding or subtracting fractions with the same denominator, we need to add or subtract just their numerators and keep the denominator as it is.

Example 11

Solve the following:

(a) $\frac{3}{8}+\frac{2}{8}$

(b) $\frac{6}{13}+\frac{5}{13}-\frac{1}{13}$

Solution

(a) $\frac{3}{8}+\frac{2}{8}=\frac{3+2}{8}=\frac{5}{8}$

(b) $\frac{6}{13}+\frac{5}{13}-\frac{1}{13}=\frac{6+5-1}{13}=\frac{10}{13}$

When adding or subtracting fractions with unlike denominators, we need to find common denominator by multiplying the denominators together and multiply each numerator with the other denominator, then add or subtract them.

Example 12

Solve the following:

(a) $\frac{2}{3}+\frac{5}{4}$

(b) $\frac{5}{2}+\frac{7}{3}-\frac{1}{4}$

Solution

(a) $\frac{2}{3}+\frac{5}{4}=\frac{(4\times 2)+(3\times 5)}{3\times 4}=\frac{8+15}{12}=\frac{23}{12}$

(b) $\frac{5}{2}+\frac{7}{3}-\frac{1}{4}=\frac{(6\times 5)+(4\times 7)-(1\times 3)}{12}=\frac{30+28-3}{12}=\frac{55}{12}$

1.1.1.7 Multiplying fractions

In multiplying fractions operation, all we have to do is to multiply numerators together and denominators together.

In fractions multiplications, we can simplify before the multiplications, if there is any, to save the calculation time.

Example 13

Solve the following:

(a) $\frac{5}{4} \times \frac{7}{3}$

(b) $\frac{5}{4} \times \frac{2}{3}$

Solution

(a) $\frac{5}{4} \times \frac{7}{3} = \frac{35}{12}$

(b) $\frac{5}{\cancel{4}2} \times \frac{\cancel{2}1}{3} = \frac{5}{6}$

Example 14

How much is three-sevenths of two-thirds?

Solution

$$\frac{\cancel{3}1}{7} \times \frac{2}{\cancel{3}1} = \frac{2}{7}$$

1.1.1.8 Dividing fractions

You need to change the division operator (—, / or ÷) between the fractions into multiplication operation (×) and flip the divided by fraction (replace the numerator with its denominator and replace the denominator with its numerator) to apply the fraction multiplications method, also called the reciprocal.

Example 15

How much is three-fourths divided by one-fifth?

Solution

$$\frac{3}{4} \div \frac{1}{5} = \frac{\frac{3}{4}}{\frac{1}{5}} = \frac{3}{4} \times \frac{5}{1} = \frac{15}{4}$$

It is essential to know the correct order of operations in which any expression can be simplified.

Rules for order of operations

1. First, parentheses
2. Second, multiplications and divisions, starting from left to right
3. Third, addition and subtraction, starting from left to right

Example 16

Perform the indicated operations.

(a) $3 + 8 \div 2$
(b) $(4 + 8) \div 4$
(c) $3 \times (4 + 1)$
(d) $5 \times 2 + 3 \times 3$
(e) $7 \times 3 + 12 \div 2 - 8 \times 5$
(f) $4 \times (3 + 2) \div 6 - 9 \times 2$

Solution

(a) $3 + (8 \div 2) = 3 + 4 = 7$
(b) $(4 + 8) \div 4 = 12 \div 4 = 3$
(c) $3 \times (4 + 1) = 3 \times 5 = 15$
(d) $(5 \times 2) + (3 \times 3) = 10 + 9 = 19$
(e) $(7 \times 3) + (12 \div 2) - (8 \times 5) = 21 + 6 - 40 = 27 - 40 = -13$
(f)

$$\begin{aligned} 4 \times (3 + 2) \div 10 - 9 \times 2 &= 4 \times 5 \div 10 - 9 \times 2 \\ &= 20 \div 10 - 9 \times 2 \\ &= 2 - 9 \times 2 \\ &= 2 - 18 = -16 \end{aligned}$$

We can use MATLAB to do basic arithmetic as the following:

```
>>  12 + 19
ans  =
    31
>>  4 – 15
ans  =
    –11
```

```
>> 5*8
ans =
    40
>> format rat
>> 35/15
ans =
    7/3
>> (8/5) + (9/3)
ans =
    23/5
>> (17/6) - (12/3)
ans =
    -7/6
>> (3/2) * (7/5)
ans =
    21/10
>> (15/4)/(12/5)
ans =
    25/16
```

We can use Maple to do basic arithmetic as the following:

```
> 11 + 19
            30
> 7 - 15
            -8
> 6*8
            48
> 45/10
            9
            -
            2
```

Continued

$> (9/5) + (7/3)$

$$\frac{62}{15}$$

$> (11/3) - (8/5)$

$$\frac{31}{15}$$

$> (2/5)(4/3)$

$$\frac{2}{5}$$

$> \frac{(6/5)}{(4/3)}$

$$\frac{9}{10}$$

1.1 EXERCISES

Write each of the improper fractions *in problems 1–4 as a* mixed number.

1. $\frac{5}{3}$

2. $\frac{7}{2}$

3. $\frac{21}{5}$

4. $\frac{39}{9}$

Write each of the mixed numbers *in problems 5–8 as an* improper fraction.

5. $1\frac{1}{2}$

6. $2\frac{3}{4}$

7. $3\frac{2}{5}$

8. $3\frac{4}{5}$

Write the fractions in simplest form as in problems 9–12.

9. $\frac{5}{15}$

10. $\frac{14}{21}$

11. $\frac{2}{12}$

12. $\frac{9}{24}$

Simplify if possible before performing the operations in problems 13–46.

13. $\frac{2}{3} \times \frac{5}{7}$

14. $\frac{1}{8} \times \frac{5}{2}$

15. $\frac{3}{2} \times \frac{4}{9}$

16. $\frac{16}{5} \times \frac{25}{64}$

17. $\frac{5}{2} \div \frac{1}{7}$

18. $\frac{6}{11} \div \frac{7}{2}$

19. $\frac{5}{3} \div \frac{5}{2}$

20. $\frac{3}{7} \div \frac{3}{11}$

21. $\frac{3}{4} \div \frac{5}{4}$

22. $\frac{4}{3} \div \frac{8}{3}$

23. $3 \times \frac{2}{5}$

24. $\frac{3}{5} \times 2$

25. $3 \times \frac{7}{3}$

26. $9 \times \frac{2}{5}$

27. $\frac{3}{4} \div 5$

28. $\frac{2}{7} \div 3$

29. $2\frac{3}{5} \div 7$

30. $3\frac{1}{2} \div 5$

31. $8 \div 5\frac{1}{3}$

32. $5 \div 3\frac{2}{7}$

33. $3\frac{1}{2} \times 5\frac{2}{7}$

34. $2\frac{3}{2} \times 7\frac{1}{5}$

35. $5\frac{3}{4} \times \frac{1}{8}$

36. $3\frac{1}{2} \times \frac{5}{4}$

37. $2\frac{1}{5} \div 3\frac{1}{10}$

38. $5\frac{1}{3} \div 2\frac{1}{7}$

39. $\frac{1}{3} \times \frac{5}{8} \times \frac{4}{20}$

40. $\frac{3}{7} \times \frac{14}{11} \times \frac{22}{15}$

41. $\frac{1}{3} + \frac{13}{3}$

42. $\frac{11}{5} + \frac{3}{5}$

43. $\frac{2}{3} + \frac{1}{7}$

44. $\frac{3}{4}+\frac{5}{3}$

45. $\frac{9}{7}-\frac{3}{7}$

46. $\frac{5}{2}-\frac{1}{3}$

47. Adam paid the following bills: house payment \$825, energy bill \$125, water bill \$50, car payment \$272, and insurance premium \$65. What is the total amount of the bills?

48. John drives 45 miles per h (mph) for 2 h and 65 mph for 4 h. Determine the average speed for the total driving time.

Hint: Average rate $= \frac{\text{Total distance}}{\text{Total time}}$.

49. A welded support base is cut into three pieces: A, B, C. Find the fractional part of the total length that each of the three pieces represents when piece A = 18 in, piece B = 6 in, and piece C = 32 in.

50. The operation time sheet for machining aluminum housing identifies 2 h for facing, 3 4/5 h for milling, 4/10 h for drilling, 6/10 h for tapping, and 3/5 h for setting up. Determine the total time allotted for this operation.

51. Find the length of the grip of the bolt shown where all dimensions are in inches.

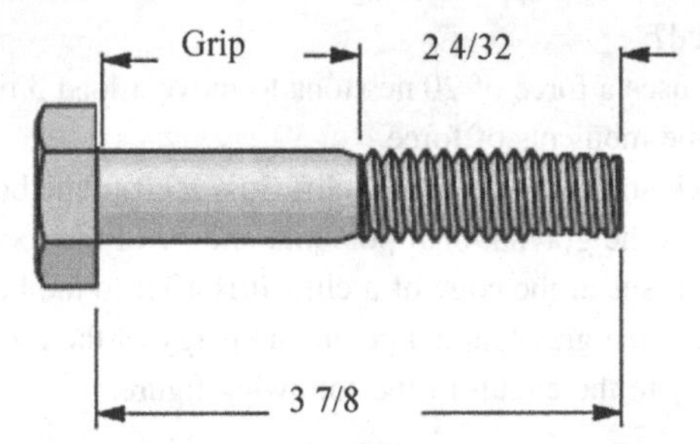

52. Calculate the distance (D) between the center lines of the first and fifth rivets connecting the two metal plates shown in the figure.

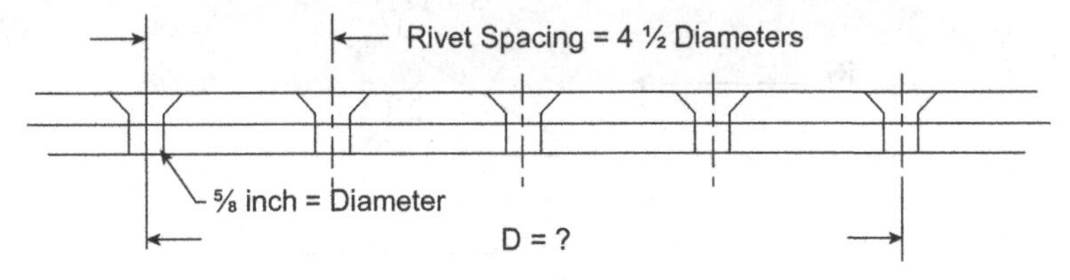

53. A steel block has the following parameters:
6 3/4 in long, 3 7/12 in wide, and 5/6 in thick. What is the volume of the block?

Hint: volume = length × width × thickness.

54. Assume that a light bulb draws 0.4 A current (I) at an input voltage (V) of 240 V.
Find the resistance (R) of the filament and the power (P) dissipated.

Hint: $P = VI$ and $R = \frac{V}{I}$

55. An electric heater is rated 1100 watts at 110 volts. Find:
1 – the current (I)
2 – the resistance (R)

56. A carpenter bought 420 board feet of pine, 35 board feet of birch lumber, and 728 board feet of oak. What was the total board feet that he bought?

57. An electrician needs 420 ft of light wire, 950 ft medium wire, 265 ft heavy wire. Determine the number of feet of wire that the electrician needs.

58. How many joules (J) of work (W) are done when a force of 200 newtons (N) is applied for a distance of 100 meters?

59. If a rider and a bicycle weighing 180 lb. travel 600 ft in 30 s, how much power (P) is developed?

60. If a rider and a bicycle with a mass of 80 kg move 400 ft in 25 s, how much power (P) is developed?

61. If a force of 40 lb is applied to move a load 15 ft, how much torque is developed?

62. An engine uses a force of 20 newtons to move a load 3 meters. Calculate the moments of force.

63. A 25 lb rock sits at the edge of a cliff. It is 50 ft to the bottom of the cliff. What is the gravitational potential energy of the rock?

64. A 35 lb rock sits at the edge of a cliff. It is 40 ft to the bottom of the cliff. What is the gravitational potential energy of the rock?

65. Calculate R_T in the circuit of the following figure.

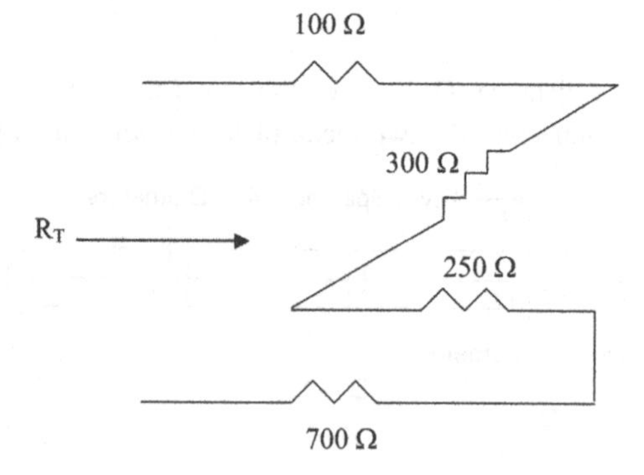

Hint: The total resistance in a series circuit is $R_T = R_1 + R_2 + R_3 + R_4$.

66. Calculate R_T in the circuit of the following Figure.

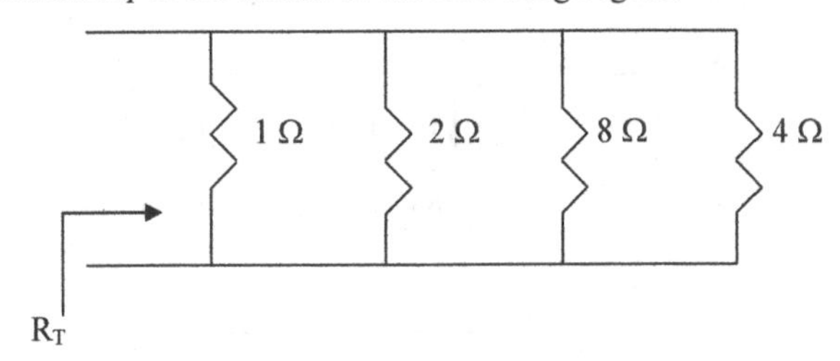

Hint: The total resistance in a parallel circuit is

$$R_T = \frac{1}{\frac{1}{R_1} + \frac{1}{R_2} + \frac{1}{R_3} + \frac{1}{R_4}}$$

67. Consider the series circuit in the Figure given below. Find:

(a) the total resistance R_T

(b) the current I

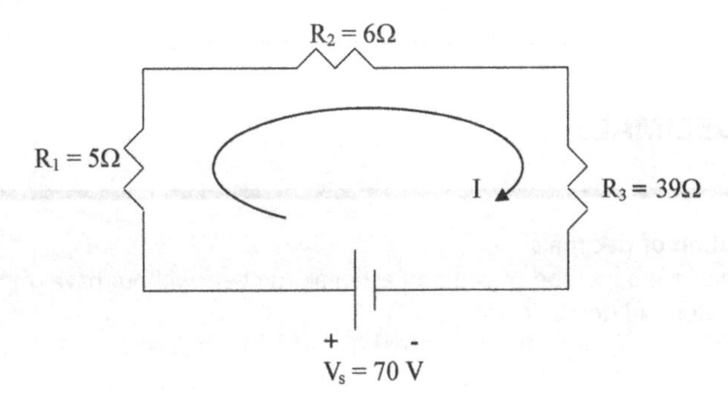

68. Consider the circuit in the following Figure. Find the voltage across each resistor.

R1 = 6 Ω
+ V1 -
+
12 V
-
+
V2
-
R2 = 2 kΩ

Hint: Using voltage divider rule:

$$V_1 = \left(\frac{R_1}{R_1 + R_2}\right)V, \quad V_2 = \left(\frac{R_2}{R_1 + R_2}\right)V$$

69. Find the total current I_t for the circuit in the following Figure.

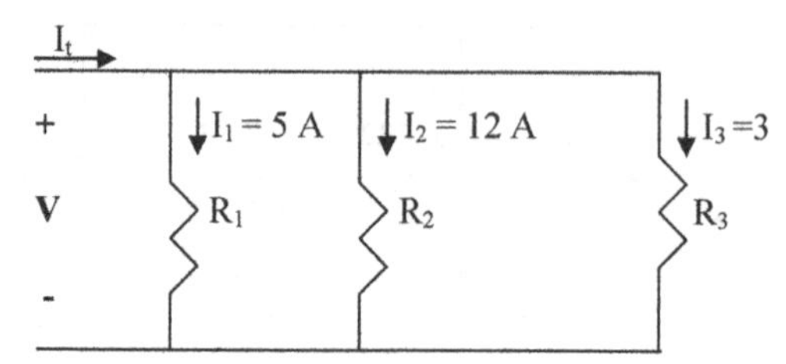

70. Find the total current I_t for the circuit in the following Figure.

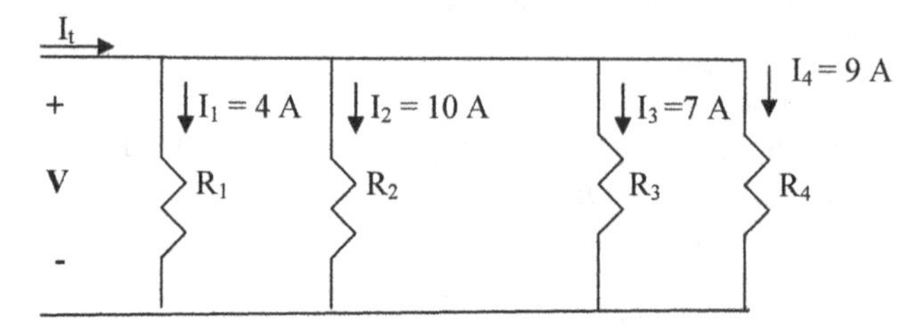

1.2 DECIMALS

Definition of decimals

Decimals are a method of writing fractional numbers without having numerator and denominator.

For example, we can write the fraction $\frac{5}{10}$ in decimal as 0.5. We use a period to indicate that the number is a decimal. Both $\frac{5}{10}$ and 0.5 are rational numbers.

It is necessary to know the value of each digit in a decimal.

Place Values names								
Numbers	•	Tenth	Hundredths	Thousandths	Ten thousandths	Hundred thousandths	Millionths

Now we can read the decimal 0.5 as five tenths or zero point five. When a decimal is less than 1, we place a zero before the decimal point, for example, we write 0.5 not .5.

In naming a decimal, we read the number from left to right as we read a whole number, and then we use the place value name for last digit of the number.

Example 1

Name the following decimals.
(a) 0.3521
(b) 0.0064
(c) 0.72345
(d) 0.000009

Solution

(a) First we write in words of 3521, and then we write the place value name for the last number, which is 1 after it: Three thousand five hundred twenty one thousandths or zero point three five two one.
(b) Sixty-four ten thousandths or zero point zero zero six four.
(c) Seventy-two thousand three hundred forty-five hundred thousandths.
(d) First we write in words of 9, and then we write the place value name for the last number, which is 9 after it: Nine millionths or zero point zero zero zero zero zero nine.

If we have a whole number and a decimal, the decimal point is written using the word "and."

Example 2

Name the following decimals.
(a) 11.725
(b) 41.93
(c) 81.13
(d) 276.5

Solution

(a) Eleven and seven hundred twenty-five thousandths.
(b) Forty-one and ninety-three hundredths.
(c) Eighty-one and thirteen hundredths.
(d) Two hundred seventy-six and five tenths.

1.2.1 Rules to rounding off decimals

The following rules are used in rounding off numbers:

1. When the number following the last digit to be retained is greater than 5, increase the last number by 1. For example, 0.016 becomes 0.02 and 0.047 becomes 0.05.
2. When the number following the last digit to be retained is less than 5, retain the last number. For example, 0.071 becomes 0.07 and 0.093 becomes 0.09.
3. When the number following the last digit to be retained is exactly 5, and the number to be retained is odd, increase the last number by 1. For example, 0.375 becomes 0.38 and 0.835 becomes 0.84.
4. When the number following the last digit to be retained is exactly 5, and the number to be retained is even, retain the last number. For example, 0.125 becomes 0.12 and 0.645 becomes 0.64

So, we can conclude that to round a decimal to a specific place value, we need to do the following steps:

1. If the digit to the right of the number needs to be rounded is 0, 1, 2, 3, or 4, the number remains the same.
2. If the digit to the right of the number needs to be rounded is 5, 6, 7, 8, or 9, we add one to the number.
3. After the rounding, all digits to the right of the rounded number are dropped.

Example 3

Round the following decimals to the given specific place value:

(a) 3.731 to the nearest one
(b) 0.25 to the nearest tenth
(c) 0.3412 to the nearest hundredth
(d) 48.6723 to the nearest hundredth
(e) 32.5468 to the nearest thousandth

Solution

(a) 4
(b) 0.3
(c) 0.34
(d) 48.67
(e) 32.547

Zeros at the end of a decimal can be affixed on the right side of the decimal point. Also, zeros can be dropped if they are located at the end of a decimal on the right side of the decimal point.

Example 4

0.75 can be written as 0.750 or 0.7500

0.560 can be written as 0.56

Now, we will go over the basic operations additions, subtraction, multiplication, and division.

We can use MATLAB to do rounding as below:

```
>> X = 0.135678;
>> Y = sprintf('%.2f',X)
Y =
0.14
>> X = 0.125678;
>> Y = sprintf('%.2f',X)
Y =
0.13
```

1.2.2 Addition of decimals

In order to add two or more decimals, we put the decimals in a column, placing the decimal points of each number in a vertical line, then adding the numbers at the same place value starting from the rightmost. If there are fewer digits in the added decimals prefix as many zeros as needed.

Example 5

Add the following decimals:
(a) 1.2 + 43.56 + 756.123
(b) 0.31 + 1.7 + 6.211

Solution

(a)

$$\begin{array}{r} 1.200 \\ 43.560 \\ +\quad 756.123 \\ \hline 800.883 \end{array}$$

(b)

$$\begin{array}{r} 0.310 \\ 1.700 \\ +\quad 6.211 \\ \hline 8.221 \end{array}$$

1.2.3 Subtraction of decimals

In order to subtract two decimals, we put the decimals in a column, placing the decimal points of each number in a vertical line, then subtracting the numbers at the same place value starting from the farthest right.

Example 6

Subtract the following decimals:
(a) 56.963 − 17.35
(b) 16.4 − 8.368

Solution

(a)

$$\begin{array}{r} 56.963 \\ -\ 17.350 \\ \hline 39.613 \end{array}$$

(b)

$$\begin{array}{r} 16.400 \\ -\ 8.368 \\ \hline 8.032 \end{array}$$

1.2.4 Multiplication of decimals

In order to multiply two decimals, we multiply the two numbers without the decimal points, and then we count the total number of digits to the right of the decimal points in the two numbers.

Example 7

Multiply the following decimals:
(a) 3.04 × 0.2
(b) 42.8 × 0.007
(c) 0.002 × 0.02
(d) 23.4 × 0.57
(e) 13.32 × 4.3

Solution

(a)

$$\begin{array}{rl} 3.04 & \longleftarrow \text{2 digits from right} \\ \times\ 0.2 & \longleftarrow \text{1 digit from right} \\ \hline 0.608 & \longleftarrow \text{3 digits from right} \end{array}$$

(b)

$$\begin{array}{rl} 42.8 & \longleftarrow \text{1 digit from right} \\ \times\ 0.007 & \longleftarrow \text{3 digits from right} \\ \hline 0.2996 & \longleftarrow \text{4 digits from right} \end{array}$$

(c)

```
        0.002  ←—— 3 digits from right
 ×       0.02  ←—— 2digits from right
 ----------------
      0.00004  ←—— 5 digits from right
```

(d)

```
         1 2
         2 2   (crossed out)
        23.4   ←—— 1 digit from right
 ×      0.57   ←—— 2 digits from right
 ----------------
 +      1638
       1170
 ----------------
      13.338   ←—— 3 digits from right
```

(e)

```
        11     (crossed out)
       13.32   ←—— 2 digits from right
 ×       4.3   ←—— 1 digit from right
 ----------------
        3996
 +     5328
 ----------------
      57.276   ←—— 3 digits from right
```

1.2.5 **Division of decimals**

The division of decimals based on two situations:

First, when a decimal is divided by a whole number, in this case, we place the decimal point in the quotient directly above the decimal point in the dividend.

Example 8

Find the value of the divisions up to the tenths digits value $\frac{135.1}{48}$

Solution

```
       2.8
 48 ) 135.1
  -    96
      ----
       390
  -    384
      ----
        6 , we ignore it .
```

Therefore, $\frac{135.1}{48} \approx 2.8$

We can use MATLAB to do decimal operations as the following:

```
>> 1.2 + 2.7
ans =
  3.9000
>> 2.2 - 1.3
ans =
  0.9000
>> 0.23 * 0.15
ans =
    0.0345
>> 0.75/0.35
ans =
      2.1429
```

We can use Maple to do decimal operations as the following:

```
> 2.6 + 5.3
          7.9
> 5.8 - 3.4
          2.4
> (6.3) (4.7)
          29.61
```

$> \frac{8.6}{2.3}$

```
          3.739130435
>
```

1.2 EXERCISES

Name the following decimals in problems 1–4.

1. 0.53
2. 0.008
3. 0.5237
4. 34.09

Perform the following operations on the decimals in problems 5–12.

5. $0.322 + 7.5 + 54$
6. $24.5 + 6.724 + 13$
7. $73.8 - 0.96$
8. $0.86 - 3.2$
9. 53.27×0.657
10. 0.32×0.08
11. $\dfrac{145.32}{0.003}$
12. $\dfrac{145.25}{55}$
13. John paid $2.88 for six pounds of bananas. How much do the bananas cost per pound?
14. If Sarah burns about 328.15 calories while walking fast on a treadmill for 40.5 min, about how many calories does she burn per minute?
15. If a car travels at a speed of 65 miles per hour for 15 min, how far will the car travel?
Hint: Distance = rate × time.
16. If one gallon of water weighs 8.25 pounds, approximately how much does a 30-gallon container of water weigh?
17. Find the decimal fraction of distance 1 inch of distance 4 inches.

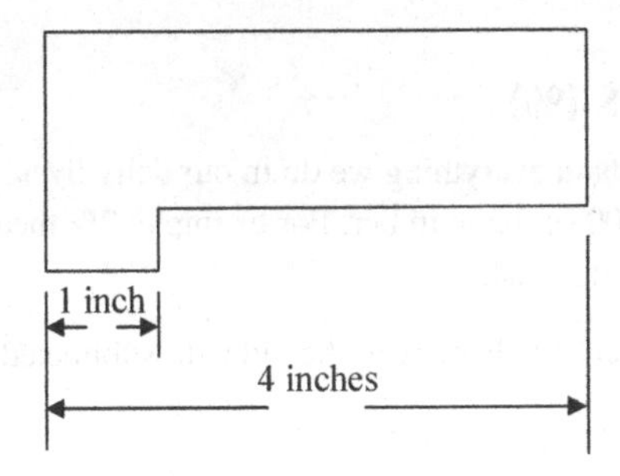

18. Find the perimeter of the following Figure.

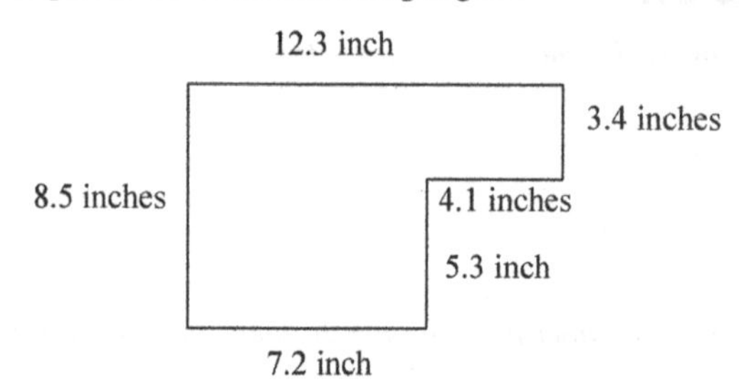

19. A charge of 6.5 C flows through an element for 0.3 s; find the amount of current through the element.
20. An electric motor delivers 50,000 J of energy (W) in 2 min. Determine the power.

    Hint: $P = \frac{W}{t}$
21. One of four 1.5-V batteries in a flashlight is put in backward. Find the voltage across the bulb.
22. Two of four 1.5-V batteries in a flashlight are put in backward. Find the voltage across the bulb.
23. Determine the number of 1.2-V batteries that must be connected in series to produce 6 V.
24. Determine the number of 1.2-V batteries that must be connected in series to produce 12 V.
25. What is the difference in power between two engines with 156.85 horsepower (hp) and 102.27 hp, respectively?
26. What is the difference in power between two engines with 126.35 hp and 101.40 hp, respectively?
27. David is given $3.25 a pay increase of per hour, what is the total amount of the pay increase per one 42-hour week?
28. Jane makes 2.5 times more money per hour than Dalia does. If Dalia earns $8.25 per hour, how much does Jane make per hour?

1.3 PERCENTS (%)

We process percents in everything we do in our daily lives. Percent (%) of a number means 1/100 of that number. For example, 7% means 7/100, which is equal to 0.07 in decimal.

Percents, like other numbers, can be added, subtracted, multiplied, or divided.

Example 9

Find the following:

(a) 7% + 4%
(b) 25% – 13%
(c) 27% / 3%
(d) 8% × 5%

Solution

(a) 7% + 4% = 11%
(b) 25% – 13% = 38%
(c) 27% ÷ 3% = 9%
(d) 8% × 5% = 40%

To convert a percent to a decimal, we remove the percent sign and move the decimal point two places to the left. For example, 35% becomes 0.35, and 3.7% becomes 0.037.

In order to convert a decimal to percent, we move the decimal places to the right then add the percent sign %. For example, the decimal number 0.56 becomes a percent number as 56%.

To convert a percent to a fraction, we divide the percent number by 100, and then we reduce it to the lowest term.

Example 10

Change 16% to a fraction in lowest terms.

Solution

16% = 16/100 = 4/25.

Example 11

2 parts of 100 parts is 2/100 = 0.02 = 2%.

Example 12

Convert 47.5% to decimal.

Solution

47.5 % ⟶ .47.5% ⟶ 0.475.

To find the percentage of a number, the percent is converted to decimal and multiplied by the number.

Example 13
What is 50% of 100?

Solution

50% ⟶ .45 % ⟶ 0.50, then $0.50 \times 100 = 50$. So 50% of $100 = 50$.

To increase a number by a percentage, like in a pay raise or sales tax, we find the percentage like in the previous example and add it to original number.

Example 14
Increase 100 by 50%.

Solution
$50\% = 0.5$
$0.5 \times 100 = 50$
$100 + 50 = 150$
So, 100 increased by 50% is 150.

Example 15
What is 8% of 75?

Solution
$8\% = 8/100 = 0.08$
$0.08 \times 75 = 6.$
So, 8% of 75 is 6.

1.3 EXERCISES

Convert the following decimal numbers to percent numbers in problems 1–4.

1. 0.34
2. 0.15
3. 0.08
4. 0.02

Convert the following percent numbers to decimal numbers in problems 5–8.

5. 43%
6. 52%
7. 7%
8. 5%

Determine the answers for problems 9–12.

9. 16% of 38

10. 15% of 2.5

11. 250 is 35% of what amount?

12. 120 is 15% of what amount?

13. An electric motor consumes 878 watts (W) and has an output of 1.14 horsepower (hp). Find the percent efficiency of the motor.
Hint: 1 hp = 746 W, percent efficiency $= \frac{\text{output}}{\text{input}} \times 100$.

14. An electric motor consumes 975 watts (W) and has an output 1.25 horsepower (hp). Find the percent efficiency of the motor.

15. In a technical mathematics class of 30 students, 75% are boys. If 65% of them are working part time, how many boys in the class are working part time?

16. If a motorcycle requires 25 units of actual work and 65 units of theoretical work, compute the mechanical efficiency.

17. Richard and Matt arrive late for a seminar and miss 5% of it. The seminar is 115 min long. How many minutes did they miss?

18. The price of gasoline drops from $3.00 per gallon to $2.35 per gallon. What is the percent of decrease?

19. Thirty percent of the people at a restaurant selected the lunch special. If 60 people did not select the special, how many people are eating at the restaurant?

20. A machine on a production line produces parts that are not acceptable by company standards 2% of the time. If the machine produces 900 parts, how many will be defective?

21. Jeff's monthly utility bill is equal to 40% of his monthly rent, which is $700 per month. How much is Jeff's utility bill each month?

22. Sam has completed 80% of his 110-page report. How many pages has he written?

CHAPTER 1 REVIEW EXERCISES

Reduce each of the following fractions in problems 1–4 to lowest terms.

1. $\frac{4}{14}$

2. $\frac{14}{22}$

3. $\frac{24}{42}$

4. $\frac{65}{135}$

Change each of the following mixed numbers in problems 5–8 to improper fractions:

5. $7\frac{2}{3}$

6. $5\frac{3}{4}$

7. $8\frac{1}{2}$

8. $6\frac{2}{5}$

Find the fraction of each portion of the following figures in problems 9 and 10.

9.

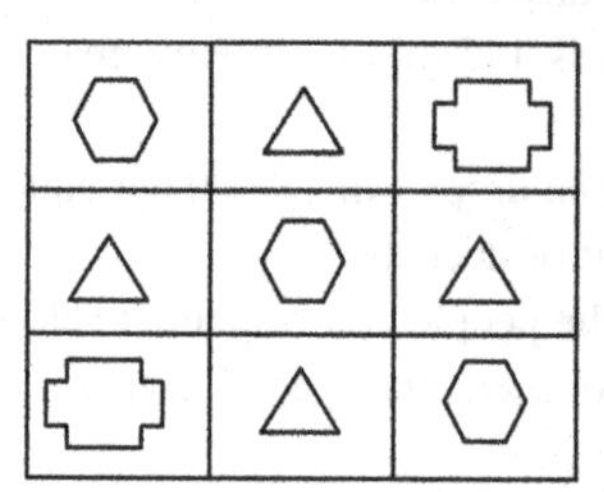

10.

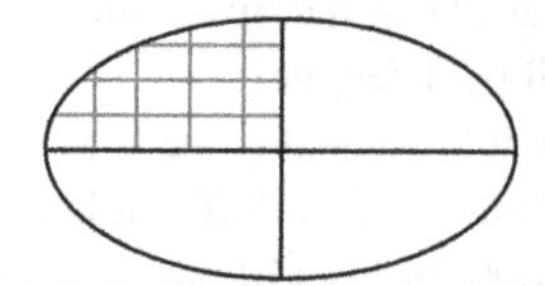

11. A batch of concrete is made by mixing 20 kg of water, 65 kg of cement, 100 kg of sand, and 225 kg of aggregate. Determine the total weight of the mixture.

12. Texas contains 268,820 square miles (mi^2), and Louisiana 52, 271 mi^2. How much larger is Texas than Louisiana?

13. The current to a lamp is 4.3 A when the line voltage is 110 V. Calculate the power dissipated in the lamp.
Hint: Power = voltage × current.

14. A gear in a machine rotates at the speed of 1603 revolutions/min. How many revolutions will it make in 6 min?

15. If 800 shares of a stock are valued at $72,000, what is the value of each share?

16. If 600 shares of a stock are valued at $42,000, what is the value of each share?

17. A charge (Q) of 90 coulombs (C) passes a given point of an electric circuit in 20 seconds (s). Find the current (I) in the circuit.

Hint: $I = \frac{Q}{t}$.

18. The power (P) dissipated in an electric circuit element is 120 W and a current (I) of 20 A is flowing through it. Determine the voltage (V) across and the resistance (R) of the element.

Hint: $P = VI$ and $R = \frac{V}{I}$.

19. Calculate the total resistance (R_T) in each of the circuits shown in the Figure given below, where
$R_1 = 1$ ohm, $R_2 = 2$ ohms, $R_3 = 3$ ohms
Hint: The total resistance in a series circuit is $R_T = R_1 + R_2 + R_3$.

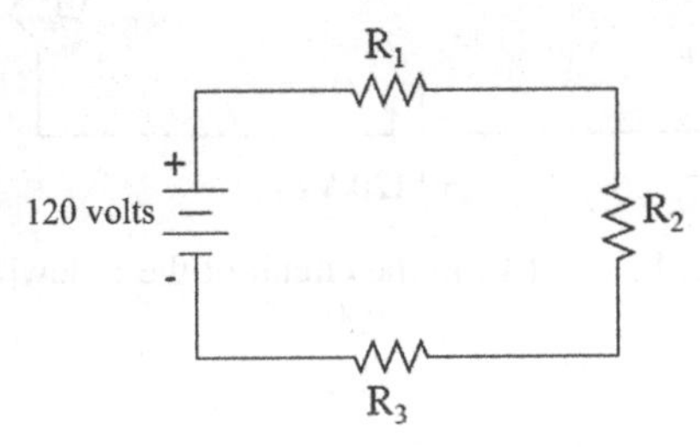

20. Calculate the total resistance (R_T) in each of the circuits shown in the Figure given below, where
$R_1 = 2$ ohm, $R_2 = 4$ ohms, $R_3 = 6$ ohms

Hint: The total resistance in a parallel circuit is $R_T = \frac{1}{\frac{1}{R_1} + \frac{1}{R_2} + \frac{1}{R_3}}$

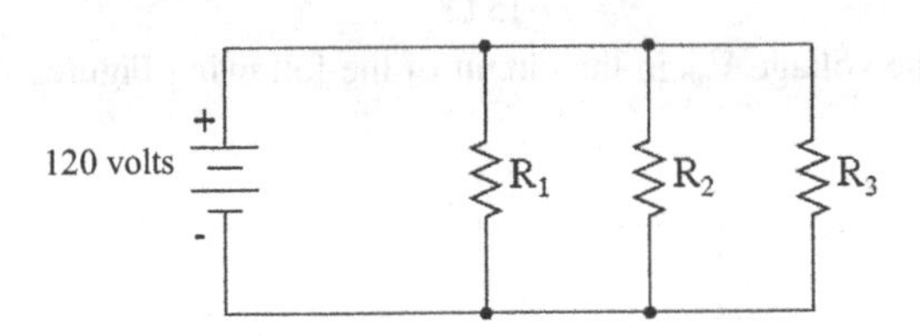

21. Determine V_1 and V_2 in the following Figure, using voltage division.

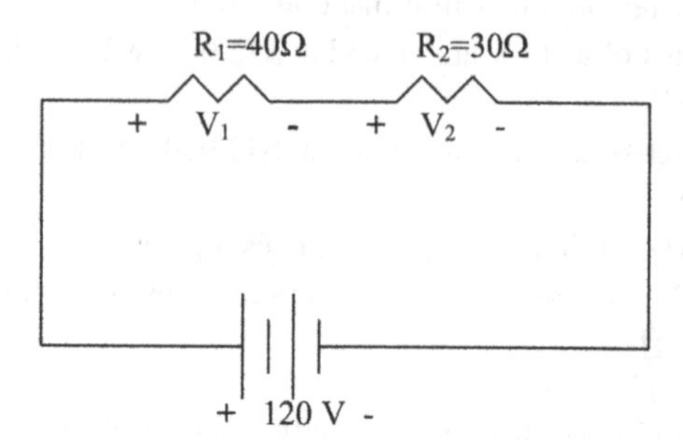

22. Find V_1, V_2 , and V_3 in the circuit of the following Figure.

R₂ =13 Ω
+ V₂ -
R₁ =7 Ω
- V₁ +
+ V₃ -
R₃ = 20Ω
+ 120 V-

23. Determine V_1, V_2 , and V_3 in the circuit of the following Figure.

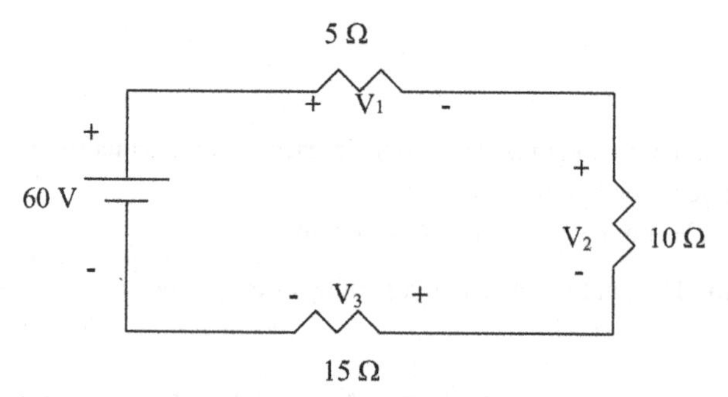

24. Find the voltage V_{ab} in the circuit of the following Figure.

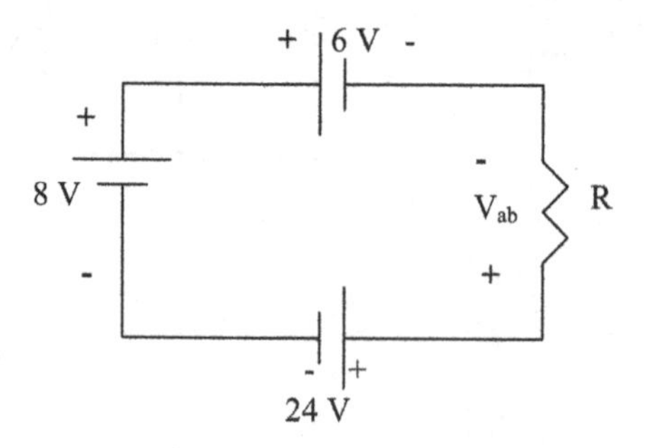

25. Find the voltage V_{ab} in the circuit of the following Figure.

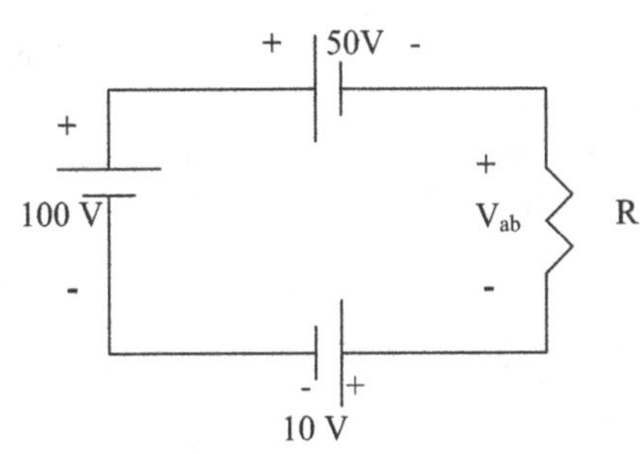

Hint: When voltage sources are connected in series, the total voltage is the sum of the voltage of the individual sources.

26. A board $9\frac{1}{2}$ ft long has a piece $2\frac{3}{4}$ ft long cut from it. How long is the remaining board?

27. How many inches are left in a 16-in welding rod after $3\frac{1}{2}$ in have been burned up?

28. John drove 87 miles and Jane drove 48 miles. What fraction of the entire trip did Jane drive?

29. A water tank, when filled, holds 825,000 L. If there are 386,000 L of liquid in the tank, what fraction represents the unfilled portion of the tank?

30. Two electrical resistors have resistances of $3\frac{1}{7}$ ohms and $2\frac{5}{14}$ ohms. What is the sum of their resistances?

31. A carpenter needs $7\frac{1}{2}$ ft of one type of trim and $95\frac{1}{4}$ ft of another type. What is the total feet of trim that he needs?

32. A mechanic has 21 bolts, each $3\frac{5}{7}$ in long. If he placed the bolts end to end, how long a string of bolts would be formed?

33. If a piece of metal trim$21\frac{5}{7}$ in long is cut into 6 pieces of equal length, what will be the length of each piece?

34. The maximum continuous load on a circuit breaker device is limited to 85% of the device rating. If the circuit breaker device is rated 40 A, what is the maximum continuous load permitted on the device?

35. A fuse must be sized no less than 120% of the continuous load. If the load is 75 A, what will be the smallest-sized fuse?

36. Increase 55 by 30%.

37. Increase 35 by 25%.

38. Sales tax is 8.25%. If an item costs $10.00, how much will it be after tax?
39. A car travels 2470 km and uses 190 L of gasoline. Find its mileage in kilometers per liter.
40. A car travels 1350 miles and uses 50 gallons of gasoline. Find its mileage in miles per gallon.

Chapter 2

Introduction to Algebra

When solving problem, dig at the roots instead of just hacking at the leaves.

Anthony J. D'Angelo

■ James Prescott Joule (1784—1858), an English physicist who studied the nature of heat and established its relationship to mechanical work. He therefore laid the foundation for the theory of conservation of energy, which later influenced the First Law of Thermodynamics. He also formulated the Joule's laws which deal with the transfer of energy.

INTRODUCTION

In this chapter, we will discuss useful and important basic topics of introductory algebra that are essential for easier understanding of technical and applied mathematics.

2.1 INTRODUCTION TO ALGEBRA

In algebra, some values are known but others are either unknown or not specified, which can be represented by letters or variables. Algebra performs mathematical operations on numbers and variables (unknown numbers), in order to find the values of variables.

Fundamentals of Technical Mathematics. http://dx.doi.org/10.1016/B978-0-12-801987-0.00002-2

Algebra uses signs to represent operations. The study of algebra involves the use of equations to solve various problems. Today, technicians and engineers are required to learn algebra to solve general and specific problems related to their fields and new technologies.

In algebra, an expression is a combination of letters that represent numbers known as variables, separated by operations creating **terms**. For example, $3x - 1$ is an expression with two terms. The number or constant in front of the variable of an expression is called a **coefficient**. For the above example, 3 is the coefficient of x. An expression is a mathematical statement that does not contain an equal sign (=), but it can contain numbers, variables, and operators ($+$, $-$, $\times$, $\div$). Equations are expressions followed by an equal sign.

In algebra, we use formulas that are shortened rules for employing letters as symbols to represent quantities. For example, we use the letter l to represent the length, the letter w to represent the width, and the letter A for area. We can write the formula for the area of a rectangle as $A = l \times w$.

Example 1

Find the coefficients of the following expressions:

(a) $25x - y$
(b) $7t + 3z$

Solution

(a) 25 is the coefficient of x and -1 is the coefficient of y
(b) 7 is the coefficient of t and 3 is the coefficient of z

Some expressions contains more than one grouping of symbols in parentheses ( ) or brackets [ ]. In this case, you need to compute the innermost grouping first and work outward.

When terms involve the same letter with different coefficients in an expression, we call them *like terms*. For example, the expression $5x - 2y + 7x + 4y$ have two like terms $5x$ with $7x$ and $-2y$ with $4y$. Usually an expression can be simplified by adding and subtracting the like terms (apples to apples and oranges to oranges).

Example 2

Compute the following

(a) $x(a + 3a) - 2$
(b) $5[x(a + 3a) - 2]$

Solution

(a) First we solve the operation in parentheses, which is $a + 3a = 4a$, then we multiply the result $4a$ by x, which is $4ax$, then we subtract 2 from $4ax$ to give us $4ax - 2$.

(b) First we solve the operation in parentheses from $a + 3a = 4a$, then we multiply the result, $4a$, by x, which is $4ax$. Then we subtract 2 from $4ax$ to give us $4ax - 2$. Next, we multiply by 5 to give us $20ax - 10$.

Example 3

Simplify the expression $8x - 3y - 2x + 15y$.

Solution

First, we collect the like terms for x, which are $8x - 2x = 6x$, and the same for the y as $-3y + 15y = 12y$. Therefore, the expression $8x - 3y - 2x + 15y$ can be simplified to $6x + 12y$.

Substitution is used in algebra to find a numerical value of a variable in an expression. It can be done by assigning a particular number value to a variable.

Example 4

Let $y = x - 2$ and $x = 5$, find the value of y.

Solution

We need to substitute 5 for x to get the value of y as $y = 5 - 2 = 3$.

2.1.1 Basic principles of addition

We can summarize the basic principles of addition as in the table below considering a, b, and c are real numbers.

Property	**Symbol**
Commutative	$a + b = b + a$
Associative	$a + (b + c) = (a + b) + c$
Distributive	$a(b + c) = ab + ac$
Identity is 0	$0 + a = a$
Inverse (opposite)	$a + (-a) = 0$

We will apply the above principles on algebra as in the below examples. We use x and y to represent variables.

Example 1

(a) $5x + 2y = 2y + 5x$
(b) $7x + (3y + 9) = (7x + 3y) + 9$
(c) $2y + 3y = y\,(2 + 3)$
(d) $0 + 6x = 6x$
(e) $7x + (-7x) = 0$

2.1.2 Basic principles of multiplication

The basic principles of multiplication are summarized as in the below table with consideration that a, b, and c are real numbers.

Properties of multiplication

Property	Symbol
Commutative	$ab = ba$
Associative	$a(bc) = (ab)c$
Distributive	$(b + c)a + ba + ca$
Identity is 1	$a \cdot 1 = a$
Inverse (reciprocal)	$a\left(\frac{1}{a}\right) = 1, a \neq 0$

Again, we applied these principles to algebra as in the below examples. It is important to know that $ab = a \times b = a \cdot b$.

Example 1

(a) $(5x)(2y) = (2y)(5x) = 10xy$
(b) $7x\,[(3y)(9)] = [(7x)(3y)]9 = 189xy$
(c) $5[t + 3] = 5t + 15$
(d) $1 \cdot 6x = 6x$
(e) $7x\,(1/7x) = 1$
(f) $x \cdot 6 = 6x$
(g) $(-9t)t = -9t^2$

We summarized some of the most basic algebraic operations and the way they simplified as in the below table with consideration that a, b, c, and d are real numbers.

Operation	**Simplified to**
$a + (-b)$	$a - b$
$a \times \frac{1}{b}, b \neq 0$	$\frac{a}{b}$
$\frac{a}{a}, a \neq 0$	1
$a(-b)$	$-(ab)$
$(-a)b$	$-(ab)$
$(-a)(-b)$	ab
$-(-a)$	a
$\frac{a}{-b}$	$\frac{-a}{b}$ or $-\frac{a}{b}$
$\frac{-a}{-b}$	$\frac{a}{b}$
$ac = bc, c \neq 0$	$a = b$
$\frac{ac}{bc}, b \neq 0$ and $c \neq 0$	$\frac{a}{b}$
If $ab = 0$	$a = 0$ or $b = 0$, or $a = b = 0$
$\frac{a}{b} + \frac{c}{d}, b \neq 0$ and $d \neq 0$	$\frac{ad + bc}{bd}$
$\frac{a}{b} \times \frac{c}{d}$	$\frac{ac}{bd}$
$\frac{(\frac{a}{b})}{(\frac{c}{d})}, b \neq 0, d \neq 0$, and $c \neq 0$	$\frac{a}{b} \times \frac{d}{c} = \frac{ad}{bc}$
$a - (b + c)$	$a - b - c$
$a - (b - c)$	$a - b + c$

When several operations exist in a single expression, then we need to follow the order of operations as we start first with parentheses, second with exponent, third with multiplication or division, left to right, and fourth with addition or subtraction, left to right.

Example 2

Simplify the following:

(a) $55 - (6 - 2) \times 3^2$

(b) $7[5 - 2(a + b)]$

(c) $x + (-y)$

(d) $x \times \frac{1}{y}$

(e) $x \times 0$

(f) $0/x$

(g) $\frac{y}{y}$

(h) $x(-y)$

(i) $(-x)y$

(j) $(-x)(-y)$

(k) $-(-x)$

(l) $\frac{x}{-y}$

(m) $\frac{-x}{-y}$

(n) $3x = 3y$

(o) $\frac{7x}{7y}$

(p) $a(b = 0)$

(q) $\frac{x}{3} + \frac{y}{4}$

(r) $\frac{7}{x} \times \frac{2}{y}$

(s) $\frac{\left(\frac{2}{x}\right)}{\left(\frac{9}{y}\right)}$

(t) $x - (3z + t)$

(u) $7z - (y - x)$

Solution

(a) $55 - (6 - 2) \times 3^2 = 55 - 4 \times 3^2$, we start first with parentheses
$=55 - 4 \times 9$, then exponent
$=55 - 36$, multiplication
$=19$, last subtraction

(b) $7[5 - (a + b)] = 7[5 - a - b] = 35 - 7a - 7b$

(c) $x + (-y) = x - y$

(d) $x \times \frac{1}{y} = \frac{x}{y}$

(e) $x \times 0 = 0$

(f) $\frac{0}{x} = 0$

(g) $\frac{y}{y} = 1$

(h) $x(-y) = -(xy) = -xy$

(i) $(-x)y = -(xy) = -xy$

(j) $(-x)(-y) = xy$

(k) $-(-x) = x$

(l) $\frac{x}{-y} = \frac{-x}{y} = -\frac{x}{y}$

(m) $\frac{-x}{-y} = \frac{\cancel{-}x}{\cancel{-}y} = \frac{x}{y}$

(n) $3x = 3y \Rightarrow \frac{\cancel{3}x}{\cancel{3}} = \frac{\cancel{3}y}{\cancel{3}} \Rightarrow x = y$

(o) $\frac{7x}{7y} = \frac{\cancel{7}x}{\cancel{7}y} = \frac{x}{y}$

(p) $a(b = 0) = a \times 0 = 0$

(q) $\frac{x}{3} + \frac{y}{4} = \frac{4 \times x + 3 \times y}{3 \times 4} = \frac{4x + 3y}{12}$

(r) $\frac{7}{x} \times \frac{2}{y} = \frac{7 \times 2}{x \times y} = \frac{14}{xy}$

(s) $\frac{\left(\frac{2}{x}\right)}{\left(\frac{9}{y}\right)} = \frac{2}{x} \times \frac{y}{9} = \frac{2y}{9x}$

(t) $x - (3z + t) = x - 3z - t$

(u) $7z - (y - x) = 7z - y + x$

Example 3

Simplify the following:

(a) $\frac{-2}{3} + \frac{5}{2}$

(b) $\frac{-9x^2}{-3x}$

(c) $\frac{5}{3} \cdot \frac{11}{2}$

(d) $\frac{12/5}{3/7}$

(e) $3x\,(2-7x)$

(f) $6x-2-3x)$

(g) $5y-(-5x+2y)$

(h) $(-7)(3x)$

(i) $(-2x)(-3y)$

(j) $(3x-2)(3x+2)$

(k) $5\times\dfrac{1}{6}+3$

Solution

(a) $\dfrac{-2}{3}+\dfrac{5}{2}=\dfrac{-4+15}{6}=\dfrac{11}{6}$

(b) $\dfrac{-9x^2}{-3x}=3x$

(c) $\dfrac{5}{3}\cdot\dfrac{11}{2}=\dfrac{55}{6}$

(d) $\dfrac{12/5}{3/7}=\dfrac{12}{5}\cdot\dfrac{7}{3}=\dfrac{28}{5}$

(e) $3x\,(2-7x)=6x-21x^2$

(f) $6x-(2-3x)=9x-2$

(g) $5y-(-5x+2y)=3y+5x$

(h) $(-7)(3x)=-21x$

(i) $(-2x)(-3y)=6xy$

(j) $(3x-2)(3x+2)=9x^2-4$

(k) $5\times\dfrac{1}{6}+3=\dfrac{5+18}{6}=\dfrac{23}{6}$

We summarized some of the most common expansions and factors and the way they simplified as in the below table with consideration that a and b are variables.

Expansion and factors

1. $a^2-b^2=(a-b)(a+b)$
2. $a^3-b^3=(a-b)(a^2+ab+b^2)$
3. $a^n-b^n=(a-b)(a^{n-1}+a^{n-2}b+....+b^{n-1})$, for n an odd integer
4. $a^n+b^n=(a+b)(a^{n-1}-a^{n-2}b+....-b^{n-1})$, for n an even integer

5. $a^3 + b^3 = (a + b)(a^2 - ab + b^2)$
6. $(a \pm b)^2 = (a \pm b)(a \pm b) = a^2 \pm 2ab + b^2$
7. $(a \pm b)^3 = a^3 \pm 3a^2b + 3ab^2 \pm b^3$

Example 4

(a) $(x - y)(x + y) = x^2 - y^2$

(b) $(x - z)(x^2 + xz + z^2) = x^3 - z^3$

(c) $(x + y)(x^2 - xy + y^2) = x^3 + y^3$

(d) $(x + y)^2 = (x + y)(x + y)$

2.1.3 **Exponent and radicals**

Definition of exponent
If n is a natural number, then $a^n = a \cdot a \cdot a \cdots a$, where a appears as a factor n times.

The expression a^n is a combination of the power n, which is the *exponent*, and a which is the *base*.

So, a *power* or *exponent* indicates how many times the base is multiplied by itself. The power is written in exponent form to represent the number of factors.

For example, 5×5 is called the second power and can be written in exponent form as 5^2.

We read 5^2 as 5 power of 2. The 5 is called the base and the 2 is called the exponent. When no exponent is written with a number, it is assumed to be one. For example, $5 = 5^1$.

Example 1
Find the value of 3^4.

Solution
$3^4 = 3 \times 3 \times 3 \times 3 = 81$.

Based on the definition of exponent, we can define the exponent a^n when n is zero or negative values.

Definition of zero and negative exponents

If a is a nonzero real number and n is a positive integer, then

$a^0 = 1$ and $a^{-n} = \frac{1}{a^n}$

We can summarize the rules of power as:

Properties of exponents

For any integers n and m, and any real numbers a and b, the following properties are true:

1. $a^n \times a^m = a^{n+m}$
2. $(a^n)^m = a^{n \times m}$
3. $\dfrac{a^n}{a^m} = a^{n-m}$
4. $(ab)^n = a^n \times b^n$
5. $\left(\dfrac{a}{b}\right)^n = \dfrac{a^n}{b^n}$

We summarize the operations with zero as in the below table with consideration that a is a real number.

Zero properties

1. $a - a = 0$
2. $a \times 0 = 0$
3. $\dfrac{0}{a} = 0$, if $a \neq 0$
4. $a^0 = 1$, if $a \neq 0$
5. $0^a = 0$
6. $\dfrac{a}{0} =$ undefined
7. $0^0 =$ undefined

Definition of radicals
If n is an even or odd natural number and $a > 0$, then $\sqrt[n]{a} = a^{\frac{1}{n}}$.

For a and b are real numbers, and m and n are natural numbers.

We summarize the common properties of square root as in the below table.

Square root properties	
$\sqrt[n]{a}$, a: real number, n: integer number	$a^{\frac{1}{n}}$
$\sqrt[n]{a^m}$ a: real number, n and m: integer numbers	$a^{\frac{m}{n}}$
$\sqrt[n]{a}\sqrt[n]{b}$	$\sqrt[n]{ab}$
$\frac{\sqrt[n]{a}}{\sqrt[n]{b}}$, $b \neq 0$	$\sqrt[n]{\frac{a}{b}}$
$\sqrt[m]{\sqrt[n]{a}}$	$\sqrt[mn]{a}$

Example 2

(a) $\sqrt[2]{x} = \sqrt{x} = x^{\frac{1}{2}}$

(b) $\sqrt[6]{y} = y^{\frac{1}{6}}$

(c) $\sqrt[7]{x^3} = x^{\frac{3}{7}}$

(d) $\sqrt[5]{x}\sqrt[5]{y} = \sqrt[5]{xy}$

(e) $\frac{\sqrt[3]{x}}{\sqrt[3]{y}} = \sqrt[3]{\frac{x}{y}}$

(f) $\sqrt[5]{\sqrt[3]{x}} = \sqrt[15]{x}$

2.1 EXERCISES

In problems 1–6, perform the indicated operations using exponents, and write each answer in a simple form.

1. 3×3
2. $5 \times 5 \times 5$

3. 2×16

4. 3×9

5. $mn \times mn \times mn \times mn$

6. $vt \times vt \times vt$

In problems 7–22, find the value for n for which each statement is true.

7. $2^4 \times 2^5 = 2^n$

8. $2^4 \times 2^6 \times 2^8 = 4^n$

9. $3^n \times 3^5 = 3^{11}$

10. $3^n \times 3^9 \times 3^4 = 3^{21}$

11. $7(7^2 \times 7^3) = 7^n$

12. $5^8(5^4 \times 5) = 5^n$

13. $2^n \times 2^n = 1$

14. $2^n \times 2^{3n} = 1$

15. $3^{n-1} = 3$

16. $3^{2n+1} = 3^4$

17. $2^5 \times 3^5 = 6^n$

18. $2^7 \times 3^7 = 6^{n-1}$

19. $2^{10} \times 5^7 = 2^3 \times 10^{n+1}$

20. $3^{12} \times 5^8 = 3^4 \times 15^{n-1}$

21. $4^5 \times 3^5 = n^5$

22. $2^7 \times 9^7 = n^7$

In problems 23–28, perform each of the indicated operations.

23. 2^7

24. 2^8

25. $(2^3)^2$

26. $(3^2)^3$

27. $(2 \cdot 3^2)^2$

28. $(5^3 \times 3^2)^0$

In problems 29–36, write each rational expression in lowest terms.

29. $\dfrac{3t^2}{9t}$

30. $\dfrac{21r^3}{7r}$

31. $\frac{2z - 12}{6z - 36}$

32. $\frac{5p + 15}{7p + 21}$

33. $\frac{5(p + 11)}{(p + 11)(p - 5)}$

34. $\frac{2(p + 4)}{(p + 4)(2p - 4)}$

35. $\frac{9m^2 + 18m}{3m}$

36. $\frac{4m^2 + 32m}{2m}$

2.2 ADDITION, SUBTRACTION, MULTIPLICATION, AND DIVISION OF MONOMIALS

A monomial is one variable that is the product of a constant (a letter that stands just for one number) and a variable with a nonnegative integer power. Recall, integer is a set of {..., −2, −1, 0, 1, 2,...}

A monomial form $= cx^k, k \geq 0$

$c =$ a constant (coefficient of the monomial)
$x =$ a variable
$k =$ an integer (degree of monomial)

Exponent (power) is a small number above the variable or number (called base) to indicate the number of times a number is multiplied by itself.

For $x^k = x \cdot x \cdot x \cdots x$ ⟵ k factors

$x =$ base
$k =$ exponent

Example 1

(a) $x^0 = 1$

(b) $x^1 = x$

(c) $x^2 = x \cdot x$

(d) $x^{-9} = \frac{1}{x^9}$

(e) $x^4 = \frac{1}{x^{-4}}$

Example 2

$7x^3$ is a monomial with a coefficient 7 and of a degree 3.

Note that $x^{-k} = \frac{1}{x^k}$, if $k \neq 0$

Two monomials can use addition and subtraction between them only when they are like terms (same variable with same degree).

Example 3

(a) $3x^2 + 5x^2 = (3 + 5)x^2 = 8x^2$

(b) $3x^2 + 5x \neq 8x^2$

(c) $-3V + 9V = (-3 + 9)V = 6V$

(d) $5I^3 - 13I^3 = (5 - 13)I^3 = -8I^3$

A *term* is an expression, for example, $6t^5$, is a term with number 6 as coefficient and t as variable, and 5 as the exponent. The expression t^5 means $t \cdot t \cdot t \cdot t \cdot t$. There are two types of terms: first, *like terms*, which are terms having the same variable and same exponent; second, *unlike terms*, which are terms that do not have the same variable and the same exponent. For example, z and z^2 are unlike terms, whereas, $8z^5$ and $3z^5$ are like terms.

2.2.1 Rules for multiplication of monomials

The rules for multiplication of monomials are summarized in the below table.

$c_1a^m \cdot c_2a^n = c_1c_2a^{m+n}$, m, n are integers
$(a^m)^n = a^{mn}$
$(ca)^n = c^na^n$

Example 1

(a) $3x^2 \cdot 2x^5 = (3 \cdot 2)x^{2+5} = 6x^7$

(b) $(x^2)^5 = x^{2 \cdot 5} = x^{10}$

(c) $(2x^2)^3 = 2^3x^{2 \cdot 3} = 8x^6$

2.2.2 Rules for division of monomials

The rules for multiplication of monomials are summarized in the below table.

$\left(\frac{a^n}{a^m}\right) = a^{n-m}$, m, n are integers
$\left(\frac{a}{c}\right)^m = \left(\frac{a^m}{c^m}\right)$
$\frac{1}{a^n} = a^{-n}$

Example 1

Simplify the following

(a) $\left(\frac{x^5}{x^2}\right)$

(b) $\left(\frac{x}{4}\right)^2$

(c) $\frac{1}{x^7}$

(d) $\frac{5}{x^4}$

(e) $19x^{-3}$

Solution

(a) $\left(\frac{x^5}{x^2}\right) = x^5 \cdot x^{-2} = x^3$ or by cancellation $\left(\frac{\not{x}^{5} x^3}{\not{x}^2}\right) = x^3$

(b) $\left(\frac{x}{4}\right)^2 = \left(\frac{x^2}{4^2}\right) = \frac{x^2}{16}$

(c) $\frac{1}{x^7} = x^{-7}$

(d) $\frac{5}{x^4} = 5x^{-4}$

(e) $19x^{-3} = \frac{19}{x^3}$

A *binomial* is the sum or difference of two monomials that are not like terms.

For example, $x + x^2$ is a binomial with two terms.

2.2.3 Addition, subtraction, multiplication, and division of polynomials

A *polynomial* is a monomial or a finite sum of monomials. In other words, a polynomial is a term or a finite sum of terms in which all variables have exponents of whole numbers and no variables with negative exponents.

For example, the following are polynomials:

-5, $3n$, and $6y^3 + 2y^2 + x$.

The general form of a polynomial is

$a_n x^n + a_{n-1} x^{n-1} + \ldots + a_1 x + a_0$

Where

$a_n, a_{n-1}, \ldots a_1 x, a_0 =$ constants (coefficients of the polynomial)

n: an integer ($n \geq 0$) and is the degree of the polynomial

x: a variable.

For example, $5z^3 + 3z - 1$ is polynomial with coefficients 5, 3, -1 with degree of 3. Note that $\frac{1}{t^2}$ is not a polynomial.

A polynomial can have more than one variable. A term (monomials that make up a polynomial) with more than one variable has a degree of the sum of all the exponents of the variables.

The degree of a polynomial in more than one variable is equal to the greatest degree of any term in the polynomial.

For example, $5z^3x^6$ has a degree of 9, and $z^3x + 3xy - 1$ has a degree of 4.

There are four types of polynomials:

1. Trinomial: a polynomial containing exactly three terms
2. Binomial: a polynomial containing exactly two terms
3. Monomial: a single term polynomial
4. Just polynomial

Example 1

1. $z^3x + 3xy - 1$ is polynomial of type trinomial
2. $z^3x + 3x$ is polynomial of type binomial
3. $3xy$ is polynomial of type monomial
4. $z^3x + 3xy + 7t - 17n$ is polynomial

Polynomials are added and subtracted by combining the like terms.

Example 2

Add or subtract the following polynomials:

(a) $(2z^3 + 4z^3 - 3z) - (z^3 + 6z^2 - z) + 17$

(b) $-3(z^2 - 4z - 3) + (z^2 + z)$

Solution

(a) Distribution of the minus sign and then combining like terms.

$(2z^3 + 4z^3 - 3z) - (z^3 + 6z^2 - z) + 17$
$= 2z^3 + 4z^3 - 3z - z^3 - 6z^2 + z + 17$
$= (2z^3 + 4z^3 - z^3) - 6z^2 + (-3z + z) + 17$
$= 5z^3 - 6z^2 - 2z + 17.$

(b) Distribution of the negative 3 and then combining like terms.

$-3(z^2 - 4z - 3) + (z^2 + z)$
$= -3z^2 + 12z + 9 + z^2 + z$
$= (-3z^2 + z^2) + (12z + z) + 9$
$= -2z^2 + 13z + 9.$

Products of polynomials are found by using the associative, distributive, and exponent properties.

Example 3

Multiply the polynomials

(a) $5V(2V - 3)$

(b) $(t + 3)(2t^2 - t + 1)$

(c) $(3I - 5)(I + 2)(I + 1)$

Solution

(a) $5V(2V - 3) = 5V(2V) - 5V(3) = 10V^2 - 15V$

(b) $(t + 3)(2t^2 - t + 1) = t(2t^2 - t + 1) +$
$3(2t^2 - t + 1) = (2t^3 - t^2 + 1) + (6t^2 - 3t + 3)$
$= 2t^3 + 5t^2 - 2t + 3$

(c) $(3I - 5)(I + 2)(I + 1)$
$= (3I - 5)[(I + 2)(I + 1)]$
$= (3I - 5)[I^2 + 3I + 2]$
$= 3I[I^2 + 3I + 2] - 5[I^2 + 3I + 2]$
$= 3I^3 + 9I^2 + 6I - 5I^2 - 15I - 10$
$= 3I^3 + 4I^2 - 9I - 10$

The division of two polynomials can be found using long division, similar to the long division of whole numbers.

Example 4

Divide $2x^3 + 3x^2 + x + 6$ by $x^2 - 1$

Solution

$$\begin{array}{r l}
 & 2x+3 \quad \leftarrow \text{Quotient} \\
\text{Divisor} \rightarrow x^2-1 \,\big)\!\! & 2x^3+3x^2+x+6 \quad \leftarrow \text{Dividend} \\
\text{Subtract} \rightarrow - & 2x^3 - 2x \\
\hline
\text{Subtract} \rightarrow - & 3x^2+3x+6 \\
 & 3x^2 - 3 \\
\hline
 & 3x+9 \quad \leftarrow \text{Remainder}
\end{array}$$

So

$$\frac{2x^3 + 3x^2 + x + 6}{x^2 - 1} = (2x + 3) + \frac{3x + 9}{x^2 - 1}$$

Note that, (Quotient) (Divisor) + Remainder = Dividend.

2.2 EXERCISES

Perform the indicated algebraic operations.

1. $(-V^2 - 5V + 1) + (2V^2 + 6V - 1)$
2. $(9V^3 + 3V^2 - 7V) - (2V^3 - 6V^2 - V - 3)$
3. $3(2I^2 - 5I + 7) + 2(-I^2 + 2I - 3)$
4. $-4(I^2 + 5I - 1) - 2(-I^2 - 3I + 6)$
5. $-2t(3t^3 + 4t^2 - 5t - 6)$
6. $3t(2t^{1/2} - 4t^2 + 3t - 5)$
7. $(5q + 2)(3q + 4)$
8. $(q - 1)(2q + 3)$
9. $(2r + t)^2$
10. $(r - 2t)^2$
11. $(2k + 1)(3k^2 + 4k - 1)$
12. $(k - 1)(2k^2 - 3k + 8)$
13. $(p + m + n)(3p - 4m - 2n)$
14. $(p + 2m - n)(p - 3m - 6n)$
15. $(p + 1)(2p + 3)(p - 2)$
16. $(4p + 5)(3p - 2)(2p - 1)$
17. $\dfrac{z^3 + 2z^2 + z + 6}{z + 1}$
18. $\dfrac{5z^3 - 2z^2 - z + 9}{z^2 + 1}$

2.3 RATIO, PROPORTION, AND VARIATION

2.3.1 Ratio

A ratio (:) is the relation between two like numbers. For example, the ratio 5:7 can be written as a fraction, 5/7, or as division 5 ÷ 7. We read the ratio 5:7 as 5 to 7 or 5 per 7.

Definition of ratio
A ratio is a comparison of two numbers.

A ratio can be expressed by a fraction $\left(\frac{a}{b}\right)$ or by a colon (:). So, the ratio of two numbers a and b can be written as a:b which is the same as the fraction $\frac{a}{b}$, where $b \neq 0$.

Example 1
Reduce the ratio 20:35 to the lowest term.

Solution
20:35 = 4/7.

Example 2
Reduce the ratio 3/5:2/7 to the lowest term.

Solution
3/5:2/7 = 3/5 ÷ 2/7 = 3/5 × 7/2 = 21/10.

Example 3
Express the following statement in a ratio form: "5 out of every 17 students in the class have a new car."

Solution
The ratio is $\frac{5}{17}$, same as 5 to 17 or 5:17.

Example 4
Find the ratio of the following:

(a) 3 to 9
(b) 7 to 13
(c) 11 to 51
(d) 17 to 99

Solution

(a) $\frac{3}{9} = \frac{1}{3}$

(b) $\frac{7}{13}$

(c) $\frac{11}{51}$

(d) $\frac{17}{99}$

Ratios can be used to find unit price or how much a service item costs.

Example 5

If 6 apples cost \$3.00, find the unit price.

Solution

This is done by putting the quality on the top and the price on the bottom, then simplify, and the quality becomes 1.

$$\frac{6 \text{ apples}}{3 \text{ dollars}} = \frac{2 \text{ apples}}{1 \text{ dollar}} = \frac{2 \text{ apples}}{\$1.00} = \frac{1 \text{ apple}}{\$0.50}.$$

So the unit price of these apples is 50 cents.

2.3.2 Proportion

A proportion (:: or $\propto$ or $=$) is a statement for two ratios that are equal. It can be written in two ways:

Definition of proportion

A proportion is an equality of two ratios.

(1) 2/3 :: 4/6, that is read, 2 is to 3 as 4 is to 6,

(2) 2:3 = 4:6 or 2/3 = 4/6, that is read, 2/3 equal 4/6.

For example, the statement $\frac{1}{5} = \frac{5}{25}$ is a proportion.

It is necessary to use the basic rules of proportions for $\frac{a}{b} = \frac{c}{d}$ as below:

1. $a \times d = b \times c$: cross multiplication, still makes them equal
2. $\frac{b}{a} = \frac{d}{c}$: by inversing of both sides, still makes them equal
3. $\frac{a}{c} = \frac{b}{d}$: by replacing the denominator of the first with the numerator of the second, still makes them equal.

4. $\dfrac{a+b}{b} = \dfrac{c+d}{d}$:

5. $\dfrac{a-b}{b} = \dfrac{c-d}{d}$:

Example 1

Find the value of V for the following proportions:

(a) $\dfrac{V}{3} = \dfrac{2}{5}$

(b) $\dfrac{5}{7} = \dfrac{V}{6}$

(c) $\dfrac{5}{V} = \dfrac{8}{2}$

(d) $\dfrac{2}{9} = \dfrac{3}{V}$

Solution

(a) $\dfrac{V}{3} = \dfrac{2}{5}$

Using cross multiplications rule $\Rightarrow 5 \times V = 2 \times 3$

$$5V = 6, \text{ Divide both sides by } 5$$

$$\frac{\cancel{5}V}{\cancel{5}} = \frac{6}{5}$$

$$V = \frac{6}{5}$$

(b) $\dfrac{5}{7} = \dfrac{V}{6}$

Using cross multiplications rule $\Rightarrow 5 \times 6 = 7 \times V$

$$30 = 7V, \text{ Divide both sides by } 7$$

$$\frac{30}{7} = \frac{\cancel{7}V}{\cancel{7}}$$

$$V = \frac{30}{7}$$

(c) $\dfrac{5}{V} = \dfrac{8}{2}$

Using cross multiplications rule $\Rightarrow 5 \times 2 = 8 \times V$

$$10 = 8V, \text{ Divide both sides by 8}$$

$$\frac{\not{1}0^5}{\not{8}^4} = \frac{\not{8}V}{\not{8}}$$

$$V = \frac{5}{4}$$

(d) $\dfrac{2}{9} = \dfrac{3}{V}$

Using cross multiplications rule $\Rightarrow 2 \times V = 3 \times 9$

$$2V = 27, \text{ Divide both sides by 2}$$

$$\frac{\not{2}V}{\not{2}} = \frac{27}{2}$$

$$V = \frac{27}{2}.$$

Example 2

Solve the following proportions.

(a) $\dfrac{9}{5} = \dfrac{t+2}{3}$

(b) $\dfrac{t-1}{2} = \dfrac{11}{6}$

Solution

(a) $\dfrac{9}{5} = \dfrac{t+2}{3}$

Using cross multiplications rule $\Rightarrow 9 \times 3 = 5 \times (t+2)$

$$27 = 5t + 10$$

$$27 - 10 = 5t + \not{10} - \not{10}$$

$$\frac{17}{5} = \frac{\not{5}t}{\not{5}}$$

$$t = \frac{17}{5}$$

(b) $\dfrac{t-1}{2} = \dfrac{11}{6}$

$$\text{Using cross multiplications rule} \Rightarrow 6 \times (t-1) = 2 \times 11$$
$$6t - 6 = 22$$
$$6t - 6 + 6 = 22 + 6$$
$$\frac{6t}{6} = \frac{28^{14}}{6^{3}}$$
$$t = \frac{14}{3}.$$

There are two kinds of proportion:

1. *Direct proportion*: is when two quantities are related such that when one of them increases this causes the other to increase or when one of them decreases this causes the other to decrease. For example, the greater the length, the greater the area.
2. *Inverse proportion*: is when two quantities are related such that when one of them increases this causes the other to decrease or when one of them decreases this causes the other to increase. For example, the greater the volume, the less the density.

2.3.3 Variation

Definition of variation
A variation is a relationship between two variables in which one is a constant multiple of the other.

Variation has two types, direct and inverse variations.

Definition of direct variation
A direct variation is a relationship between two variables in which one is a constant multiple of the other. When one variable is changed, the other changes in proportion to the first.

If y is directly proportional to x written as $y\alpha x$, then $y = kx$, where k is called the constant of variation.

Example 1
Show that in equation $y = 3x$, the variable y is directly proportional to x with variation constant equal to 3.

Solution

x	y
1	3
2	6
3	9
4	12

We see that in the table every time we increase x, we note that y increases 3 times. Doubling x causes y to double, and tripling x causes y to triple.

Example 2

If y varies directly with respect to x and $y = 5$, when $x = -3$, find y when $x = 12$.

Solution

Let $y_1 = 5$, $x_1 = -3$, $x_2 = 12$, and we want to find y_2. We can use the proportion based on what is given $\frac{y_1}{x_1} = \frac{y_2}{x_2}$. Because y varies directly with respect to x, that can lead to two equations, first, $y_1 = kx_1$, where k is a constant and $k = \frac{y_2}{x_2}$. Second, $y_2 = kx_2$, where $k = \frac{y_1}{x_1}$.

So substituting the values in $\frac{y_1}{x_1} = \frac{y_2}{x_2}$ can lead to the proportions $\frac{5}{-3} = \frac{y_2}{12}$.
$\frac{5}{-3} = \frac{y_2}{12}$, Multiplying both side by 12

$$\cancel{12}^{4} \times \left(\frac{5}{\cancel{-3}^{-1}}\right) = \left(\frac{y_2}{\cancel{12}}\right) \times \cancel{12} \Rightarrow y_2 = -20.$$

This means that when $x = 12$, $y = -20$.

Definition of joint variation
A joint variation is a relationship between one variable and a set of variables in which the one variable is directly proportional to each variable taken one at a time.

If z is directly proportional to x and y written as $z \alpha xy$, then $z = kxy$, where k is called the constant of variation.

Example 3

Show that in equation $z = 2xy$, the variable z is jointly proportional to x and y with variation constant equal to 2.

Solution

x	y	z
1	1	2
2	1	4
1	2	4
2	2	8

We see that in the table every time we double x, that z will double. Also, doubling y causes z to double, and doubling both x and y causes z to increase four times as great as x or y.

Definition of inverse variation

A direct variation is a relationship between two variables in which the product is a constant. When one variable increases the other decreases in proportion so that the product is unchanged.

If y is inversely proportional to x written as $y \alpha \frac{1}{x}$, then $y = \frac{k}{x}$, where k is called constant of variation.

Example 4

Show that in equation $y = \frac{2}{x}$, the variable y is inversely proportional to x with variation constant equal to 2.

Solution

x	y
1	2
2	1
3	$\frac{2}{3}$
4	$\frac{1}{2}$

We see that in the table every time we increase x, y decreases with unchangeable variation constant equal to 2.

2.3 EXERCISES

Reduce each ratio to the lowest terms for problems 1–4.

1. 15:30
2. 6/3
3. ½ ::3/5
4. 3/7:4/6

For problems 5–10, find the ratio of the following.

5. 3 to 11
6. 2 to 9
7. 4 to 14
8. 5 to 15
9. 17 to 5
10. 13 to 2

For problems 11–22, find the value of V in the proportions.

11. $\frac{7}{3} = \frac{V}{2}$
12. $\frac{1}{4} = \frac{V}{3}$
13. $\frac{V}{2} = \frac{3}{5}$
14. $\frac{V}{3} = \frac{6}{11}$
15. $\frac{3}{2} = \frac{9}{V}$
16. $\frac{13}{5} = \frac{2}{V}$
17. $\frac{15}{V} = \frac{2}{7}$
18. $\frac{5}{V} = \frac{9}{2}$
19. $\frac{V+3}{2} = \frac{1}{5}$
20. $\frac{V-2}{5} = \frac{7}{4}$

21. $V{:}4 = 5{:}13$
22. $V{:}5 = 7{:}3$
23. Suppose 145 gallons of oil flow through a feeder pipe in 5 min. Find the flow rate in gallons per minute.
24. Suppose 180 gallons of oil flow through a feeder pipe in 8 min. Find the flow rate in gallons per minute.
25. A transformer has a voltage of 6 volts in the primary circuit and 3850 volts in the secondary circuit. Determine the ratio of the primary voltage to the secondary voltage.
26. A transformer has 65 turns in the primary coil and 640 in the secondary coil. Determine the ratio of the secondary turns to primary turns.

Identify the direct proportion and inverse proportion between the following for problems 27–30.

27. Speed and time
28. Speed and distance
29. Temperature of gas and volume
30. Volume and density
31. If $\frac{x}{y} = \frac{z}{t}$, $x + y = 40$, $z = 5$, and $t = 4$, find y.
    Hint: rule number 4, or by rule 1 and adding $x + y$ both sides, then factoring.
32. If y varies directly with respect to the square of x and $y = 7$ when $x = -1$, find y when $x = 6$.
33. If y varies inversely with respect to x and $y = 3$ when $x = -2$, find y when $x = 2$.
34. If y varies inversely with respect to x and $y = 5$ when $x = -3$, find y when $x = 1$.

CHAPTER 2 REVIEW EXERCISES

Evaluate each expression in problems 1–4.

1. $x + 4y + 3z$ if $x = 1$, $y = -1$, and $z = 1/3$.
2. $-5x - 2y - z^2$ if $x = 3$, $y = 2$, and $z = 5$.
3. $\frac{x}{2} - y + \frac{t}{3} - 1$ if $x = 2$, $y = 2$, $t = 3$.
4. $x^3 - z_1 + k$ if $x = -2$, $z_1 = 10$, $k = -5$.

Simplify each expression in problems 5–14.

5. $\dfrac{\left(\frac{x}{t}\right)^3}{t^{-2}}$

6. t^3t^8

7. $(-3z^5)(2z^3)$

8. $(3x^2y^5z)^3$

9. $3(z+2)-7-(8y+1)$

10. $-3x^2(2x^3-1)$

11. $\dfrac{1}{3}(12x-9)$

12. $(2xy^2z^3)(-3x^{-3}y^{-7}z^{-4})$

13. $\dfrac{-4x^3y^{-1}}{2x^{11}y^{-4}}$

14. $\dfrac{9x^5y^{11}}{3x^2y^3}$

In problems 15–20, classify each expression as a monomial, binomial, trinomial, or neither. Then find the degree (order) of the polynomial.

15. $4x^2y^5-7z^9$

16. $z^3x+3xy-4$

17. $z^3x+3xtz$

18. xyz

19. $z^3+xy+xt-2n$

20. $z^2-\dfrac{2}{t}+1$

In problems 21 and 22, express the following expression in a single polynomial.

21. $(z^2-t)+(2z^2-3t)$

22. $(z^2-t)(2z^2-3t)$

In problems 23–26, perform each division.

23. $\dfrac{2x^2z^3}{2xz}$

24. $\dfrac{2x^2+7x-1}{x+1}$

25. $\dfrac{x - 2x^2}{(1 - 2x)}$

26. $\dfrac{x^2 - 1}{(x - 1)}$

In problems 27–30, write the following expressions in their simplest forms.

27. $7x + 11x$

28. $3t - t$

29. $6z - 2z + 4z + z$

30. $11m - 2n + 3m - 12n$

In problems 31 and 32, when $t = 3$, $w = 1$ find the numerical values of the expressions.

31. $2t - 1$

32. $3w + t + 5$

For problems 33–38, write the following in their simplest forms.

33. $2a \times 4$

34. $3x \times 5y$

35. $x^2 \times x^9$

36. $(2x)^3$

37. $5t \times 4t^4$

38. $3zt \times 8z^2t$

For problems 39 and 40, find the numerical values.

39. $3x^2 \times 4x$ when $x = 2$

40. $(4tz)^2 - t$ when $t = 1$, $z = 2$

Identify the direct proportion and inverse proportion between the following for problems 41 and 42.

41. The current that flows in a circuit and the applied voltage of the circuit.

42. The current that flows in a circuit and the resistance of that circuit.

Chapter 3

Equations, Inequalities, and Modeling

The more you have, the more you're occupied; the less you have, the more free you are.

Mother Teresa

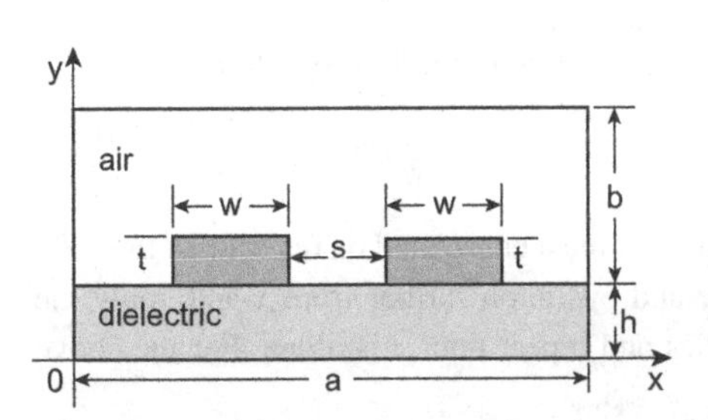

Microstrip lines are the most commonly used in all planar circuits despite the frequency ranges of the applied signals. Quasi-static analysis of microstrip lines involves evaluating them as parallel plate transmission lines, supporting a pure "TEM" mode. Advances in microwave solid-state devices have stimulated interest in the integration of microwave circuits. Today, microstrip transmission lines have attracted great attention and interest in microwave integrated circuit applications. This creates the demand for accurate modeling and simulation of microstrip transmission lines. The above model used to calculate the capacitance coupling of double-strip shielded transmission lines uses COMSOL, a finite element software.

INTRODUCTION

In this chapter, we introduce the concept of equations and their solution with application to modeling. The ability to use equations and formula is essential in all types of technical works.

Fundamentals of Technical Mathematics. http://dx.doi.org/10.1016/B978-0-12-801987-0.00003-4

3.1 EQUATIONS

Most of the time building an equation can help in finding a solution quickly and clearly.

Definition of equation
An equation is a statement that the expressions on either side of the equals sign (=) are equal.

Example 1
$5x - 8 = 0$ is an equation because the $5x - 8$ is same as 0.

Definition of the solution of equation
A solution of equation is the numbers that produce true statement for the equation.

To find the solution of an equation x must be isolated on one side of the "="; to do so, identify the number and operation further from x and apply the opposite operation to both sides and repeat until x is alone. For the above example, the solution of x is $x = 8/5$.

Definition of linear equation
A linear equation with one variable (first degree equation) is an equation that can be written in form: $ax + b = 0$, where a, and b are real numbers and $a \neq 0$.

Example 2
Solve the following equations:

(a) $x - 3 = 1$
(b) $x + 2 = 5$
(c) $x + 7 = 2$
(d) $\frac{x}{3} = 8$
(e) $\frac{5}{2}x = 15$
(f) $\frac{6}{7}x - 2 = 5$
(g) $2(x - 3) = 5(x + 2)$

Solution

(a) $x - 3 = 1$

$x = 1 + 3$

$x = 4$

(b) $x + 2 = 5$

$x = 5 - 2$

$x = 3$

(c) $x + 7 = 2$

$x = 2 - 7$

$x = -5$

(d) $\frac{x}{3} = 8$

$x = 8 \times 3$

$x = 24$

(e) $\frac{5}{2}x = 15$

$x = \cancel{15}^{3} \times \frac{2}{\cancel{5}^{1}}$

$x = 6$

(f) $\frac{6}{7}x - 2 = 5$

$\frac{6}{7}x = 5 + 2 = 7$

$x = 7 \times \frac{7}{6}$

$x = \frac{49}{6}$

(g) $2(x - 3) = 5(x + 2)$

$2x - 6 = 5x + 10$

$-10 - 6 = 5x - 2x$

$-16 = 3x$

$x = \frac{-16}{3}$

3.1.1 Constructing models to solve problems with one variable

Mathematical models are based on a process of solving applied problems. Most applied problems are described verbally and need to be translated to mathematical expressions.

A helpful process to solve technical problems is:

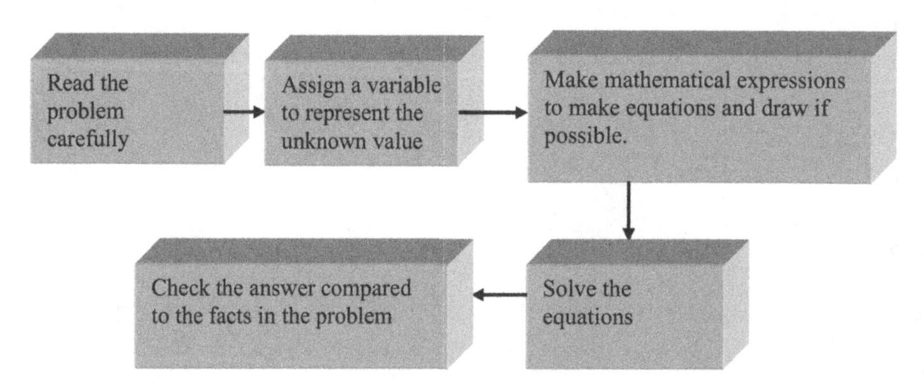

Example 1

Seven minus two more than a number is three. What is the number?

Solution

Let $x =$ the number.

Two more than a number $\longrightarrow 2 + x$

Seven minus two more than a number $\longrightarrow 7 - (2 + x)$

Seven minus two more than a number is three $\longrightarrow 7 - (2 + x) = 3$

What is the number? $\longrightarrow$ Find x

Solve $7 - (2 + x) = 3$

$7 - 2 - x = 3 \longrightarrow x = 2$

Check $\longrightarrow 7 - (2 + 2) = 3$

Example 2

An optical plate is in a rectangular shape with a length of 3 inches more than its width and has a perimeter of 14 inches. Find the dimensions.

Solution

Let $x =$ the width of the rectangle

The length $= x + 3$

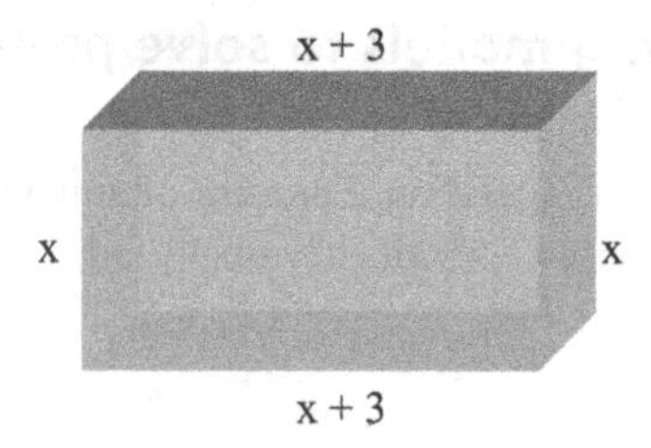

Remember that the perimeter of a rectangle is $2 \times$ length $+ 2 \times$ width.
Thus, the perimeter of the rectangle $= 2(x) + 2(x + 3) = 14$
Find the dimensions = find width (x) and length ($x + 3$).
Solve $2(x) + 2(x + 3) = 14 \longrightarrow 2x + 2x + 6 = 14$
$\longrightarrow 4x = 8 \longrightarrow x = 8/4 = 2$ inches.
The width of the rectangle = 2 inches.
The length = width + 3 = 5 inches.

3.1.2 **Equations with two variables**

Here we extend the equations from one variable to two variables.

Definition of equation with two variables
An equation with two variables is a statement that has two expressions with two variables and the expressions are equal.

Example 1
Solve the following equations:

(a) $2y - 3x = y + 9$

(b) $7\left(\frac{y}{3} + 5x\right) = 5y - 1$

Solution
(a) $2y - 3x = y + 9$

$2y - y = 3x + 9$

$y = 3x + 9$

(b) $7\left(\frac{y}{3} + 5x\right) = 5y - 1$

First we distribute 7 inside the parentheses:

$\frac{7}{3}y + 35x = 5y - 1$

Solve for y, by putting y on one side of the equation and the x and constants on the other side:

$\frac{7}{3}y - 5y = 35x - 1$

Factor y:

$$y\left(\frac{7}{3}-5\right) = 35x - 1$$

Simplify the fraction:

$$y\left(\frac{-8}{3}\right) = 35x - 1$$

Multiply both sides by $-3/8$.

$$\frac{-3}{8}\left[y\left(\frac{-8}{3}\right) = 35x - 1\right]$$

$$y = \frac{-105}{8}x + \frac{3}{8} = \frac{-105x + 3}{8}$$

3.1.3 Quadratic equations

Definition of quadratic equation

A quadratic equation is a second order equation written as $ax^2 + bx + c = 0$ where a, b, and c are coefficients of real numbers and $a \neq 0$.

Definition of quadratic formula

The quadratic formula is a general formula used for solving the quadratic equation:

$$x = \frac{-b \pm \sqrt{b^2 - 4ac}}{2a}.$$

Note that: if $x^2 = k$, where $k \geq 0$, then $x = \sqrt{k}$ or $-\sqrt{x}$ same as $x = \pm\sqrt{k}$, where $\pm$ means "plus or minus."

Example 1

Solve the following equations:

(a) $2x^2 + 3x - 4 = 0$

(b) $x^2 + 5x - 3 = 0$

Solution

(a) $2x^2 + 3x - 4 = 0$

The coefficients of the equations are:

$a = 2, b = 3, c = -4$

$$\text{So, } x = \frac{-3 \pm \sqrt{3^2 - 4\cdot2\cdot(-4)}}{2\cdot2} = \frac{-3 \pm \sqrt{9+32}}{4} = \frac{-3 \pm \sqrt{41}}{4}$$

(b) $x^2 + 5x - 3 = 0$

The coefficients of the equations are:

$a = 1, b = 5, c = -3.$

$$\text{So, } x = \frac{-5 \pm \sqrt{5^2 - 4\cdot1\cdot(-3)}}{2\cdot1} = \frac{-5 \pm \sqrt{25+12}}{2}$$

$$= \frac{-5 \pm \sqrt{37}}{2}.$$

3.1.4 Equations with rational, radical, and absolute value

3.1.4.1 Rational equation

Definition of rational equation
A rational equation is an equation that contains a rational expression for one or more terms.

Example 1
Solve the following equations:

(a) $\dfrac{4}{x-3} = 5$

(b) $\dfrac{2x}{x+6} = 5$

(c) $\dfrac{2x}{x+1} + \dfrac{3}{x} = 2$

(d) $\dfrac{x+1}{2} + \dfrac{3x}{x-2} = 4$

(e) $\dfrac{x+1}{3} - \dfrac{5}{x} = 0$

Solution

(a) $(x-3)\times\left[\dfrac{4}{x-3}=5\right]\Rightarrow(x-3)\times\dfrac{4}{x-3}=(x-3)\times 5$

$$\cancel{(x-3)}\times\frac{4}{\cancel{(x-3)}}=5x-15$$

$$4=5x-15\Rightarrow 5x=19$$

$$x=\frac{19}{5}$$

(b) $\dfrac{2x}{x+6}=5$

$$(x+6)\times\left[\frac{2x}{x+6}=5\right]$$

$$\cancel{(x+6)}\times\frac{2x}{\cancel{(x+6)}}=(x+6)\times 5$$

$$2x=5x+30$$

$$-30=5x-2x$$

$$-30=3x$$

$$x=-\frac{30}{3}=-10$$

(c) $\dfrac{2x}{x+1}+\dfrac{3}{x}=2$

$$(x+1)x\times\left[\frac{2x}{x+1}+\frac{3}{x}=2\right]$$

$$\cancel{(x+1)}x\times\frac{2x}{\cancel{(x+1)}}+(x+1)\cancel{x}\times\frac{3}{\cancel{x}}=(x+1)x\times 2$$

$$2x^2+3x+3=2x^2+2x$$

$$\cancel{2x^2}+3x-\cancel{2x^2}-2x=-3$$

$$x=-3$$

(d) $\dfrac{x+1}{2}+\dfrac{3x}{x-2}=4$

$$2(x-2)\times\left[\frac{x+1}{2}+\frac{3x}{x-2}=4\right]=(x+2)(x+1)+6x=8(x+2)$$

$$\Rightarrow x^2+3x+2+6x=8x+16\Rightarrow x^2+x-14=0$$

$$x=\frac{-1\pm\sqrt{1-4\cdot1\cdot(-14)}}{2\cdot1}=\frac{-1\pm\sqrt{1+56}}{2}=\frac{-1\pm\sqrt{57}}{2}$$

(e) $\frac{x+1}{3} - \frac{5}{x} = 0$

Multiply both sides of the equation by $3x$:

$$3x\left(\frac{x+1}{3} - \frac{5}{x} = 0\right) \quad \Rightarrow \quad x^2 + x - 15 = 0$$

Solve the quadratic equation for x:
$a = 1, b = 1, c = -15$

$$x = \frac{-1 \pm \sqrt{1 - 4 \cdot 1 \cdot (-15)}}{2 \cdot 1} = \frac{-1 \pm \sqrt{121}}{2}$$

3.1.4.2 Radical equation

Equations can have unknowns in a radical form.

Definition of a radical equation
A radical equation is an equation that contains variables in a square root, cube root, and so on.

Example 1
Find the real solutions of the following equations:

(a) $\sqrt{x+1} = 3$

(b) $\sqrt{x-2} - 3 = 0$

(c) $\sqrt{\frac{2}{3}x - 1} + 5 = 0$

Solution

(a) $\sqrt{x+1} = 3$

We rewrite the square root in fraction form:

$$(x+1)^{\frac{1}{2}} = 3$$

We remove the square root from x to keep it in first order by squaring both sides of the equation:

$$\left[(x+1)^{\frac{1}{2}}\right]^2 = 3^2$$

$x + 1 = 9$

$x = 8$

(b) $\sqrt{x-2} - 3 = 0$

First we put x on one side and the constant on the other side:

$$\sqrt{x-2} = 3$$

We rewrite the square root in fraction form:

$$(x-2)^{1/2} = 3$$

We remove the square root from x to keep it in first order by squaring both sides of the equation:

$$[(x-2)^{1/2}]^2 = [3]^2 \quad \Rightarrow \quad (x-2) = 9$$

Solve for x:

$$(x-2) = 9 \quad \Rightarrow \quad x = 11$$

(c) $\sqrt{\frac{2}{3}x - 1} + 5 = 0$

First we put x on one side and the constant on the other side:

$$\sqrt{\frac{2}{3}x - 1} = -5$$

We rewrite the square root in fraction form:

$$\left(\frac{2}{3}x - 1\right)^{1/2} = (-5)$$

We remove the square root from x to keep it in first order by squaring both sides of the equation:

$$\left(\frac{2}{3}x - 1\right)^{\frac{1}{2}\cdot 2} = (-5)^2 \quad \Rightarrow \quad \frac{2}{3}x - 1 = 25$$

Solve for x:

$$3\left(\frac{2}{3}x - 1 = 25\right) \quad \Rightarrow \quad 2x - 3 = 75 \quad \Rightarrow \quad 2x = 75 + 3$$

$$x = \frac{78}{2} = 39$$

3.1.4.3 Absolute value equation

Definition of absolute value
An absolute value equation is an equation that contains an absolute value expression for one or more terms. If a is a positive real number and u is any algebraic expression where the absolute value of u is $|u|$ equal to a, then $|u| = a \leftrightarrow u = a$ or $u = -a$.

$\leftrightarrow$: Equivalent to

Example 1

Solve the following equations:

(a) $|x - 3| = 7$

(b) $|x^2 - 9| = 6$

Solution

(a) There are two possible solutions:

$$x - 3 = 7 \quad \text{or} \quad x - 3 = -7$$

$$x = 10 \quad \text{or} \quad x = -4$$

So the solution set for x is $\{-4,10\}$.

(b) There are two possible solutions:

$$x^2 - 9 = 6 \qquad x^2 - 9 = -6$$

$$x^2 = 15 \quad \text{or} \quad x^2 = 3$$

$$x = \pm\sqrt{15} \qquad x = \pm\sqrt{3}$$

So the solution set for x is $\{-\sqrt{15}, -\sqrt{3}, \sqrt{3}, \sqrt{15}\}$.

3.1 EXERCISES

For problems 1–18, solve the following equations.

1. $x - 7 = 2$

2. $y + 9 = -9$

3. $6 - x = 11$

4. $\frac{x}{2} = 5$

5. $3x + 1 = 13$

6. $\frac{5}{3}x - 1 = 5$

7. $6 - 2x = 5x$

8. $x + 5 = 9 - 3x$

9. $4x = x + 11$

10. $5x = 3x + 10$

11. $t - 5 = -2 + 13t$

12. $\frac{m}{2} + 3 = 4m$

13. $5 - 4p = 5p - 13$

14. $2(v - 7) = 3$

15. $\frac{3}{2}c - 1 = 5$

16. $18 = 4(t + 3)$

17. $3(f - 1) - 4(3f + 7) + 10 = 0$

18. $3(2g - 4) - 7 = 0$

For problems 19–22, write each phrase or sentence in algebraic symbols.

19. A number decreased by five
20. A number increased by nine
21. A number times seven
22. A number divided by two
23. If a current (I) in an electric circuit is 4 A, find the voltage (V) across a 8 Ω resistor (R) in the circuit.
    Hint: V = IR.
24. If a round table has a circumference of 50 inches, what is the diameter of the table?
    Hint: Circumference = $2\pi r$.
25. A fence needs to have a perimeter of 24 yards and the width needs to be 2 yards shorter than the length. Find the side of the fence.
26. A rectangular box with square ends has its length 20 cm greater than its breadth, and the total length of its edges is 200 cm. Find the width of the box and its volume.
27. If twice a certain number is increased by 7, the result is equal to 4 less than triple the number. Find the number.
28. If twice a certain number is increased by 11, the result is equal to 5 less than triple the number. Find the number.

29. Nine times a certain number is 71. What is the number?

30. Determine the length of a side of a square if the perimeter is 36 ft.

31. A 8.5 inch piece is cut from a 14-inch length of bar stock. Determine the length of the unused piece (no allowance for thickness of the cut).

32. The sum of two angles equals 120°. One angle is twice as large as the other. Determine the size of the smaller angle.

33. The power we use at home has a frequency of 60 Hz (50 Hz in Europe). Determine the period of this sine wave.

Hint: $T = \frac{1}{f}$.

34. The period of a signal is 200 ms. What is its frequency in kilohertz?

For problems 35–56, solve the equations.

35. $4y + 5x = y + 7$

36. $9y - 3x = y + 7$

37. $3\left(\frac{y}{3} + 2x\right) = 2y - 3$

38. $2\left(\frac{y}{3} + x\right) = y - 1$

39. $x^2 - 2x + 1 = 0$

40. $x^2 + 2x - 6 = 0$

41. $x^2 + 4x - 5 = 0$

42. $3x^2 - 2x + 1 = 0$

43. $\frac{2}{x - 5} = 1$

44. $\frac{x}{x + 3} = 2$

45. $\frac{x}{x + 1} + \frac{2}{x} = 4$

46. $\frac{x + 1}{3} + \frac{2x}{x - 3} = 6$

47. $\frac{x + 1}{3} - \frac{5}{x} = 0$

48. $\frac{x + 2}{2} - \frac{1}{x} = 0$

49. $6\sqrt{t + 6} = 4\sqrt{6t - 24}$

50. $\sqrt{t + 3} - 3 = 5$

51. $\sqrt{m - 3} - 8 = 0$

52. $\sqrt{m} - 7 = 18$

53. $|x-2| = 4$

54. $|x+6| = 9$

55. $|x^2 - 24| = 1$

56. $|x^2 + 4| = 2$

3.2 INEQUALITY EQUATIONS AND INTERVALS

We illustrate here how to solve inequality equations and apply intervals.

Definition of inequality equation

An inequality equation is an equation containing an expression in which one side is less than (<), less than or equal to (≤), greater than (>), greater than or equal to (≥).

Definition of interval

An interval is a part of a number line (a correspondence between the set of points of a line and the set of real numbers).

Interval notations (describing intervals using parentheses for no equal sign and brackets for equal sign included):

Let a (left endpoint) and b (right endpoint) be two real numbers where $a < b$.

1. A close interval, denoted by $[a,b]$, this is an interval notation, consists of all (∀) real numbers x for which $a \leq x \leq b$ (does include a and b), this is in an inequality notation as:

   In graph notation $-\infty$ ← a ● —— x —— ● b → ∞

   ∞ (Infinity): is not real number but a notation used to indicate unboundedness in the positive direction.
   −∞ (Minus infinity): is not real number but a notation used to indicate unboundedness in the negative direction.
2. An open interval, denoted by (a,b), this is an interval notation, consists of all (∀) real numbers x for which $a < x < b$ (does not include either a or b), this is an inequality notation.

   In graph notation $-\infty$ ← a ○ —— x —— ○ b → ∞

3. Half-open interval, or half-closed interval denoted by $(a,b]$ or $[a,b)$, this is an interval notation, consists of all ($\forall$) real numbers x for which $a < x \leq b$ (b is included but not a) or $a \leq x < b$ (a is included but not b), respectively; this is an inequality notation.

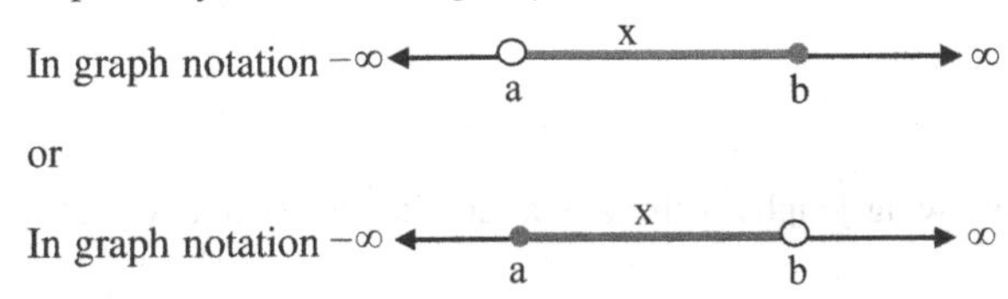

Summary table

Interval notation	Inequality notation	Graph notation
$(-\infty, \infty)$	$-\infty < x < \infty$ (all real numbers)	−∞ ← X → ∞
$[a,b]$	$a \leq x \leq b$	−∞ ← a ● X ● b → ∞
(a,b)	$a < x < b$	−∞ ← a ○ X ○ b → ∞
$(a,b]$	$a < x \leq b$	−∞ ← a ○ X ● b → ∞
$[a,b)$	$a \leq x < b$	−∞ ← a ● X ○ b → ∞
$[a,\infty)$	$x \geq a$	−∞ ← a ● X → ∞
(a,∞)	$x > a$	−∞ ← a ○ X → ∞
$(-\infty,b]$	$x \leq b$	−∞ ← X ● b → ∞
$(-\infty,b)$	$x < b$	−∞ ← X ○ b → ∞

Note: $-\infty$ ←————→ ∞ : is called number line

Note: Set notation is based on the inequality notation. A set is a collection of elements enclosed in braces { }.

Example 1

$x \geq 2$ can be written in set notation as $\{x|x \in R, x \geq 2\}$ a set of x such that x belongs to (contains) real number R where x is greater or equal to 2.

$|$: such that

$\in$: belongs

Example 2

Describe the following graphs with interval and inequality notations.

(a)

(b)

Solution

(a) interval notation, $[4,\infty)$; inequality notation, $x \geq 4$

(b) interval notation, $(-1,1]$; inequality notation, $-1 < x \leq 1$

Example 3

Write each inequality notation using interval notation.

(a) $-2 < x \leq 5$

(b) $7 \leq x \leq 11$

(c) $x \geq 2$

Solution

(a) $(-2,5]$

(b) $[7,11]$

(c) $[2,\infty)$

Properties of inequalities
a, *b*, and *c* are real numbers

Property	Symbol
Nonnegative	$a^2 \geq 0$
Addition and subtraction	If $a \leq b$, then $a + c \leq b + c$
	If $a \leq b$, then $a - c \leq b - c$
	If $a \geq b$, then $a + c \geq b + c$
	If $a \geq b$, then $a - c \geq b - c$

Multiplication and division	If $a \leq b, c > 0$, then $ac \leq bc$.
	If $a \geq b, c > 0$, then $ac \geq bc$.
	If $a \leq b, c < 0$, then $ac \geq bc$.
	If $a \geq b, c < 0$, then $ac \leq bc$.
	If $a \leq b, c > 0$, then $a/c \leq b/c$.
	If $a \geq b, c > 0$, then $a/c \geq b/c$.
	If $a \leq b, c < 0$, then $a/c \geq b/c$.
	If $a \geq b, c < 0$, then $a/c \leq b/c$.

Note: All these properties are true when the inequality $\leq$ is replaced by $<$, and when the inequality $\geq$ is replaced by $>$.

Example 4

If $-2x < 1$, then $x > -1/2$

Note: If we multiply or divide by a negative, we change the direction of the inequality sign.

3.2.1 **Solving linear inequality**

To solve a linear inequality equation, we define it as:

Definition of linear inequality

A linear inequality in one variable is an equality that can be written as:

$$ax + b \begin{pmatrix} > \\ \text{or} \\ \geq \\ \text{or} \\ < \\ \text{or} \\ \leq \end{pmatrix} 0$$

where a and b are real numbers with $a \neq 0$.

Example 1

Solve the inequality $5(x - 3) \geq 20$.

Solution

$$5(x-3) \geq 20 \longrightarrow (x-3) \geq 20/5 \longrightarrow x \geq 7$$

Example 2

Solve the inequality $2x + 3 \leq 5x - 1$, and graph the solution and write it in interval notation.

Solution

$2x + 3 \leq 5x - 1$

$3 + 1 \leq 5x - 2x \rightarrow 4 \leq 3x \rightarrow 4/3 \leq x \rightarrow x \geq 4/3$

The graph of the solution:

Interval notation is [4/3,∞).

Example 3

Solve the inequality $-3 < 6x - 4 < 2$, and draw the solution on the number line.

Solution

$-3 < 6x - 4 < 2 \rightarrow -3 + 4 < 6x < 2 + 4 \rightarrow 1 < 6x < 6 \rightarrow 1/6 < x < 1$ or $(1/6, 1)$

The graph:

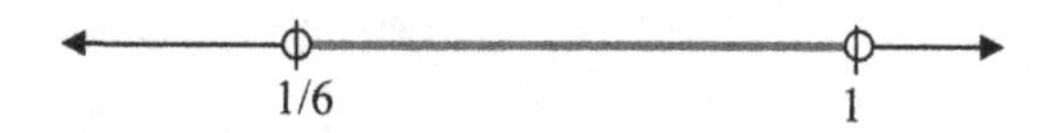

3.2.1.1 Inequality involving absolute value (| |)

Absolute values are used in equality equations.

Let $a > 0$ and u be any algebraic expression, then

1. $|u| \leq a \Leftrightarrow -a \leq u \leq a \Leftrightarrow (u \geq -a \text{ and } u \leq a)$
2. $|u| < a \Leftrightarrow -a < u < a \Leftrightarrow (u > -a \text{ and } u < a)$
3. $|u| \geq a \Leftrightarrow u \leq -a \text{ or } u \geq a$
4. $|u| > a \Leftrightarrow u < -a \text{ or } u > a$

Note: $\Leftrightarrow$: equivalent to

Example 1

Solve the inequality $|5x - 1| \leq 9$, graph the solution on the number line, and write the solution in interval form.

Solution

$|5x - 1| \leq 9 \rightarrow -9 \leq 5x - 1 \leq 9 \rightarrow -8 \leq 5x \leq 10 \rightarrow -8/5 \leq x \leq 2$

Graph form:

Interval form: $[-8/5, 2]$.

Example 2

Solve the inequality $|2x + 1| > 7$, and graph the solution on the number line, and write the solution in interval form.

Solution

$|2x + 1| > 7 \Leftrightarrow 2x + 1 > 7$ or $2x + 1 < -7$.
$2x + 1 > 7 \rightarrow 2x > 6 \rightarrow x > 3$ or $2x + 1 < -7 \rightarrow 2x < -8 \rightarrow x < -4$.

Graph form:

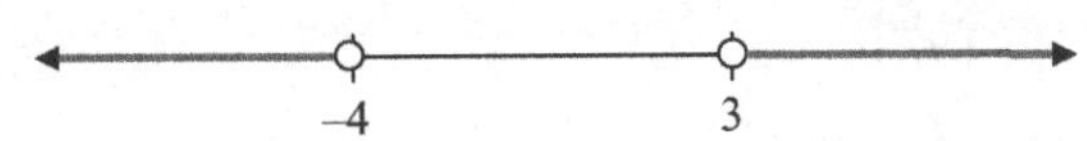

Interval form: $(-\infty, -4) \cup (3, \infty)$.

3.2 EXERCISES

For exercises 1–8, draw the number line for each interval. State whether the interval is an open, closed, half-open, or half-closed interval.

1. $[2,5]$
2. $[-4,-1]$
3. $(2,6)$
4. $(-1,-5)$
5. $(-4,0]$
6. $(2,6]$
7. $[3,5)$
8. $[1,3)$

For exercises 9–12, draw the number line for the equations.

9. $x \leq -5$

10. $x > 2$

11. $x < 1$

12. $x \geq 3$

For exercises 13–38, solve the equations. Give the answer using interval notation, and show the solution on a number line.

13. $y - 5 < 2$

14. $y + 2 < 7$

15. $1 - 3t \leq 9$

16. $6 - 2m \leq 8$

17. $2n - 9 > 11$

18. $3v + 2 > 8$

19. $3p + 23 \leq p - 5$

20. $4v + 8 \leq 10v - 6$

21. $2x + 4 < x + 6$

22. $-2x + 15 \geq 4x - 10$

23. $\frac{-4t}{3} - 2 < \frac{-t}{3}$

24. $-3x < -5x + 16$

25. $5m - 6 \leq 2m + 1$

26. $n + 2 \leq 2n - 7$

27. $|t| < 5$

28. $|t| < 15$

29. $|u| > 3$

30. $|u| > 11$

31. $|2x| < 12$

32. $|3x| < 24$

33. $|2x - 1| \leq 7$

34. $|x + 1| \leq 4$

35. $|x + 1| > 5$

36. $|x - 3| > 9$

37. $|x - 1| + 2 > 9$

38. $|x - 3| - 1 > -5$

39. Express the fact that T differs from 6 by less than 1 as an equality involving an absolute value. Solve for T.

40. Express the fact that T differs from 7 by more than 2 as an equality involving an absolute value. Solve for T.

3.3 COMPLEX NUMBERS

Complex numbers are used in engineering, science, and other technology applications, especially, in control engineering.

Definition of complex number

A complex number is a number that can be written in standard form as $a + bi$, where a and b are real numbers and i is an imaginary unit ($i = \sqrt{-1}$).

We call a the real part of the complex number and bi the imaginary part of the complex number. The imaginary unit (number) i is sometimes denoted by j.

For a complex number, z has the rectangular form as $z = x + yi$, where x and y are real numbers,

$x = \text{Re } z$ (real part of z),

$y = \text{Im } z$ (imaginary part of z)

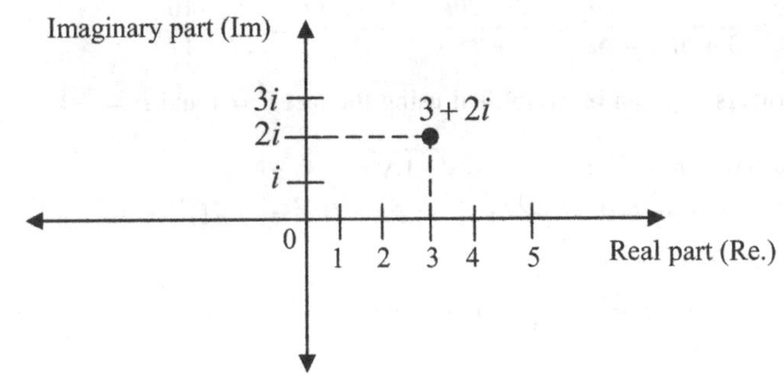

Plane diagram of complex number in rectangular form

For example, the following are complex numbers:

$$0+0i = 0 \quad 0+2i = 2i \quad 7+0i \quad 1+3i \quad 2-9i \quad 5+\frac{3}{8}i \quad \frac{2}{3}-i$$

To use complex numbers, we need to know their properties and rules.

The following are basic properties and rules for complex numbers:

1. $i^2 = -1 \rightarrow i = \sqrt{-1}$

2. $\sqrt{-c} = (\sqrt{c})i$, when $c > 0$.

3. $a+bi = c+di$, if and only if $a = c$ and $b = d$ (equality)

4. $(a+bi)+(c+di) = (a+c)+(b+d)i$ (addition)

5. $(a+bi)-(c+di) = (a-c)+(b-d)i$ (subtraction)

6. $(a+bi)(c+di) = ac+adi+bci+bdi^2$

$$= ac+adi+bci-bd$$

$$= (ac-bd)+(ad+bc)i \qquad \text{(multiplication)}$$

7. Conjugate of $a+bi$ is $a-bi$

8. $(a+bi)(a-bi) = a^2+b^2$

9. The reciprocal of $a+bi$ is $\frac{1}{a+bi}$, where $a+bi \neq 0$

To simplify the reciprocal of a nonzero complex number to a standard form, multiply the numerator and denominator by the denominator conjugate.

For example, $\frac{4}{3+5i}$ can be simplified as:

$$\frac{4}{3+5i} = \frac{4}{3+5i}\cdot\frac{3-5i}{3-5i} = \frac{12-20i}{9+25} = \frac{12-20i}{34} = \frac{6}{17}-\frac{10}{17}i$$

The powers of i can be simplified using the fact $i^1 = i$ and $i^2 = -1$.

$$i^1 = \sqrt{-1} = i \qquad i^2 = i.i = \sqrt{-1}.\sqrt{-1} = -1$$

$$i^3 = i^2\cdot i = (-1)\cdot i = -i \qquad i^4 = (i^2)^2 = (-1)^2 = 1$$

$$i^5 = (i^4)(i) = (i^2)^2(i) = (-1)^2 i = i$$

Note: $(-1)^{\text{even number}} = 1$; $(-1)^{\text{odd number}} = -1$.

Example 1

Simplify the following in standard form:

(a) $(5+3i)-(3+i)$

(b) $\sqrt{25}$

(c) $\sqrt{-9}$

(d) $\sqrt{-7}$

(e) $\sqrt{\dfrac{-1}{5}}$

(f) $-3(2-5i)$

(g) $(4+3i)+(-1+6i)$

(h) $(3-4i)(2+3i)$

(i) $\dfrac{2+i}{3i}$

(j) $\dfrac{5}{3+2i}$

(k) $\dfrac{3-2i}{5-7i}$

(l) i^{12}

(m) $\dfrac{1}{i^5}$

Solution

(a) $(5+3i)-(3+i)=(5-3)+(3-1)i=2+2i$

(b) $\sqrt{25}=5$

(c) $\sqrt{-9}=3i$

(d) $\sqrt{-7}=(\sqrt{7})i$

(e) $\sqrt{\dfrac{-1}{5}}=\dfrac{\sqrt{-1}}{\sqrt{5}}=\dfrac{i}{\sqrt{5}}=\dfrac{i}{\sqrt{5}}\cdot\dfrac{\sqrt{5}}{\sqrt{5}}=\dfrac{i\sqrt{5}}{5}$

(f) $-3(2-5i)=-6+15i$

(g) $(4+3i)+(-1+6i)=[4+(-1)]+(3+6)i=3+9i$

(h) $(3-4i)(2+3i)=3\cdot(2+3i)-4i(2+3i)=6+9i-8i-12i^2$

$$=6+i-12(-1)=6+i+12=18+i$$

(i) $\dfrac{2+i}{3i}=\dfrac{2+i}{3i}\cdot\dfrac{i}{i}=\dfrac{2i+i^2}{3i^2}=\dfrac{2i+(-1)}{3(-1)}=\dfrac{2i-1}{-3}=\dfrac{1}{3}-\dfrac{2}{3}i$

(j) $\frac{5}{3+2i} = \frac{5}{3+2i} \cdot \frac{3-2i}{3-2i} = \frac{15-20i}{3+4} = \frac{15}{7} - \frac{20}{7}i$

(k) $\frac{3-2i}{5-7i} = \frac{3-2i}{5-7i} \cdot \frac{5+7i}{5+7i} = \frac{15+21i-10i-14i^2}{25+49}$

$= \frac{29+11i}{74} = \frac{29}{74} + \frac{11}{74}i$

(l) $i^{12} = (i^2)^6 = (-1)^6 = 1$

(m) $\frac{1}{i^5} = i^{-5} = i^{-6} \cdot i = (i^2)^{-3} \cdot i = (-1)^{-3} \cdot i = -i$

3.3 EXERCISES

In problems 1–26, perform the indicated operations, and write each answer in standard form.

1. $(1 + 3i) + (2 + 6i)$

2. $(5 + i) + (7 + 2i)$

3. $(12 - 2i) + (-3 + 4i)$

4. $(-9 - 3i) + (17 - 8i)$

5. $(5 - 2i) - (7 + 3i)$

6. $(-13 - 2i) - (6 - 5i)$

7. $4i + (6 - 3i)$

8. $7 - (1 - 4i)$

9. $(3i)(9i)$

10. $(4i)(6i)$

11. $-3i(5 + 2i)$

12. $-2i(-1 - 5i)$

13. $(1 + 3i)(2 + i)$

14. $(2 - 5i)(4 - 3i)$

15. $(3 - i)(2 + 6i)$

16. $(6 + 7i)(-4 - 5i)$

17. $\dfrac{1}{1+3i}$

18. $\dfrac{7}{3+5i}$

19. $\dfrac{5}{2-4i}$

20. $\dfrac{4}{7-2i}$

21. $\dfrac{i}{2+3i}$

22. $\dfrac{5i}{1-2i}$

23. $\dfrac{3+i}{2+4i}$

24. $\dfrac{1-2i}{3-5i}$

25. $\dfrac{(3-\sqrt{-9})}{(2+\sqrt{-25})}$

26. $\dfrac{(7+\sqrt{-4})}{(1-\sqrt{-16})}$

Write problems 27–34 in standard form.

27. $\dfrac{1}{2i}$

28. $\dfrac{3}{4i}$

29. $\dfrac{1+5i}{3i}$

30. $\dfrac{7-i}{2i}$

31. $(1+3i)+2(3-i)^2+11$

32. $3(1-3i)+(2-5i)^2-7$

33. $(2-i)^2-3(4-i)-8$

34. $(3-2i)^2-4(5-3i)-9$

In problems 35–38, evaluate the expressions.

35. $z^2 + 3z - 1$ for $z = 1 + i$

36. $z^2 - 2z + 3$ for $z = 2 - i$

37. $2z^2 - z + 4$ for $z = 1 - i$

38. $3z^2 - 2z - 5$ for $z = 1 + 2i$

In problems 39–42, simplify the complex numbers.

39. i^{16}, i^{28}, i^{42}

40. i^{19}, i^{31}, i^{63}

41. $\frac{1}{i^{22}}, \frac{3}{i^{36}}, \frac{5}{i^{54}}$

42. $\frac{1}{i^{21}}, \frac{2}{i^{33}}, \frac{4}{i^{57}}$

In problems 43–48, what are the real values of x and y to make the following equations to be true?

43. $(3x + 1) + (2y - 5)i = 7 + 8i$

44. $(x + 1) + (4y + 2)i = 6 + i$

45. $(2x - 4) - (3y - 7)i = 9 - 5i$

46. $(x - 5) - (2y - 6)i = 1 - 3i$

47. $2x - (3y + 5)i = (4 + 3x) + (9y - 5)i$

48. $3x + (y - 7)i = (5 - 2x) - (4y + 6)i$

In problems 49–56, what are the real values of z to make the following expressions an imaginary number (unit)?

49. $\sqrt{2 - z}$

50. $\sqrt{7 - z}$

51. $\sqrt{5 + z}$

52. $\sqrt{8 + z}$

53. $\sqrt{2 - 4z}$

54. $\sqrt{5 - 3z}$

55. $\sqrt{11 + 4z}$

56. $\sqrt{10 + 3z}$

CHAPTER 3 REVIEW EXERCISES

Solve the following equations in problems 1–14.

1. $x + 5 = 13$

2. $z - 6 = 3$

3. $3y + 1 - 5y - 7 = 7y + 8 + y - 12$

4. $-5z = 11$

5. $4\frac{y}{2} = 3$

6. $\frac{z}{-3} = -11$

7. $7z - 3 = 2$

8. $5 - \frac{t}{3} = 9$

9. $\frac{4t}{7} - 1 = \frac{8t}{3} + 5$

10. $\frac{9s}{2} - 11 = s + 3$

11. $6(s - 4) = 3$

12. $7(t + 1) - 5 = 14$

13. $\frac{3z - 2}{5} = z + 1$

14. $\frac{3x}{5} + \frac{6x}{7} = 2$

Solve the following quadratic equations by factoring in problems 15–18.

15. $x^2 = 16$

16. $x^2 = 281$

17. $10z^2 - 100 = 60$

18. $9x^2 + 5x = 0$

Solve the following quadratic equations using the quadratic formula in problems 19–22.

19. $4x^2 + 9x - 11 = 0$

20. $y^2 - 4y - 5 = 0$

21. $3z^2 + 7z = 1$

22. $2z^2 + 5z = 0$

Solve the following equations in problems 23–30.

23. $\sqrt{x-6} = 2$

24. $\sqrt{x^2+1} = 3$

25. $\sqrt{x-2} + 2x = 5$

26. $\sqrt{3x-1} = x-2$

27. $\sqrt{x-1} = 2\sqrt{x+5}$

28. $\sqrt[3]{(2x+7)^2} = 9$

29. $|x-5| = 7$

30. $|3x+1| - 2 = 11$

Describe the following graphs with interval and inequality notations in problems 31–36.

31.

32.

6 8

33.

−2 5

34.

−3

35.

−7 −3

36.

−2 6

Write each inequality notation using interval notation in problems 37–44.

37. $-1 < x \leq 3$

38. $0 \leq x < 1$

39. $1 \leq x \leq 15$

40. $-5 < x < 1$

41. $x \geq 7$

42. $x > 19$

43. $x < 43$

44. $x \leq 65$

Solve for x, graph the solution on the number line, and write the solution in interval form in problems 45–60.

45. $\frac{x-2}{3} + \frac{4}{7} \leq 5$

46. $\frac{2x}{3} - \frac{5}{6} \geq 1$

47. $(x+3)(x-2) > 0$

48. $2x^2 - 3x \leq 12$

49. $\frac{x-4}{x+1} \geq 0$

50. $(x+5)(2x-1) \geq 0$

51. $x^2 - 3x \leq 14$

52. $x^2 - 4 \geq 0$

53. $x(x+1)(x-1) \geq 0$

54. $2x^2 \leq 4x$

55. $5 \leq x - 2 < 6$

56. $-12 \leq 8x + 3 \leq -2$

57. $|2x - 8| < 9$

58. $|x + 3| > 6$

59. $|x + 4| + 3 \geq 7$

60. $|9x - 2| - 1 \leq 8$

In problems 61–64, solve each equation in the complex standard form.

61. $x^2 + 9 = 0$

62. $x^2 + 16 = 0$

63. $x^2 + 3x + 9 = 0$

64. $3x^2 + 2x + 1 = 0$

For problems 65–76, simplify each expression in the standard form.

65. $(2 + 4i) + (5 + 9i)$

66. $(3 - i) - (1 - 5i)$

67. $4i(3 - 2i)$

68. $-3i(2 - 5i)$

69. $\dfrac{i}{2 + i}$

70. $\dfrac{3 + 2i}{1 + 4i}$

71. $\dfrac{6 - 7i}{3 - i}$

72. $\dfrac{1 - 3i}{2i}$

73. $(3 - 2i)^2 + 4(5 + i) - 7$

74. $(1 - i)^2 + 3(2 + i) - 9$

75. $(3 - 2i)(-4 + 7i)$

76. $(2 + i)(3 + 2i)$

For problems 77–82, simplify the following complex numbers.

77. i^{17}

78. i^{64}

79. i^{77}

80. i^{45}

81. i^{-11}

82. i^{-13}

For problems 83 and 84, find the value of x and y in order to make the following equations true statements.

83. $7 - 11i = (3x - 4) + (y + 6)2i$

84. $(8x - 3) + (7 - 2y)i = 2x + (3y - 1)i$

Chapter 4

Graphs and Functions

To be able to look back upon one's past life with satisfaction is to live twice.

Marcus Valerius Martial

■ Albert Einstein (1879—1955), a German physicist and mathematician. He discovered the hugely important and iconic equation, $E = mc^2$, which showed that energy and matter can be converted into one another. It was Einstein's wish that people should be respected for their humanity and not for their country of origin or religion.

Data visualization is a branch of computer graphics and user interface design that is concerned with presenting data to users by means of images. This field seeks ways to help users explore, make sense of, and communicate about data. It is an active research area, drawing on theory in information graphics, computer graphics, human—computer interaction and cognitive science.

Fundamentals of Technical Mathematics. http://dx.doi.org/10.1016/B978-0-12-801987-0.00004-6

INTRODUCTION

In this chapter, we introduce the basic concepts of graphs and functions.

4.1 GRAPHS

Graphs are used in many applications for scientist, engineers, and technologists. Graphs can be created by connecting specific points (coordinates).

4.1.1 Rectangular (Cartesian) coordinates

An ordered pair of a real number is a pair written in the form (x, y) where x and y are real numbers and are the coordinates of the order pair. The coordinates x and y are used to locate a point in the coordinate plane. The coordinates x and y are called rectangular coordinates.

The coordinate plane (rectangular coordinate system) is formed when horizontal and vertical number lines are placed perpendicular to each other with respect to their zeros (origin). The horizontal number line is called the **x-axis** and the vertical number line is called the **y-axis**. The intersection between x-axis and y-axis is called the **origin**. The coordinate plane is divided into four regions called **quadrants**.

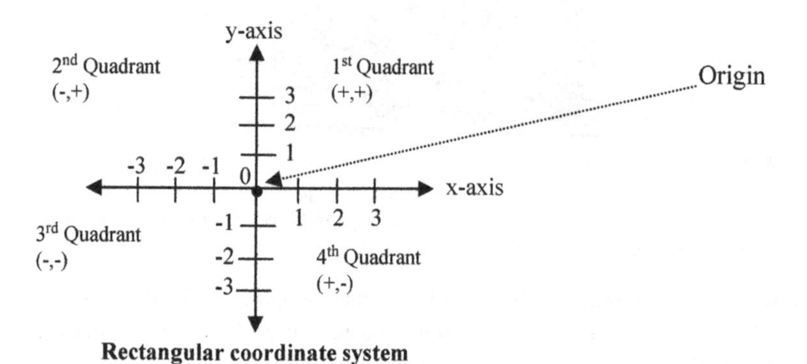

Rectangular coordinate system

Example 1

Plot the point P (2, 3).

Solution

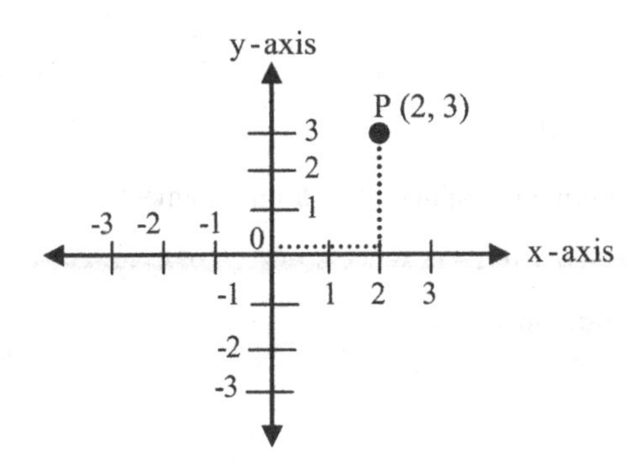

Example 2

Plot the point P (2, −3).

Solution

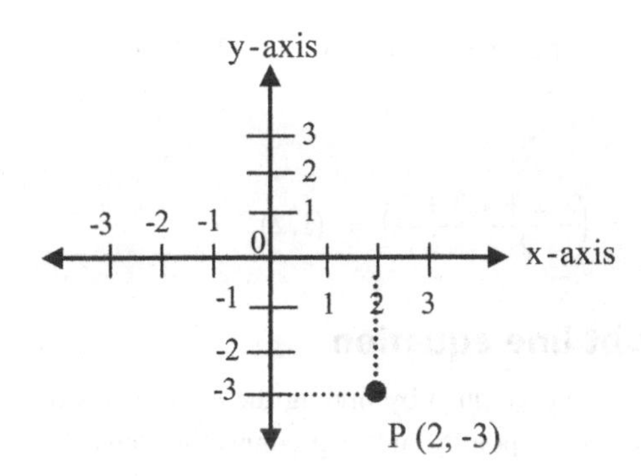

4.1.2 **Distance between points and midpoint**

Distance between points can be described in here.

> **Definition of distance between points**
> The distance d between the two points (x_1, y_1) and (x_2, y_2) is given by
>
> $$d = \sqrt{(x_2 - x_1)^2 + (y_2 - y_1)^2}$$

Example 1

Find the distance between (6, −2) and (−3, −5).

Solution

$$d = \sqrt{(-3-6)^2 + [-5-(-2)]^2} = \sqrt{(-9)^2 + (-3)^2} = \sqrt{81+9}$$
$$= \sqrt{90} = 3\sqrt{10}.$$

Now, we can describe the midpoint of a line segment.

Definition of midpoint
The midpoint $M = (x_m, y_m)$ of the line segment with endpoints (x_1, y_1) and (x_2, y_2) is given by

$$M = (x_m, y_m) = \left(\frac{x_1 + x_2}{2}, \frac{y_1 + y_2}{2}\right)$$

Example 2
Find the midpoint of a line segment from (–2, 3) to (4, 1).

Solution

$$M = (x_m, y_m) = \left(\frac{-2+4}{2}, \frac{3+1}{2}\right) = (1, 2)$$

4.1.3 Straight line equation

A straight line can be graphed by finding the coordinates of two points on the line, plotting those points, and then connecting them.

One important property of the graph of a straight line is its slope.

Definition of slope of a line
The slope m of a line passes through the points (x_1, y_1) and (x_2, y_2) is

$$m = \frac{\Delta y}{\Delta x} = \frac{y_2 - y_1}{x_2 - x_1} = \frac{\text{rise}}{\text{run}}, \text{ where } x_1 \neq x_2$$

Example 1
Find the slope of the line that passes through points (2, 7) and (8, 9).

Solution

$$\text{Slope} = m = \frac{\Delta y}{\Delta x} = \frac{y_2 - y_1}{x_2 - x_1} = \frac{9-7}{8-2} = \frac{2}{6} = \frac{1}{3}$$

Theorem 1: Straight line equation (standard form)
If A, B, and C are real numbers, and A and B are not both zero, then the graph of the equation $Ax + By = C$ is a straight line.

Also, any equation of the form $y = mx + b$ is a straight line, where m and b are real numbers. Where m is called the slope of the line; b is called the y-intercept (the point where the graph crosses the y-axis)

Example 2

Find the slope and y-intercept for the line equation $y = 3x + 1$

Solution

Slope $= m = 3$ and the y-intercept $= 1$

Theorem 2: A line equation passing a point
A line equation of a slope m that contains the point (x_1, y_1) is $y = m(x - x_1) + y_1$

Example 3

Find the line equation and its y-intercept with slope 5 that contains a point (3, 2).

Solution

$$y = m(x - x_1) + y_1$$

$$y = 5(x - 3) + 2 \longrightarrow y = 5x - 15 + 2 \longrightarrow y = 5x - 13$$

so, the y-intercept $= -13$.

A graph of an equation in two variables consists of the set of points in xy-plane whose coordinates (x, y) satisfy the equation.

Example 4

Determine the following points that are on the graph of the equation $3x + y = 4$.

(a) (0, 2)

(b) (−2, 10)

Solution

(a) Point (0, 2) means $x = 0$ and $y = 2$; substitute these values in the left expression of the equation
$3(0) + 2 = 2 \longrightarrow 2 \neq 4$ the equation is not satisfied, so the point (0, 2) is not on the graph.

(b) Point (−2, 10) means $x = -2$ and $y = 10$; substitute these values in the left expression of the equation
$3(-2) + 10 = 4 \longrightarrow 4 = 4$ the equation is satisfied, so the point (−2, 10) is on the graph.

To graph an equation by plotting points, it will be helpful to find first the x-intercept and y-intercept points.

x-intercept $\longrightarrow$ make $y = 0$
y-intercept $\longrightarrow$ make $x = 0$

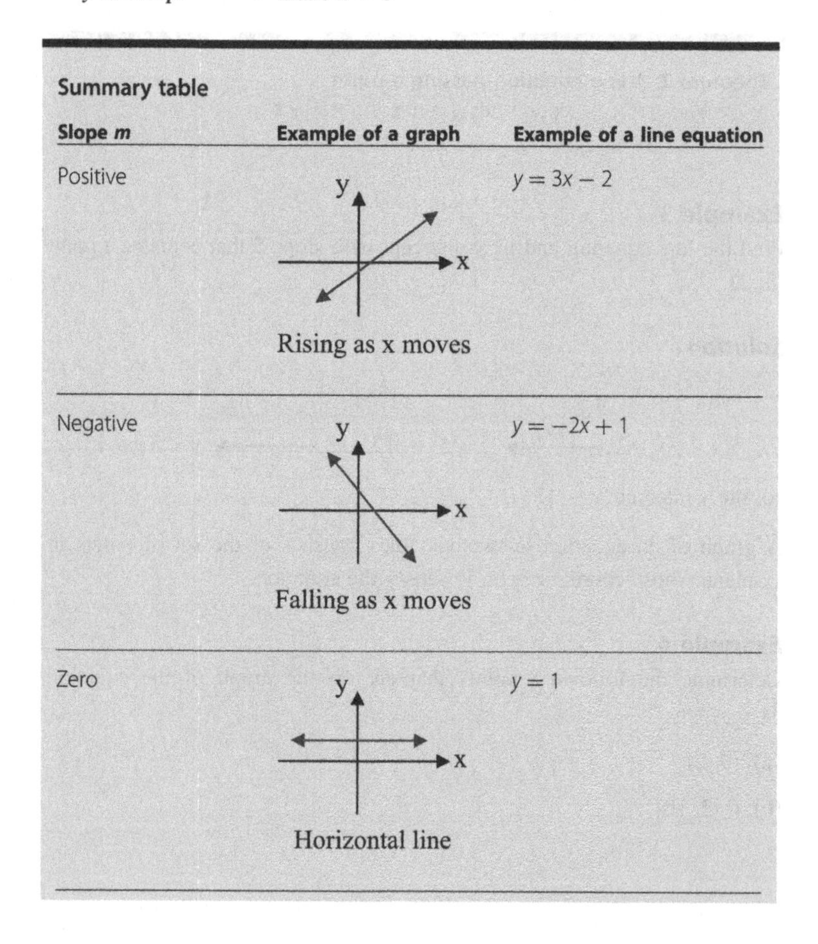

Summary table

Slope m	Example of a graph	Example of a line equation
Positive	Rising as x moves	$y = 3x - 2$
Negative	Falling as x moves	$y = -2x + 1$
Zero	Horizontal line	$y = 1$

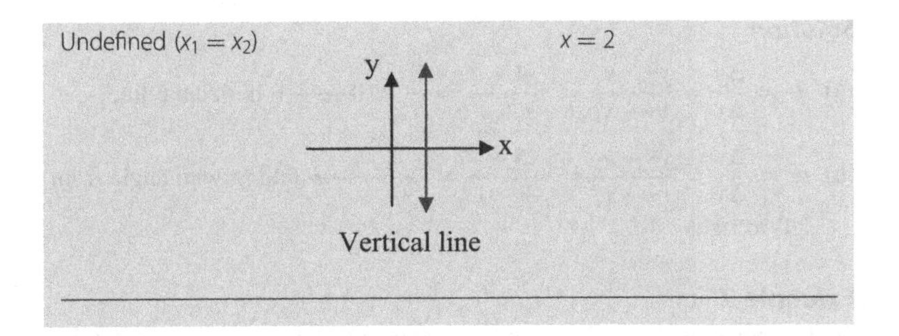

Example 5

Graph the equation $x + y = 4$.

Solution

x-intercept ⟶ make $y = 0$ ⟶ $x = 4$ ⟶ x-intercept $= 4$

y-intercept ⟶ make $x = 0$ ⟶ $y = 4$ ⟶ y-intercept $= 4$

if x is 1, then $y = 4 - 1 = 3$

x	y
4	0
0	4
1	3

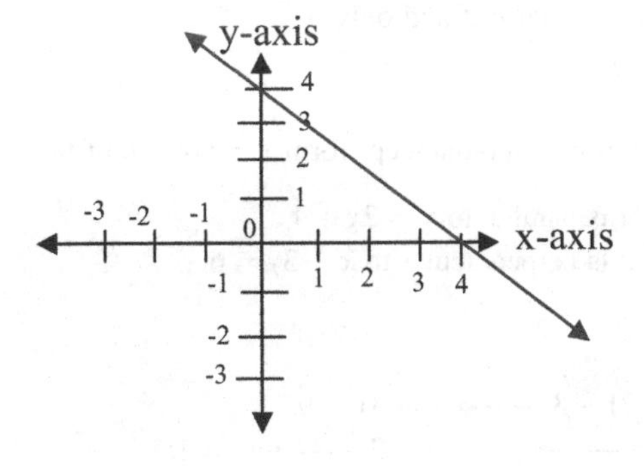

Example 6

Find the slope m of the line through the given points.

(a) $(-4, -4), (3, -4)$

(b) $(0, 3), (3, -5)$

Solution

(a) $m = \dfrac{\Delta y}{\Delta x} = \dfrac{y_2 - y_1}{x_2 - x_1} = \dfrac{-4+4}{3+4} = \dfrac{0}{7} = 0 \longrightarrow$ horizontal line

(b) $m = \dfrac{\Delta y}{\Delta x} = \dfrac{y_2 - y_1}{x_2 - x_1} = \dfrac{-5-3}{3-0} = \dfrac{-8}{3} \longrightarrow$ falling with angle from left to right

Example 7

Find an equation for the line that has a slope 3 and passes through the point (−1, 2). Write the equation in the standard form $Ax + By = C$.

Solution

$$y = m(x - x_1) + y_1 = 3(x+1) + 2 = 3x + 5 \rightarrow y - 3x = 5$$

$$\longrightarrow y = 3x + 5$$

Theorem 3: Parallel and perpendicular lines

Two nonvertical lines L_1 and L_2 with slopes m_1 and m_2, respectively, then

1. $L_1 \parallel L_2 \leftrightarrow m_1 = m_2$
2. $L_1 \perp L_2 \leftrightarrow m_1 m_2 = -1$

Note: Symbol $\|$ means: parallel
Symbol $\perp$ means: perpendicular
Symbol $\leftrightarrow$ means: if and only if

Example 8

Find the equation in slope-intercept form $y = mx + b$ of the line through

(a) (2, −3) that is parallel to $y = 2x + 1$
(b) (−1, 2) that is perpendicular to $x - 3y = 6$

Solution

(a) $y = 3(x-2) - 3 \longrightarrow y = 3x - 9$
(b) $x - 3y = 6 \longrightarrow y = x/3 - 2 \longrightarrow m = 1/3$
perpendicular $\longrightarrow m_2 = -3$
$y = -3(x+1) + 2 \longrightarrow y = -3x - 1$

Example 9

Let L_1: $y = -2x + 1$ and L_2: $y = (3 - k)x + 1$, find k if

(a) $L_1 \perp L_2$

(b) $L_1 \parallel L_2$.

Solution

(a) line 1 is perpendicular to line 2, $L_1 \perp L_2$, if and only if, the slope of line 1 multiplied by the slope of line 2 will give negative one, $m_1 \times m_2 = -1$.

$-2(3 - k) = -1 \longrightarrow -6 + 2k = -1 \longrightarrow 2k = 5 \longrightarrow k = 5/2$.

(b) line 1 is parallel to line 2, $L_1 \parallel L_2$, if and only if , the slope of line 1 is equal to the slope of line 2, $m_1 = m_2$.

$-2 = 3 - k \longrightarrow k = 5$.

4.1 EXERCISES

Plot each point in the xy-plane (rectangular plane) and tell in which quadrant each point lies for problems 1–8.

1. $P_1 = (5, -2)$

2. $P_2 = (1, 4)$

3. $P_3 = (-3, 1)$

4. $P_4 = (5, -3)$

5. $P_5 = (1, -6)$

6. $P_6 = (0, 6)$

7. $P_7 = (-4, 0)$

8. $P_8 = (-4, -5)$

Find the distance d and midpoint M between points for problems 9–14.

9. A = (2, 4) and B = (8, 9)

10. A = (3, 1) and B = (5, 0)

11. A = (5, 6) and B = (3, 7)

12. A = (2, −1) and B = (0, −3)

13. A = (−1, 6) and B = (−2, −1)

14. A = (8, 3) and B = (1, 5)

Find the slope of each straight line for problems 15–22.

15. rise = 8; run = 4

16. rise = 12; run = 3

17. rise = −6; run = 6

18. rise = −15; run = −5

19. connecting pairs of points (5, 2) and (1, 0)

20. connecting pairs of points (2, 4) and (1, 3)

21. connecting pairs of points (−3, 3) and (1, −1)

22. connecting pairs of points (7, 4) and (6, 1)

Write the equation of each straight line in slope-intercept form, and make a graph for problems 23–28.

23. slope = 5; y-intercept = −2

24. slope = 5; y-intercept = −2

25. slope = 5; y-intercept = −2

26. slope = 5; y-intercept = −2

27. slope = 5; y-intercept = −2

28. slope = 5; y-intercept = −2

Find the slope and the y-intercept for each equation, and make a graph for problems 29–34.

29. $y = 2x - 7$

30. $y = 5x - 3$

31. $y = 6x + 1$

32. $y = 3x + 4$

33. $y = -2x + 3$

34. $y = -5x + 2$

Find the coordinates of the y-intercepts and the x-intercepts of each equation for problems 35–40.

35. $y = 5x - 5$

36. $y = 3x - 2$

37. $2x = y + 12$

38. $3x = y + 9$

39. $2y - 4x - 6 = 0$

40. $5y - 10x - 15 = 0$

Find the equation of the line for problems 41–68.

41. slope = 2; containing the point (−3, 2)

42. slope = 3; containing the point (0, 5)

43. slope $= -\frac{3}{4}$; containing the point (2, 1)

44. slope $= \frac{1}{2}$; containing the point (−3, 2)

45. x-intercept = 3; y-intercept = −1

46. x-intercept = 4; y-intercept = 6

47. x-intercept = −5; y-intercept = 5

48. x-intercept = −3; y-intercept = −4

49. slope undefined; containing the point (3, 6)

50. slope undefined; containing the point (1, 2)

51. slope undefined; containing the point (2, 5)

52. slope undefined; containing the point (4, 9)

53. Parallel to the line $y = 3x$; containing the point (−2, 3)

54. Parallel to the line $y = 5x$; containing the point (−1, 1)

55. Parallel to the line $y = -4x$; containing the point (−1, 1)

56. Parallel to the line $y = -5x$; containing the point (2, −1)

57. Parallel to the line $y = -2x$; containing the point (−3, −1)

58. Parallel to the line $y = 3x + 3$; containing the point (0, 0)

59. Parallel to the line $y = 2x + 1$; containing the point (0, 0)

60. Parallel to the line $x = 4$; containing the point (6, 3)

61. Parallel to the line $y = 4$; containing the point (6, 3)

62. Perpendicular to the line $y = 2x + 6$; containing the point (−1, 2)

63. Perpendicular to the line $y = 3x - 4$; containing the point (2, 1)

64. Perpendicular to the line $y = 3x + 3$; containing the point (−4, 0)

65. Perpendicular to the line $y = 2x - 7$; containing the point (0, 3)

66. Perpendicular to the line $y = 6x - 1$; containing the point (0, 1)

67. Perpendicular to the line $x = 4$; containing the point (2, 3)

68. Perpendicular to the line $y = 7$; containing the point (4, 5)

4.2 FUNCTIONS

Definition of function
A function is a set of ordered pairs (a relation) in which no two ordered pairs have the same one component and different other components.

Functions can have domain and range.

Definition of domain
Domain of a function is the set of all first components in a function.

Definition of range
Range of a function is the set of all second components in a function.

Therefore, a function is a relation in which, for each value of the first component of the ordered pairs, there is exactly one value of the second component.

4.2.1 Theorem: Vertical line test

Theorem 1: Vertical line test
A graph is a function if no vertical line crosses the graph more than once.

Example 1
Determine which of the graphs in the figures below are graphs of functions.

(a)

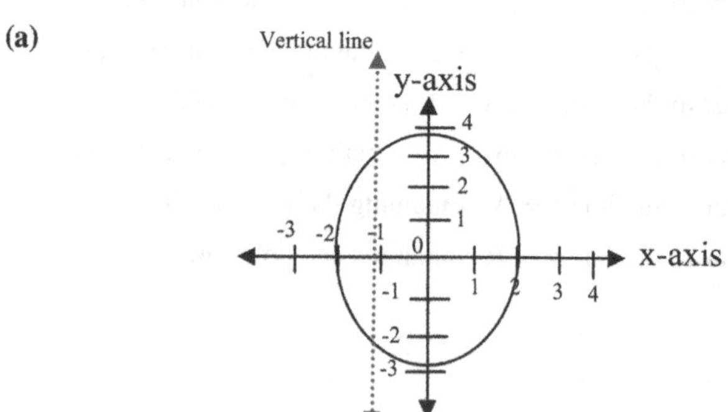

(b)

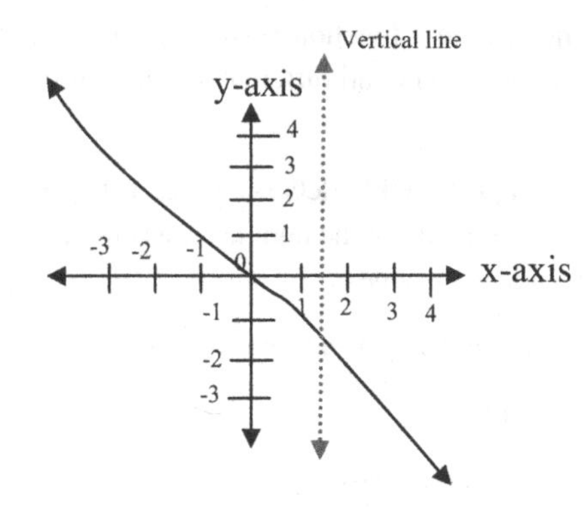

(c)

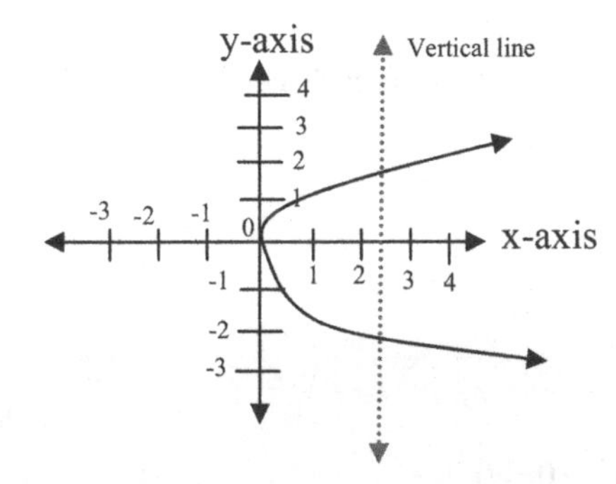

Solution

(a) The graph is not function because the vertical line crosses the graph two times.

(b) The graph is a function because the vertical line crosses the graph only once.

(c) The graph is not a function because the vertical line crosses the graph two times.

Example 2

Determine whether each set of relations is a function. State the set of domain and range of the function.

(a) A = {(1, 2), (2, 3), (4, 3), (5, 2)}

(b) B = {(1, 2), (1, 4), (2, 5), (5, 3)}

Solution

(a) Set A defines a function because no two ordered pairs have the same first component and different second components.

Domain = {1, 2, 4, 5}

Range = {2, 3}

(b) Set B does not define a function because there are ordered pairs with the same first component and different second components [(1, 2) and (1, 4)].

Functions notation are denoted by letters such as f, *f*, *g*, *F*, and *G*. The symbol $f(x)$ refers to the value of f at the number x, where f a dependent function and x is an independent variable.

x values represent the domain of the function $f(x)$

$f(x)$ values represent the range

Sometimes a function can be as $y = f(x)$

Example 3

For the function $f(x) = 3x^2 - x$, evaluate:

(a) $f(0)$

(b) $f(1)$

(c) $f(-x)$

(d) $-f(x)$

Solution

(a) $f(0) = 3(0)^2 - 0 = 0$

(b) $f(1) = 3(1)^2 - 1 = 2$

(c) $f(-x) = 3(-x)^2 - (-x) = 3x^2 + x$

Theorem 2: Even and odd functions

1. The function $f(x)$ is even if and only if $f(-x) = f(x)$.
2. The function $f(x)$ is odd if and only if $f(-x) = -f(x)$.

Example 4

Determine if the following functions are even, odd, or neither.

(a) $f(x) = 2x^2$

(b) $f(x) = 5x^3$

(c) $f(x) = 5x^2 - x$

Solution

(a) $f(-x) = 2(-x)^2 = 2x^2 = f(x)$, therefore it is Even function

(b) $f(-x) = 5(-x)^3 = -5x^3 = -f(x)$, therefore it is Odd function

(c) $f(-x) = 5(-x)^2 - (-x) = 5x^2 + x \neq f(x)$ and $\neq -f(x)$, therefore it is neither.

Definition of one-to-one function
One-to-one function is a function that has no two ordered pairs with different first coordinates and the same second coordinate.

For example, let's look to a function f that has domain D = {2, 4, 6, 8} and range R = {1, 3, 5, 7}.

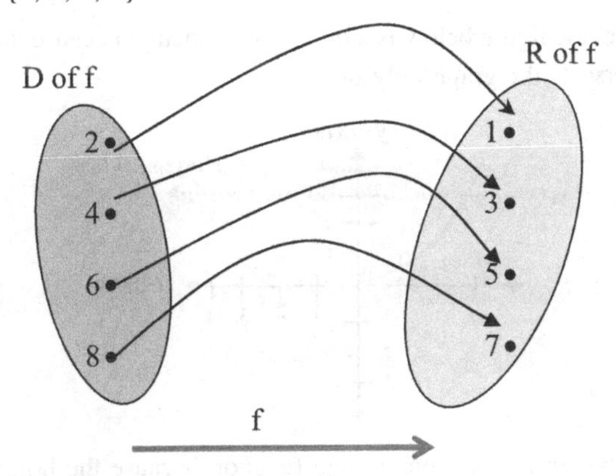

f in the figure is a one-to-one function.

For example, if function g has domain D = {2, 4, 6, 8} and range R = {3, 5, 7}.

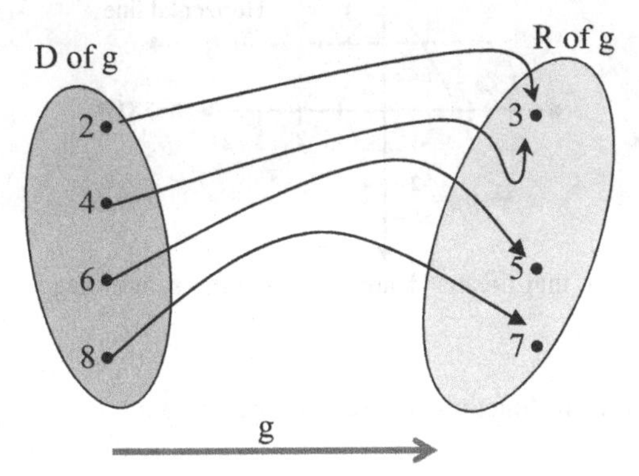

Then, g is not a one-to-one function because elements 2 and 4 have the same image 3.

In other words, a one-to-one function is a function that has no two elements of its domain that have the same image (range), that is $f(x_1) \neq f(x_2)$, where $x_1 \neq x_2$, this means, if $f(x_1) = f(x_2)$, then $x_1 = x_2$.

Theory 3: Horizontal line test
A function is one-to-one if and only if a horizontal line intersects its graph only one time.

Therefore, a horizontal line test is used to determine whether a function in one-to-one.

For example, the figure below is a one-to-one function because the horizontal line intersects the graph only once.

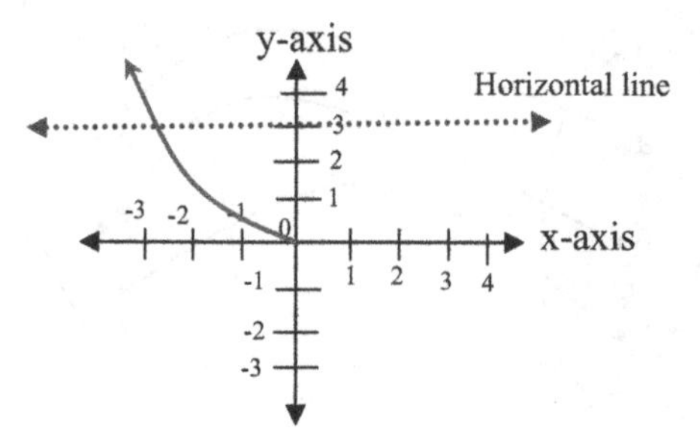

The figure below is not a one-to-one function because the horizontal line intersects the graph more than once.

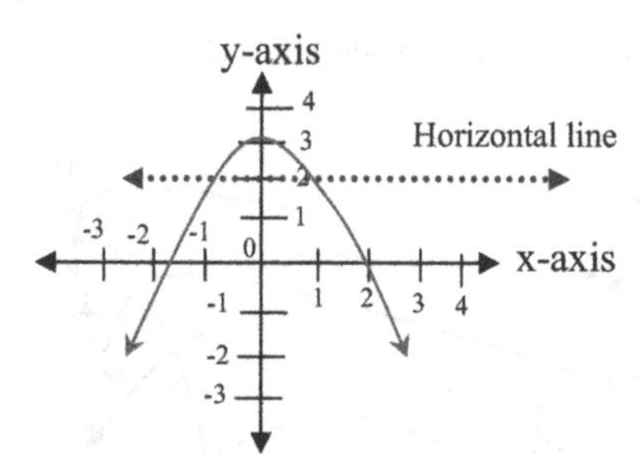

$f(1) = 2 = f(-1)$, that means, 1 and -1 have the same image.

Example 5

Show the following functions are one-to-one or not.

(a) $f(x) = x^5$

(b) $f(x) = 2x + 3$

Solution

(a) If two numbers x_1 and x_2 are not equal, $x_1 \neq x_2$, then $x_1^5 \neq x_2^5$, therefore, $f(x) = x^5$ is a one-to-one function.

(b) If a function of two numbers x_1 and x_2 are equal, $f(x_1) = f(x_2)$, then

$2x_1 + 3 = 2x_2 + 3$

$2x_1 = 2x_2$

$x_1 = x_2$

Therefore, f(x) is a one-to-one function.

4.2.2 Inverse of function $f^{-1}(x)$

Definition of inverse function $f^{-1}(x)$

Let f be a one-to-one function with domain D and range R. Then its inverse f^{-1} has domain R and range D, that is,

$f(x) = y \Leftrightarrow f^{-1}(y) = x$, for any y in R and x in D.

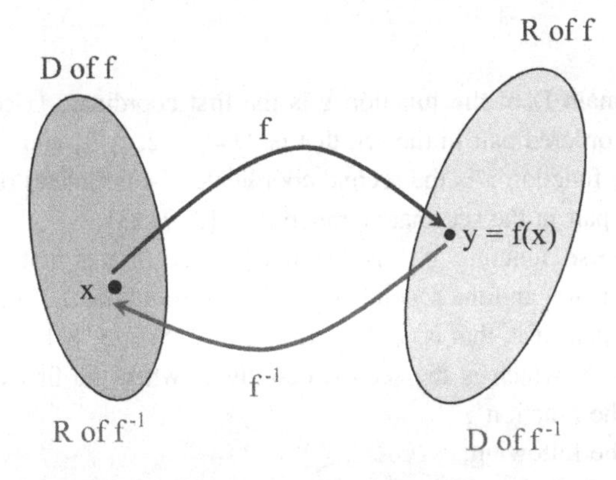

The above figure illustrates the relation between the function f and its inverse f^{-1}.

This means, the function f maps the domain value x into the range value f(x).

Domain of f = Range of f^{-1}

Domain of f^{-1} = Range of f

Note:

1. $f^{-1}(x) \neq 1/f(x)$.
2. $\Leftrightarrow$, means the statement, if and only if.

Example 1

Determine whether each of the following functions is invertible. If it is invertible, find its inverse.

(a) $f = \{(0, 2), (3, 2), (5, 6), (7, 13)\}$

(b) $g = \{(-4, 7), (5, 9), (8, 15), (10, 19)\}$

Solution

(a) The function f is not one-to-one because of the ordered pairs (0, 2) and (3, 2), therefore f is not invertible.

(b) The function g is not one-to-one, therefore g is invertible, where $g^{-1} = \{(7, -4), (9, 5), (15, 8), (19, 10)\}$.

Example 2

Let $g = \{(-2, 3), (5, 7), (8, 13)\}$, find the following:

(a) Domain and range of g

(b) g^{-1}

(c) $g^{-1}(7)$

(d) Domain and range of g^{-1}

Solution

(a) The domain D of the function g is the first coordinate (x-coordinate) of each ordered pair in the set, that is, $D = \{-2, 5, 8\}$ and the range R of the function g is the second coordinate (y-coordinate) of each ordered pair in the set, that is, range $R = \{3, 7, 13\}$.

(b) The inverse function g^{-1} is the interchange of the first coordinate (x-coordinate) and the second coordinate (y-coordinate) of each ordered pair of g, that is, $g^{-1} = \{(3, -2), (7, 5), (13, 8)\}$.

(c) $g^{-1}(7) = 5$, which is the second coordinate when the first coordinate is 7 in the function g^{-1}.

(d) Recall the following:
Domain D of g^{-1} = Range R of g, that is, $\{3, 7, 13\}$.
Range R of g^{-1} = Domain D of g, that is, $\{-2, 5, 8\}$.

Example 3

Identify and explain why the following figures represent not a function, not one-to-one function, and one-to-one function.

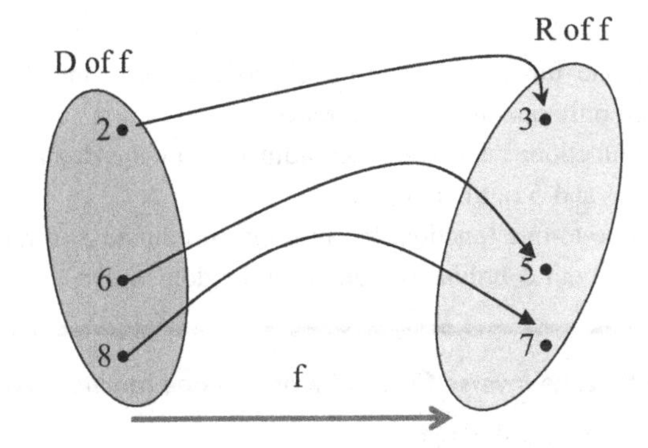

(b)

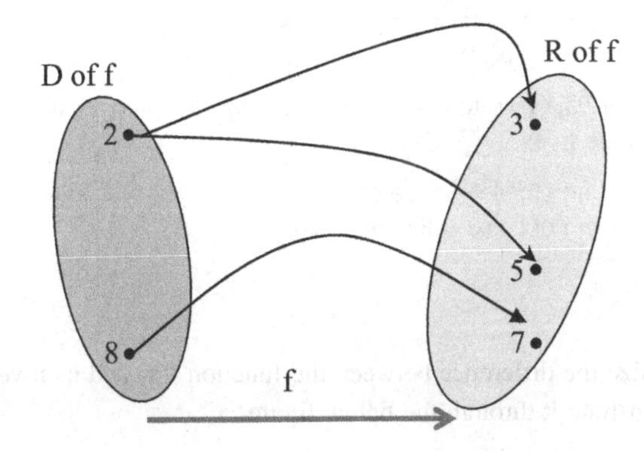

(c)

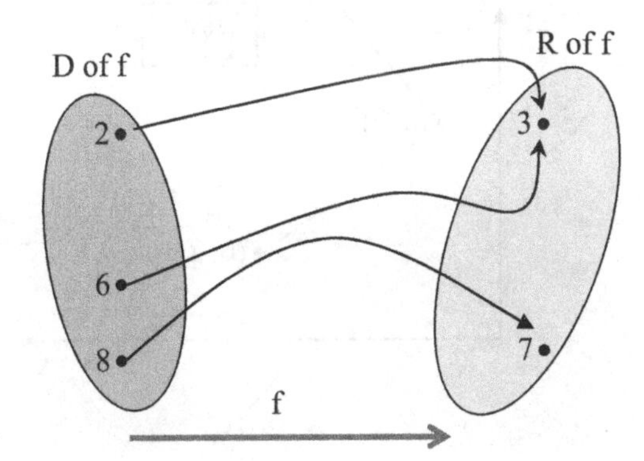

Solution

(a) One-to-one function, because each coordinate in the domain D has one and only one image in the range R.
(b) Not a function, because the coordinate 2 in the domain D has two images 3 and 5 in the range R.
(c) Not a one-to-one function, because the coordinate 3 of range R is the image of both coordinates 2 and 6 of the domain D.

Steps to find the inverse $f^{-1}(x)$ of a one-to-one function f(x)

1. Replace f(x) by y, that is, f(x) = y.
2. Interchange x and y.
3. Solve the equation for y in terms of x.
4. Replace y by $f^{-1}(x)$, that is, $f^{-1}(x) = y$.

Note: it is always nice to check after you finish the steps that the following are true:

1. the Domain of f(x) = Range of $f^{-1}(x)$
2. the Domain of $f^{-1}(x)$ = Range of f(x)

To visualize the difference between the function f(x) and its inverse $f^{-1}(x)$, we demonstrate it through the below figure.

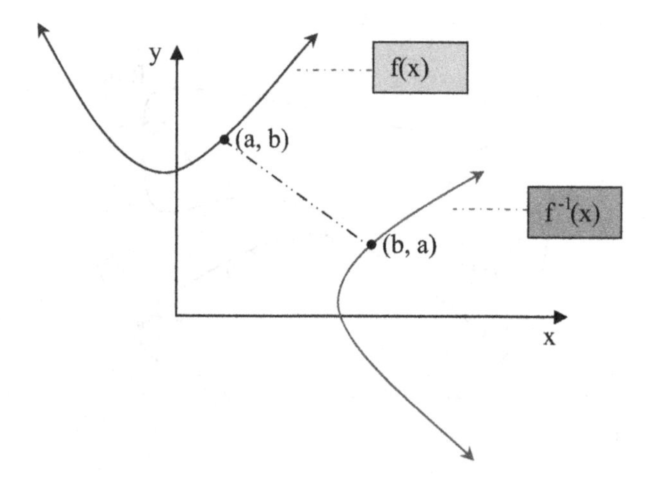

The inverse $f^{-1}(x)$ is a reflection of the function f(x). Let us assume a point (a, b) on the graph of f, then the point (b, a) is on the graph of the inverse function $f^{-1}(x)$.

We can identify the domain and range of a function f(x) from its graph.

Example 4

Graph the function $f(x) = x - 2$ and find its domain and range.

Solution

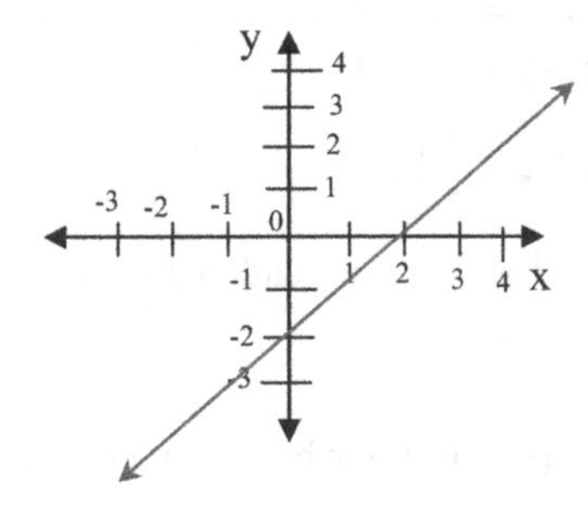

Domain D of $f(x) = (-\infty, \infty)$

Range R of $f(x) = (-\infty, \infty)$

Example 5

Graph the function $f(x) = \sqrt{x}$; find its domain and range.

Solution

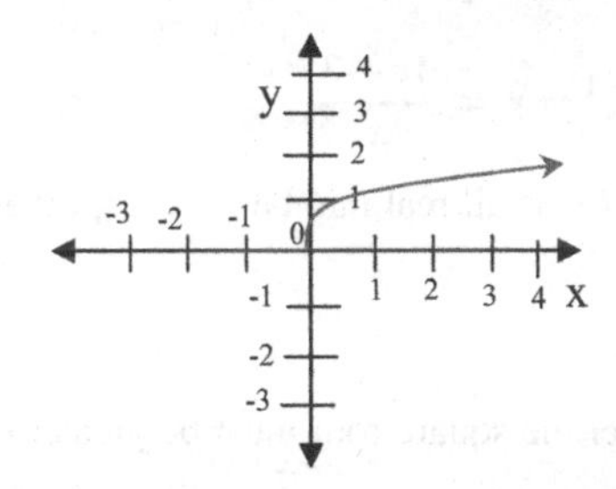

Domain D of $f(x) = [0, \infty)$

Range R of $f(x) = [0, \infty)$

Example 6

Find the domain and range of each function f(x) and $f^{-1}(x)$:

(a) $f(x) = 2x + 1$

(b) $f(x) = \dfrac{3}{x - 4}$

(c) $f(x) = \sqrt{7 - 5x}$

Solution

(a) $f(x) = 2x + 1$

The domain D of f is $(-\infty, \infty)$ and that is the range R of f^{-1}.

Note: $(-\infty, \infty) = \{x|x \in \mathbb{R}\}$.

To find the f^{-1}:

Replace f(x) with y: $y = 2x + 1$

Interchange x and y: $x = 2y + 1$

Solve for y: $y = f^{-1} = \dfrac{x-1}{2}$

The domain D of f^{-1} is $(-\infty, \infty)$ and that is the range R of f.

(b) $f(x) = \dfrac{3}{x-4}$

The domain D of f is all real numbers $\mathbb{R}$ except 4 and that is the range R of f^{-1}.

Note: The domain D of f $= \{x|x \in \mathbb{R}, \text{except when } x = 4\}$.

To find the f^{-1}:

Replace f(x) with y: $y = \dfrac{3}{x-4}$

Interchange x and y: $x = \dfrac{3}{y-4}$

Solve for y: $y = f^{-1} = y = \dfrac{4x+3}{x}$

The domain D of f^{-1} is all real number $\mathbb{R}$ except 0 and that is the range R of f.

(c) $f(x) = \sqrt{7-5x}$

The equation under the square root must be greater or equal to zero.

By solving $7 - 5x \geq 0$, we get $x \leq \frac{7}{5}$.

Therefore, the domain D of f is the interval $\left(-\infty, \frac{7}{5}\right]$ and that is the range R of f^{-1}.

Note: $\left(-\infty, \frac{7}{5}\right] = \left\{x|x \leq \frac{7}{5}\right\}$.

To find the f^{-1}:

Replace f(x) with y: $y = \sqrt{7-5x}$

Interchange x and y: $x = \sqrt{7-5y}$

Solve for y:

$$x^2 = 7 - 5y \longrightarrow 5y = 7 - x^2 \longrightarrow f^{-1} = y = \frac{-x^2 + 7}{5}$$

The domain D of f^{-1} is all real numbers $\mathbb{R}$, $(-\infty, \infty)$, and that is the range R of f.

4.2 EXERCISES

Find the domain and range of each function $f(x)$ and $f^{-1}(x)$ for problems 1–10.

1. $f(x) = 2x - 1$
2. $y = x + 4$
3. $f(x) = \dfrac{1}{x+3}$
4. $y = \dfrac{x}{x-7}$
5. $f(x) = \sqrt{x+9}$
6. $f(x) = \sqrt{x-5}$
7. $y = \sqrt{x} - 9$
8. $y = 3x^2$
9. $f(x) = \dfrac{3}{(x-1)(x+2)}$
10. $f(x) = \dfrac{5}{(x-5)(x+3)}$

Determine which of the graphs in the figures below are graphs of functions, explain the reason, and determine the domain and range of the function for problems 11 and 12.

11.

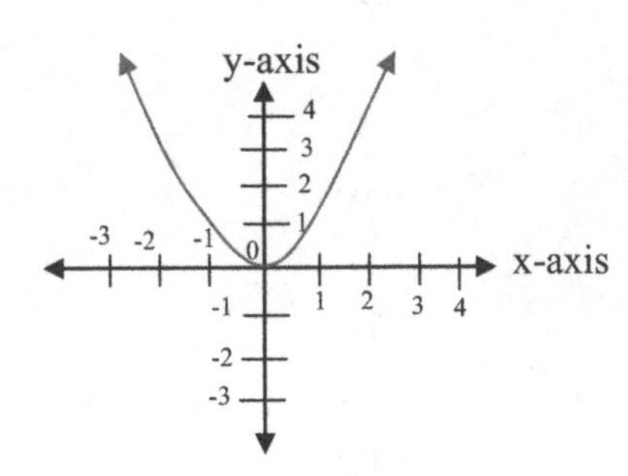

12.

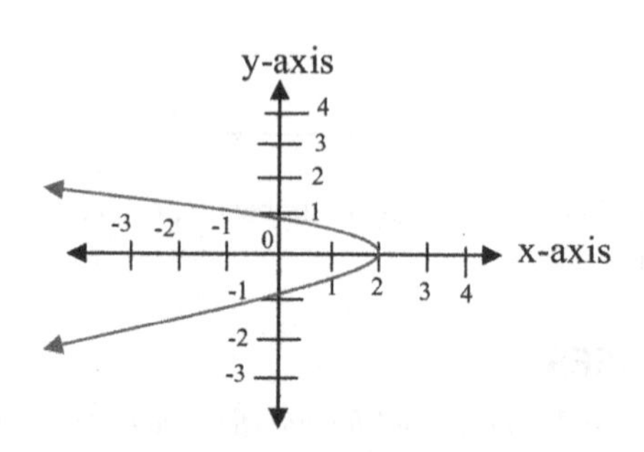

Indicate whether each set defines a function. Find the domain and range of each function for problems 13–16.

13. A = {(2, 2), (3, 4), (5, 8)}

14. B = {(2, 2), (2, 4), (6, 7)}

15. C = {(1, 3), (2, 3), (4, 11)}

16. D = {(3, 8), (1, 7), (2, 7)}

Use the horizontal line test to determine whether the functions are one-to-one for problems 17 and 18.

17. $f(x) = |x| + 1$

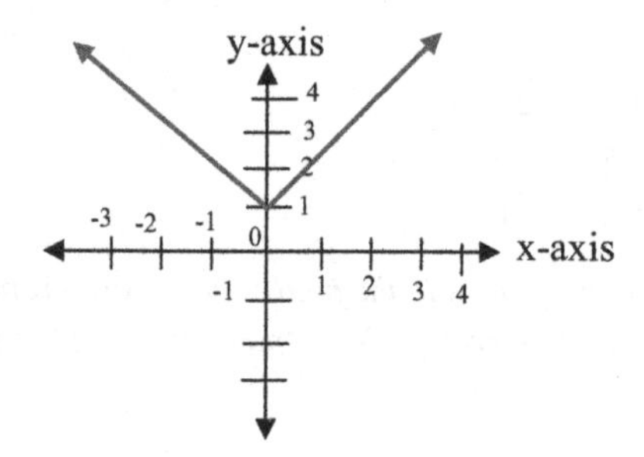

18. $f(x) = 1/x$.

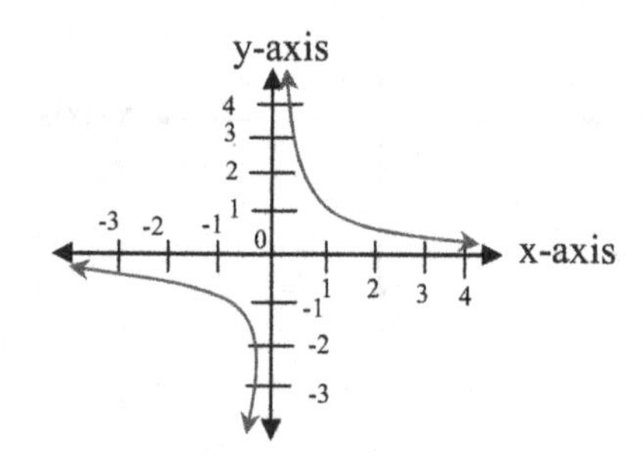

Find the domain and range of the following graphs for problems 19 and 20.

19. $f(x) = x + 1$

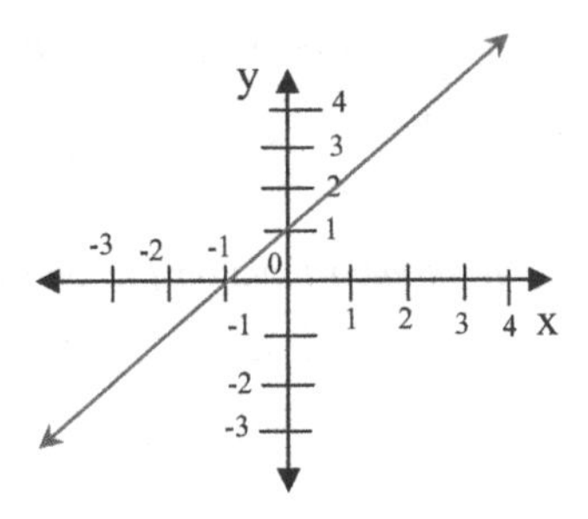

20. $f(x) = x + 2$

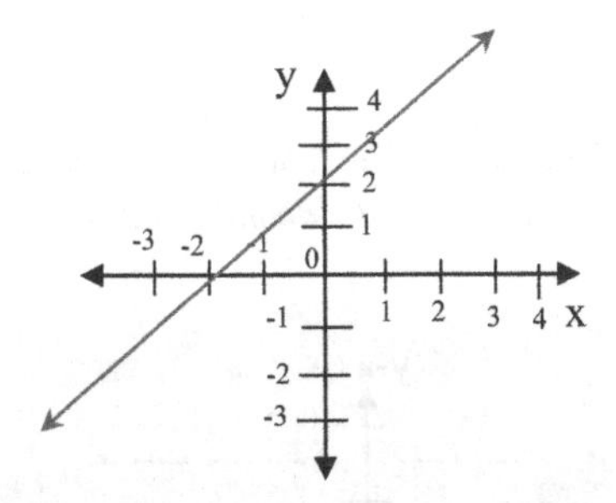

Find the inverse of the following functions for problems 21–26.

21. $f(x) = \sqrt{x-4}$

22. $f(x) = \sqrt{x+3}$

23. $f(x) = 5x$

24. $f(x) = x^5$

25. $f(x) = \dfrac{4x}{6x-1}$,

Hint: $\dfrac{y}{y} = 1$.

26. $f(x) = \dfrac{x^2-9}{x+3}$

CHAPTER 4 REVIEW EXERCISES

Plot each point in the xy-plane (rectangular plane) for problems 1–4.

1. $P = (-4, 1)$

2. $K = (0, 3)$

3. D = (−1, 0)
4. L = (−3, −2)

Find the distance d and midpoint M between points for problems 5 and 6.

5. A = (3, −5) and B = (7, 2)
6. A = (5, −1) and B = (−3, −6)

Find the domain and range of each function f(x) and $f^{-1}(x)$ for problems 7–10.

7. $f(x) = 5x - 7$
8. $f(x) = \dfrac{5}{x-9}$
9. $f(x) = \sqrt{x+4}$
10. $f(x) = \sqrt{x-3}$

Determine which of the graphs in the figures below are graphs of functions, explain the reason, and determine the domain and range of the function for problems 11 and 12.

11.

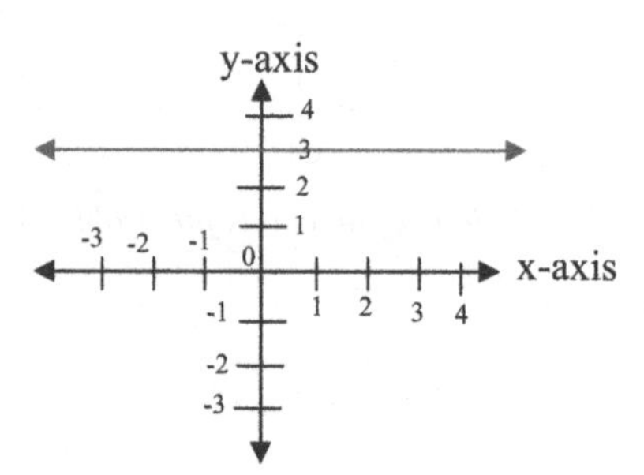

12.

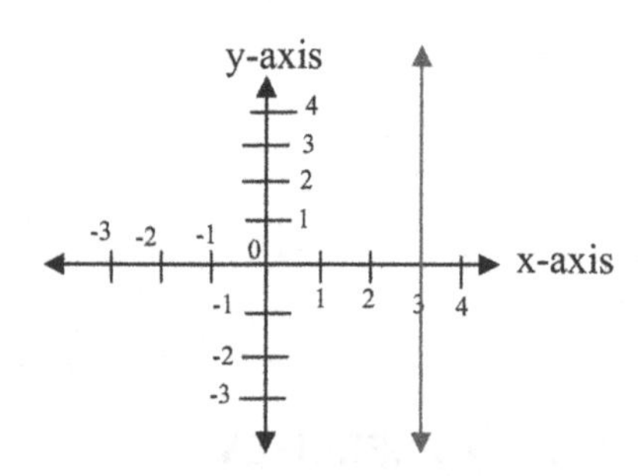

Indicate whether each set defines a function. Find the domain and range of each function for problems 13–16.

13. A = {(2, 8), (1, 12), (5, 14)}
14. B = {(3, 2), (3, 4), (6, 9)}

15. C = {(1, 2), (3, 2), (4, 5)}

16. D = {(1, 9), (3, 5), (4, 5)}

Find the domain and range of the following graphs for problems 17 and 18.

17. $f(x) = x^2$ for $x \geq 0$

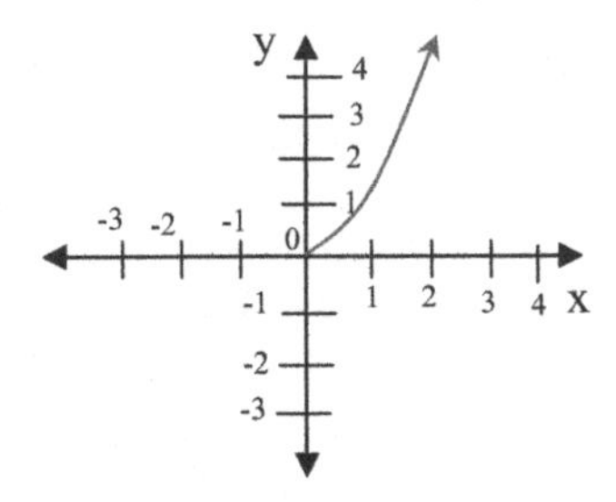

18.

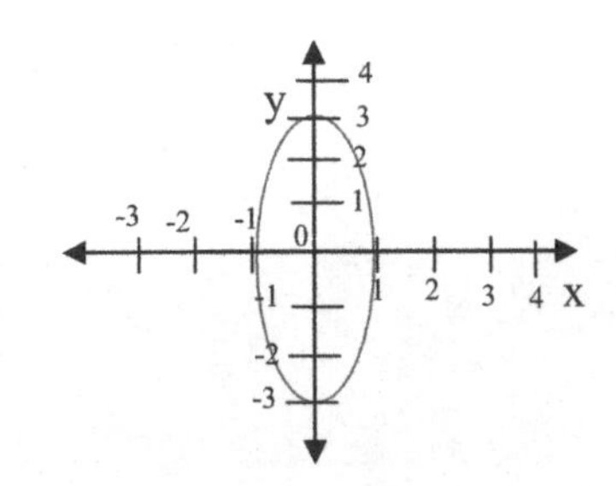

Find the inverse of the following functions for problems 19–24.

19. $f(x) = \sqrt{x-2}$

20. $f(x) = \sqrt{x+1}$

21. $f(x) = 3x$

22. $f(x) = x^3$

23. $f(x) = \dfrac{2x}{3x-1}$,

Hint: $\dfrac{y}{y} = 1$.

24. $f(x) = \dfrac{x^2-4}{x+2}$

Chapter 5

Measurement

We shall never know all the good that a simple smile can do.
Mother Teresa

■ A ground-level view of the huge Apollo 17 space vehicle on its way to Pad A, Launch Complex 39, Kennedy Space Center (KSC). The Saturn V stack and its mobile launch tower are atop a mammoth crawler-transporter. The mission launched on December 7, 1972.

INTRODUCTION

Recording measurements is an essential tool in many jobs and professions.

Measurement is the collection of quantitative data. A number with a unit, an assignment to represent its size or magnitude, is called a *measure*. Today, two different systems of measurement are used in America: the metric system or the Standard International Metric system (SI) and the U.S. customary system of units.

Fundamentals of Technical Mathematics. http://dx.doi.org/10.1016/B978-0-12-801987-0.00005-8

The metric system of measurement is used by various people around the world in numerous fields such as computer technology, science, business, and medicine. In this chapter, we focus on the U.S. customary system and metric system and their unit conversions.

When writing units after a numerical value, there are two recommendations for expressing those units:

1. The product of two or more units is indicated by a dot (·); we would write V·s, but not Vs.
2. The division units can be written for example as N/m, or $N \cdot m^{-1}$, or $\frac{N}{m}$ and W/m^2 or $W \cdot m^{-2}$.

5.1 LENGTH MEASUREMENTS

5.1.1 The metric system

In the metric system measurement of length, the *meter* is the basic unit of length. A prefix is used in front of the basic unit (meter) to identify the smaller and larger units. Table 5.1 shows the most popular metric measurements of length.

Example 1

Change the following values of units to *centi*meter (cm).

(a) 3 m
(b) 5.5 m
(c) 4 mm
(d) 3.5 mm

Solution

(a) $3\,\cancel{m} \times \frac{100\text{ cm}}{1\,\cancel{m}} = 300\text{ cm}$, therefore, 3 m = 300 cm

(b) $\frac{55\,\cancel{m}}{1\cancel{0}} \times \frac{10\cancel{0}\text{ cm}}{1\,\cancel{m}} = 550\text{ cm}$, therefore, 5.5 m = 550 cm

(c) $4\,\cancel{mm} \times \frac{0.001\,\cancel{m}}{1\,\cancel{mm}} \times \frac{100\text{ cm}}{1\,\cancel{m}} = 0.4\text{ cm}$, therefore, 4 mm = 0.4 cm

(d) $\frac{35\,\cancel{mm}}{10} \times \frac{0.001\,\cancel{m}}{1\,\cancel{mm}} \times \frac{100\text{ cm}}{1\,\cancel{m}} = 0.35\text{ cm}$, therefore, 3.5 mm = 0.35 cm

Table 5.1 Metric Units of Length

1 meter (m) = 1.0 meter (is the basic unit of length)
1 *milli*meter (mm) = 0.001 meter (m) ⇒ 1 m = 1000 mm
1 *centi*meter (cm) = 0.01 meter (m) ⇒ 1 m = 100 cm
1 *kilo*meter (km) = 1000 meters (m) ⇒ 1 km = 1000 m

5.1.2 The U.S. customary system

In the U.S. customary system, the units are not related as in the metric system. Table 5.2 shows the most popular and basic units used in the U.S. customary system for length.

Table 5.2 U.S. Customary Units of Length

1 yard (yd) = 3 feet (ft or ′)
1 yard (yd) = 36 inches (in. or ″)
1 foot (ft) = 12 inches (in.)
1 mile (mi) = 1760 yards (yd)
1 mile (mi) = 5280 feet
1 rod (rd) = 5.50 yards
1 rod (rd) = 16.5′

Example 1

Convert the following units of U.S. customary system to feet (ft).

(a) 7 yd

(b) 9.5 yd

(c) 5 mi

(d) 6.5 mi

(e) 16 in.

(f) 8.5 in.

Solution

(a) $7\ \cancel{\text{yd}} \times \dfrac{3\ \text{ft}}{1\ \cancel{\text{yd}}} = 21\ \text{ft}$

(b) $9.5\ \cancel{\text{yd}} \times \dfrac{3\ \text{ft}}{1\ \cancel{\text{yd}}} = 28.5\ \text{ft}$

(c) $5\ \cancel{\text{mi}} \times \dfrac{1760\ \cancel{\text{yd}}}{1\ \cancel{\text{mi}}} \times \dfrac{3\ \text{ft}}{1\ \cancel{\text{yd}}} = 26400\ \text{ft}$

(d) $6.5\ \cancel{\text{mi}} \times \dfrac{1760\ \cancel{\text{yd}}}{1\ \cancel{\text{mi}}} \times \dfrac{3\ \text{ft}}{1\ \cancel{\text{yd}}} = 34320\ \text{ft}$

(e) $16\ \cancel{\text{in}}. \times \dfrac{1\ \text{ft}}{12\ \cancel{\text{in}}.} = 1.333\ \text{ft}$

(f) $8.5\ \cancel{\text{in}}. \times \dfrac{1\ \text{ft}}{12\ \cancel{\text{in}}.} = 0.708\ \text{ft}$

Table 5.3 Metric System and U.S. Customary System

Metric to U.S.	U.S. to metric
1 m = 3.28 ft	1 ft = 0.305 m
1 m = 1.09 yd	1 yd = 0.914 m
1 cm = 0.394 in.	1 in. = 2.54 cm
1 km = 0.62 mi	1 mi = 1.61 km

Table 5.3 shows the relationship between the metric system and U.S. customary system for units of length.

Example 2

Convert the following units:

(a) 4 ft to centimeters

(b) 13 mi to kilometers

(c) 15 m to feet

(d) 7 km to miles

(e) 7 ft to miles

Solution

(a) $4\,\not{ft} \times \frac{0.305\,\not{m}}{1\not{ft}} \times \frac{100\text{ cm}}{1\not{m}} = 122\text{ cm}$

(b) $13\,\not{mi} \times \frac{1.61\text{ km}}{1\,\not{mi}} = 20.93\text{ km}$

(c) $15\,\not{m} \times \frac{1\text{ ft}}{0.305\,\not{m}} = 49.18\text{ ft}$

(d) $7\,\not{km} \times \frac{1\text{ mi}}{1.61\not{km}} = 4.348\text{ mi}$

(e) $7\,\not{ft} \times \frac{0.305\text{ mi}}{1\,\not{ft}} = 2.135\text{ mi}$

Table 5.4 summarizes the commonly used metric prefixes and their abbreviations.

Example 3

Covert the following units:

(a) 2500 mm to meters

(b) 3500 cm to meters

(c) 4 μm to nanometers

(d) 12 km to meters

(e) 0.035 m to millimeters

Solution

(a) $2500\text{ mm} \times \dfrac{1\text{ m}}{1000\text{ mm}} = 2.5\text{ m}$

(b) $3500\text{ cm} \times \dfrac{1\text{ m}}{100\text{ cm}} = 35\text{ m}$

(c) $4\ \mu\text{m} \times \dfrac{1\text{ m}}{10^{6}\ \mu\text{m}} \times \dfrac{1\text{ nm}}{10^{-9}\text{ m}} = 4000\text{ nm}$

(d) $12\text{ km} \times \dfrac{1\text{ m}}{10^{-3}\text{ km}} = 12000\text{ m}$

(e) $0.035\text{ m} \times \dfrac{1\text{ mm}}{10^{-3}\text{ m}} = 35\text{ mm}$

Table 5.4 The International System of Units (SI) Prefixes

Power	Prefix	Symbol
10^{-24}	yocto	y
10^{-21}	zepto	z
10^{-18}	atto	a
10^{-15}	femto	f
10^{-12}	pico	p
10^{-9}	nano	n
10^{-6}	micro	μ
10^{-3}	milli	m
10^{-2}	centi	c
10^{-1}	deci	d
10^{1}	deka	da
10^{2}	hecto	h
10^{3}	kilo	k
10^{6}	mega	M
10^{9}	giga	G
10^{12}	tera	T
10^{15}	peta	P
10^{18}	exa	E
10^{21}	zetta	Z
10^{24}	yotta	Y

5.1 EXERCISES

In problems 1–22, express each of the following lengths as indicated. If necessary round the answer to two decimal places.

1. 84 in. as feet

2. 111 in. as feet

3. $2\frac{1}{2}$ ft as inches

4. $5\frac{1}{2}$ ft as inches

5. 0.6 yd as inches

6. $3\frac{1}{4}$ yd as inches

7. $\frac{1}{9}$ yd as inches

8. 0.40 yd as inches

9. 4 yd as feet

10. 7.2 yd as feet

11. 33 ft as yards

12. 54 ft as yards

13. 27 cm as millimeters

14. 12.64 cm as millimeters

15. 223.85 mm as centimeters

16. 83.94 mm as centimeters

17. 0.93 m as centimeters

18. 0.28 m as centimeters

19. 145 mm as meters

20. 768 mm as meters

21. 0.83 mm as centimeters

22. 0.0064 m as millimeters

In problems 23–32, perform the indicated operations and express the answer in the indicated unit. If necessary round the answer to two decimal places.

23. 23.54 mm + 6.4 cm = ... mm

24. 4.3 m + 87 cm = ... m

25. 39.4 cm − 53.6 mm = ... cm

26. 162.6 mm − 14.5 cm = …… mm

27. 1.08 m − 47.3 cm = …… cm

28. 0.68 m − 432.4 mm = …… mm

29. 326 mm + 78.3 cm + 0.8 m = …… m

30. 0.046 m + 8.74 cm + 75.2 mm = …… mm

31. 68.2 mm + 7.03 cm + 308.7 mm = …… mm

32. 2.736 m − 624 mm = …… m

In problems 33–46, express each of the following customary units of lengths as in the indicated metric unit of length. If necessary round the answer to two decimal places.

33. 43.00 mm as inches

34. 118.40 mm as inches

35. 13.60 cm as inches

36. 0.62 cm as inches

37. 4.60 m as inches

38. 0.07 m as inches

39. 6.00 m as feet

40. 12.40 m as feet

41. 954.00 mm as inches

42. 76.03 mm as inches

43. 6.00 m as yards

44. 4.00 m as yards

45. 85.00 cm as feet

46. 340 mm as feet

In problems 47–60, express each of the following customary units of lengths as in the indicated metric unit of length. If necessary round the answer to two decimal places.

47. 7.00 in. as millimeters

48. 0.64 in. as millimeters

49. 76.00 in. as millimeters

50. 40.65 in. as centimeters

51. 8.00 ft as meters

52. 0.65 ft as meters

53. 6.50 yd as meters

54. 2.70 yd as meters

55. 4.58 in. as millimeters

56. 0.54 in. as centimeters

57. 428.00 in. as centimeters

58. $\frac{1}{4}$ in. as millimeters

59. $6\frac{1}{2}$ in. as centimeters

60. $46\frac{5}{7}$ in. as meters

61. What is the perimeter of a city lot in the shape of polygon ABCD in meters?

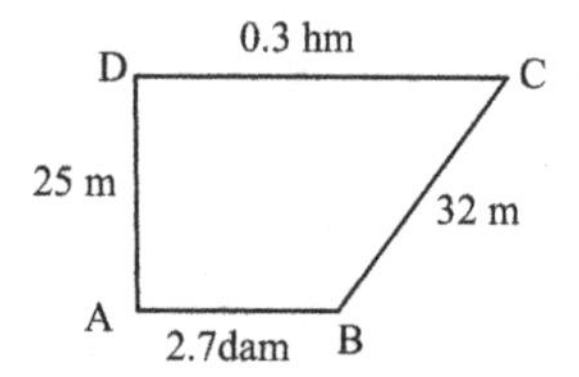

62. What is the perimeter of a city lot as in the below shape in centimeters?

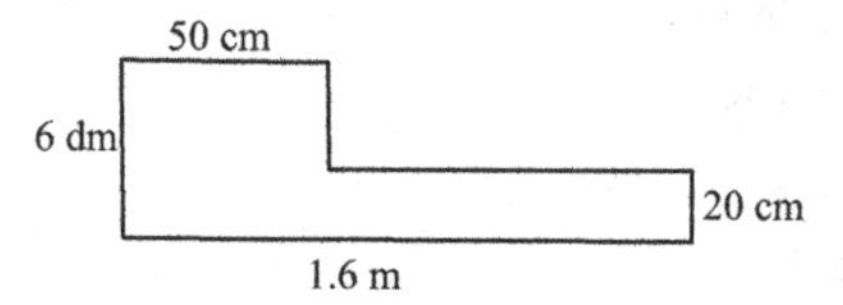

63. An aluminum slab 0.074 m thick is machined with three equal cuts; each cut is 10 mm deep. Find the finished thickness of the slab in millimeters.

64. A piece of sheet metal is 2.24 m wide. Strips each 3.4 cm wide are cut. Allow 3 mm for cutting each strip. Find the width of the waste strip in millimeters.

65. Three metal pipes are welded together as 125 mm, 40 cm, and 2.5 dm. Determine the overall length of the welded pipe in millimeters.

66. Three metal rods are welded together as 6 cm, 58 mm, and 0.8 dm. Determine the total length of the welded rod in millimeters.

5.2 MASS AND WEIGHT MEASUREMENTS

The *mass* of an object is a measure of the amount of material of which the object is composed, that is, the amount of material in an object remains constant regardless of where it is measured. The *weight* is the force that gravity exerts on the object. This force varies from place to place.

In metric system measurement of mass, the *gram* is the basic unit of weight. A prefix is used in front of the basic unit (gram) to identify the smaller and larger units. Table 5.5 shows the most popular metric measurement of weight.

Table 5.5 Metric Units of Weight (Mass)

1 gram (g) = 1.0 gram (g) (is the basic unit of mass)
1 *kilo*gram (kg) = 1000 grams (g)

In the U.S. customary system, the units are not related as in the metric system. Table 5.6 shows the most popular and basic units used in the U.S. customary system for weight.

Table 5.6 U.S. Customary Units of Weight

1 ton (T) = 2000 pounds (lb)
1 pound (lb) = 16 ounce (oz)

Table 5.7 shows the relationship between the metric system and the U.S. customary system for units of weight.

Table 5.7 Metric System and U.S. System of Weight

Metric to U.S.	U.S. to metric
1 g = 0.035 oz	1 oz = 28.35 g
1 kg = 2.205 lb	1 lb = 0.454 kg

Example 1

Convert the following weight units.

(a) 46 kg to pounds

(b) 46 kg to grams

(c) 25 grams to ounces

Solution

(a) $46 \text{ kg} \times \dfrac{2.205 \text{ lb}}{1 \text{ kg}} = 101.43 \text{ lb}$

(b) $46 \text{ kg} \times \dfrac{1000 \text{ g}}{1 \text{ kg}} = 46,000 \text{ g}$

(c) $25 \text{ g} \times \dfrac{0.035 \text{ oz}}{1 \text{ g}} = 0.875 \text{ oz}$

5.2 EXERCISES

In problems 1–18, express each of the following weights as indicated. If necessary round the answer to two decimal places.

1. 185 lb to kilograms

2. 14.5 oz to grams

3. 600 g to ounces

4. 70.6 kg to pounds

5. 645 g to kilograms

6. 148 mg to grams

7. 75 g to milligrams

8. 320 g to milligrams

9. 3.5 kg to grams

10. 5.4 kg to grams

11. 645 μg to milligrams

12. 235 μg to milligrams

13. 16 mg to micrograms

14. 7.8 mg to micrograms

15. 4.5 tons to kilograms

16. 30 tons to kilograms

17. 12,000 kg to tons

18. 335,000 kg to tons

In problems 19–24, which is heavier or equal?

19. 1 lb or 1 oz

20. 2 g or 1 kg

21. 2 kg or 4 Gg

22. 1 oz or 25 g

23. 1 ton or 200 lb

24. 12 oz or 2 lb

25. How many ounces is 5 lb of potatoes?

26. How many grams is 5 oz?

27. How many pounds is a 60 kg box?

28. How many pounds is a 25 kg box?

29. A box of 50 washers weighs 12 oz. Find the weight in grams of washers.

30. The total weight of 8 metal bars is 140 lb. Find the average weight of each bar in kilograms.

31. The weight of an aluminum casting is 8.35 kg. Determine the weight of the casting in pounds.

32. The weight of a cement step casting is 500 kg. Determine the weight of the casting in pounds.

33. How much does 2 g of water weigh in pounds at sea level?

34. How much does 7 g of water weigh in pounds at sea level?

35. How could a technician mixing chemicals express 5400 fluid ounces in gallons?

36. To mix an order of feed, the following feeds are combined: 5400, 400, 3500 and 2000 lb. Calculate the final mixture in tons.

5.3 CAPACITY MEASUREMENTS OF LIQUID

Capacity is a measure of a liquid occupies into a container. Liquid volume is used in everyday life as we see it in bottles of sodas, water, and milk. Table 5.8 indicates the metric units of liquid volume.

Table 5.8 Metric Units of Capacity for Liquid

1 liter (L) = 1.0 liter (L) (is the basic unit)
1 *milli*liter (mL) = 0.001 liter (L) → 1 L = 1000 mL
1 kiloliter (kL) = 1000 liter (L) → 1 kL = 1000 L

Example 1

How many liters (L) are there in the following units?

(a) 7 kL

(b) 9 kL

(c) 15 mL

(d) 6 mL

Solution

(a) $7 \text{ kL} \times \dfrac{1000 \text{ L}}{1 \text{ kL}} = 7000 \text{ L}$

(b) $9 \text{ kL} \times \dfrac{1000 \text{ L}}{1 \text{ kL}} = 9000 \text{ L}$

(c) $15 \text{ mL} \times \dfrac{1 \text{ L}}{1000 \text{ mL}} = 0.015 \text{ L}$

(d) $6 \text{ mL} \times \dfrac{1 \text{ L}}{1000 \text{ mL}} = 0.006 \text{ L}$

While soft drinks use metric units, milk and similar products use the customary units. Buckets and containers also use this system.

Table 5.9 shows the most popular and basic units used in the U.S. customary system for capacity for liquid.

Table 5.9 U.S. Customary Units of Capacity for Liquid

1 gallon (gal) = 4 quarts (qt)
1 quart (qt) = 2 pints (pt)
1 pint (pt) = 2 cups (c)
1 cup (c) = 8 ounces (oz)
1 tablespoon (tbs) = 3 teaspoons (tsp)

Example 2

How many gallons (gal) are there in the following units?

(a) 8 pt

(b) 5 pt

(c) 8 qt

(d) 11 qt

Solution

(a) $8 \text{ pt} \times \dfrac{1 \text{ qt}}{2 \text{ pt}} \times \dfrac{1 \text{ gal}}{4 \text{ qt}} = 1 \text{ gal}$

(b) $5 \text{ pt} \times \dfrac{1 \text{ qt}}{2 \text{ pt}} \times \dfrac{1 \text{ gal}}{4 \text{ qt}} = \dfrac{5}{8} \text{ gal} = 0.625 \text{ gal}$

(c) $8 \text{ qt} \times \dfrac{1 \text{ gal}}{4 \text{ qt}} = 2 \text{ gal}$

(d) $11 \text{ qt} \times \dfrac{1 \text{ gal}}{4 \text{ qt}} = \dfrac{11}{4} \text{ gal} = 2.75 \text{ gal}$

Table 5.10 shows the relationship between the metric system and U.S. customary system for units of capacity for liquid.

Example 3

Table 5.10 U.S. and Metric System of Capacity Equivalents

Metric to U.S.	U.S. to metric
1 L = 1.06 qt	1 qt = 0.946 L
1 L = 0.264 gal	1 gal = 3.785 L

Change 5 L to quarts.

Solution

$$5\ \text{L} \times \frac{1.06\ \text{qt}}{1\ \text{L}} = 5.30\ \text{qt}$$

Example 4

Change 4 qt to pints.

Solution

$$4\ \text{qt} \times \frac{2\ \text{pt}}{1\ \text{qt}} = 8\ \text{pt}$$

Example 5

Convert 50 fluid ounces to cups.

Solution

$$\frac{50\ \text{oz}}{1\ \text{cup}} \times \frac{1\ \text{cup}}{8\ \text{oz}} = 6.25\ \text{oz}$$

5.3 EXERCISES

In problems 1–18, express each of the following capacities as indicated. If necessary round the answer to two decimal places.

1. 47 L to quarts
2. 18.6 gal to liters
3. 14 lb to ounces
4. 6 lb to ounces

5. 30 lb to ounces
6. 140 oz to pounds
7. 16 tons to pounds
8. 700 oz to pounds
9. 15,600 lb to tons
10. 45 tons to pounds
11. 384,000 oz to tons
12. 2000 lb to tons
13. 3 tons to ounces
14. 5 tons to ounces
15. 1200 mL to liters
16. 90 mL to liters
17. 0.80 L to milliliters
18. 7 L to milliliters
19. How many quarts in 2 L of cola?
20. A 1 L bottle has how many milliliters?
21. How many liters in a 9-quart bucket?
22. How many liters in a 5-quart bucket?
23. A car gasoline tank has a capacity of 13.5 gal. Determine the capacity of the tank in liters.
24. A storage tank holds 125 gal of fuel oil. Determine the capacity of the tank in liters.

5.4 TIME MEASUREMENTS

Time is usually measured in centuries, years, weeks, days, hours, minutes, and seconds as shown in Table 5.11.

Table 5.11 Time Measurement

1 millennium = 1000 years
1 century (cr) = 100 years (yr)
1 year (yr) = 12 months (mo)
1 quarter (qtr) = 3 months
1 month = 4 weeks
1 week (wk) = 7 days (da)
1 day (da) = 24 hours (hr)
1 hour (hr) = 60 minutes (min)
1 minute (min) = 60 seconds (sec)

Example 1

Change the following times from years to months:

(a) 3 years

(b) 7 years

(c) 9 years

(d) 13 years

Solution

(a) $\text{time in months} = 3\ \cancel{\text{years}} \times \dfrac{12\ \text{months}}{1\ \cancel{\text{year}}} = 36\ \text{months}$

(b) $\text{time in months} = 7\ \cancel{\text{years}} \times \dfrac{12\ \text{months}}{1\ \cancel{\text{year}}} = 84\ \text{months}$

(c) $\text{time in months} = 9\ \cancel{\text{years}} \times \dfrac{12\ \text{months}}{1\ \cancel{\text{year}}} = 108\ \text{months}$

(d) $\text{time in months} = 13\ \cancel{\text{years}} \times \dfrac{12\ \text{months}}{1\ \cancel{\text{year}}} = 156\ \text{months}$

5.4 EXERCISES

In exercises 1–24, convert the units as indicated. If necessary round the answer to two decimal places.

1. change 6 hours to seconds

2. change 2.5 hours to seconds

3. change 14 weeks to days

4. change 17 weeks to days

5. change 20 years to decades

6. change 12 years to decades

7. change 7 hours to minutes

8. change 5 hours to minutes

9. change 325 weeks to years

10. change 120 weeks to years

11. change 2 years to days

12. change 5 years to days

13. change 33 months to quarters

14. change 17 months to quarters

15. change 16 years to months

16. change 9 years to months

17. change 10 quarters to years
18. change 6 quarters to years
19. change 8100 decades to centuries
20. change 6400 decades to centuries
21. change 4200 seconds to minutes
22. change 1600 seconds to minutes
23. change 1 day to seconds
24. change 2 days to seconds
25. How many seconds in one week?
26. How many hours in a half century?
27. If you lived a quarter century, how many years have you lived?
28. If you sleep 8 hours a day, how many days are you asleep during one week?
29. A flight from Houston, TX, to Chicago, IL, takes about 3 hours. If Richard makes two round-trip flights each month for business, how many days does he travel in a year?
30. A flight from Boston, MA, to Los Angeles, CA, takes about 6 hours. If Roger makes two round-trip flights each month for business, how many days does he travel in a year?

5.5 TEMPERATURE (T) MEASUREMENT

Temperature is a measure of the warmth or coldness of an object, substance, or environment. The metric system uses Celsius (C) to measure temperature. U.S. customary uses Fahrenheit (F) to measure temperature.

The formulas used for F and C are:

$$F = \frac{9}{5}C + 32$$

to find F from C

and

$$C = \frac{5}{9}(F - 32)$$

to find C from F

Water freezes at $T = 0°$ Celsius (C) and boils at $T = 100°$ Celsius (C) in the metric system. Water freezes at $T = 32°$ Fahrenheit (F) and boils at $T = 212°$ Fahrenheit (F) in the U.S. customary system.

Example 1

Convert the following Celsius (C) temperatures to Fahrenheit (F):

(a) 32°
(b) 41°
(c) 23°
(d) 9°

Solution

(a) We used the formula $F = \frac{9}{5}C + 32$ to find the temperature in Fahrenheit (F), $F = \frac{9}{5}(32) + 32 = \frac{288}{5} + 32 = 89.6° \Rightarrow T = 89.6\ °F$

(b) For $C = 41°$, we find the temperature in Fahrenheit (F) by $F = \frac{9}{5}(41) + 32 = \frac{369}{5} + 32 = 105.8° \Rightarrow T = 105.8\ °F$

(c) Here $C = 23°$, we find the temperature in Fahrenheit (F) by $F = \frac{9}{5}(23) + 32 = \frac{207}{5} + 32 = 73.4° \Rightarrow T = 73.4\ °F$

(d) $C = 9°$, to find the temperature in Fahrenheit (F), $F = \frac{9}{5}(9) + 32 = \frac{81}{5} + 32 = 48.2° \Rightarrow T = 48.2\ °F$

Example 2

Convert the following Fahrenheit (F) temperatures to Celsius (C):

(a) −11°
(b) −6°
(c) 39°
(d) 8°

Solution

(a) We use the formula $C = \frac{5}{9}(F - 32)$ to find C from given Fahrenheit (F), here $T = -11\ °F$, then $C = \frac{5}{9}(-11 - 32) = \frac{-215}{9} = -23.9°$ $\Rightarrow T = -23.9\ °C$

(b) Here $T = 6\ °F$, then we use $C = \frac{5}{9}(-6 - 32) = \frac{-190}{9} = -21.1°$ $\Rightarrow T = -21.1\ °C$

(c) Here T=39° F, then we use $C = \frac{5}{9}(39 - 32) = \frac{35}{9} = 3.9°$ $\Rightarrow T = 3.9\ °C$

(d) Here T = 8 °F, then we use $C = \frac{5}{9}(8 - 32) = \frac{-120}{9} =$ $-13.3° \Rightarrow T = -13.3\ °C$

5.5 EXERCISES

1. Write the formula to convert temperatures from degrees Fahrenheit to degrees Celsius.

2. Write the formula to convert temperatures from degrees Celsius to degrees Fahrenheit.

In problems 1–16, express each of the following temperature units as indicated. If necessary round the answer to two decimal places.

3. 32 °C to °F

4. 41 °C to °F

5. 23 °C to °F

6. 9 °C to °F

7. 1 °C to °F

8. 12° C to °F

9. 32 °F to °C

10. 41 °F to °C

11. 23 °F to °C

12. 9 °F to °C

13. 1 °F to °C

14. 98 °F to °C

15. 87 °F to °C

16. 65 °F to °C

17. If it is 80 °F in America, how much would it read in °C in Europe?

18. If it is 66 °F in America, how much would it read in °C in Europe?

19. When the oil temperature has preheated to 110 °C, how much has the temperature reached in °F?

20. If the beach water is about 28 °C, what is the temperature in °F?

5.6 DERIVED UNITS

Derived units are expressed in terms of base and supplementary units. Several derived units have been given special names and symbols, such as the watt with symbol W. The special names and symbols can be used to express the units of a quantity in a simpler way than by using the base units. Tables 5.12–5.14 list many quantities of derived SI units.

Table 5.12 Quantities of Units Having Special Names

Quantity	SI name	SI symbol	Expression in terms of other units
Frequency	hertz	Hz	cycle/s
Force	newton	N	kg·m/s^2
Pressure, stress	pascal	Pa	N/m^2
Work, energy	joule	J	N·m
Power	watt	W	J/s
Electric charge	coulomb	C	A·s
Electrical potential	volt	V	W/A
Capacitance	farad	F	C/V
Electric resistance	ohm	Ω	V/A
Conductance	siemens	S	A/V
Magnetic flux	weber	Wb	V·s
Magnetic flux density	tesla	T	Wb/m^2
Inductance	henry	H	Wb/A
Current	ampere	A	C/s

Table 5.13 Quantities of Units Expressed in Terms of Base and Supplementary Units

Quantity	SI name	SI symbol
Area	square meter	m^2
Volume	cubic meter	m^3
Speed, velocity	meter per second	m/s
Acceleration	meter per second squared	m/s^2
Density	kilogram per cubic meter	kg/m^3
Specific volume	cubic meter per kilogram	m^3/kg
Magnetic field strength	ampere per meter	A/m
Concentration	moles per cubic meter	mol/m^3
Luminance	candela per square meter	cd/m^2
Kinematic viscosity	square meter per second	m^2/s
Angular velocity	radian per second	rad/s
Angular acceleration	radian per second squared	rad/s^2

Table 5.14 Quantities of Units Expressed in Terms of Derived Units with Special Names

Quantity	SI name	SI symbol
Viscosity	pascal second	Pa·s
Moment of force, torque	newton meter	N·m
Surface tension	newton per meter	N/m
Heat flux density	watt per square meter	W/m^2
Entropy	joule per kelvin	J/K
Thermal conductivity	watt per meter kelvin	W/K·m
Electric field strength	volt per meter	V/m
Electric charge density	coulomb per cubic meter	C/m^3
Electric flux density	coulomb per square meter	C/m^2
Permittivity	farad per meter	F/m
Permeability	henry per meter	H/m

Example 1

The force of gravity on an object at sea level is 3000 N. Find the mass of the object.

Solution

The force of gravity on a 100 g mass is 1 N.

$$3000\ \text{N} \times \frac{100\ \text{g}}{1\ \text{N}} = 300,000\ \text{g (or 300 kg)}$$

Example 2

How much energy is required to exert a force of 20 N through a distance of 8 m?

Solution

Energy (J) = Force (N) × distance (m)

Energy = 20 N × 8 m = 160 N·m (or 160 J)

Example 3

How much power is required to expend 20 J of energy in 10 seconds?

Solution

$$\text{Power (w)} = \frac{\text{Energy (J)}}{\text{Time (s)}} = \frac{20\ \text{J}}{10\ \text{s}} = 2\ \text{watts}$$

Knowing that 1 L of water weighs 1 kg, 1 cm^3 of water weighs 1 g, and 1 ton = 1000 kg, we can demonstrate the following examples.

Example 4

A container holds 684 cm^3 of water. Find the weight of water in kilograms.

Solution

1 cm^3 of water weighs 1 g. Thus, 684 g = 0.684 kg, thus, 684 cm^3 of water weighs 0.684 kg.

Example 5

If the tangential speed of an object is 5 m/s, which is in a circular motion on a path of a radius of 2 m, compute the centripetal acceleration of the object.

Solution

The centripetal acceleration = (the tangential speed of the object)^2/radius = $(5)^2/2 = 12.5$ m/s^2.

5.6 EXERCISES

In exercise 1–4, give the symbol and the meaning of the given unit.

1. millimeter
2. microampere
3. megawatt
4. kilowatt
5. Determine the force of gravity in newtons on a mass of 900 g.
6. Determine the force of gravity in newtons on a mass of 4 kg.
7. The voltage across a 6 kΩ resistor is 18 V. Find the current through the resistor.
8. The voltage across a 3 kΩ resistor is 12 V. Find the current through the resistor.
9. An electric iron draws a current 3 A at a voltage 120 V. Determine the value of its resistance.
10. When the voltage across a resistor is 120 V, the current through it is 5 mA. Find its conductance.
11. How much power is required to expend 15 joules of energy in 6 seconds?

12. An electric circuit draws 6 A of current and expends energy at the rate of 20 kJ per minute. Determine the voltage that is necessary in this circuit.
13. A rectangular cement block has dimensions 30 cm by 14 cm by 2 dm. Calculate the mass of the block.
Hint: cement has a mass of 3 g/cm^3.
14. A rectangular cement block has dimensions 25 cm by 12 cm by 1 dm. Calculate the mass of the block.
Hint: cement has a mass of 3 g/cm^3.
15. A force of 12 N is extracted on a square plate of side 15 cm. Determine the pressure on the plate in pascals.
16. A force of 15 N is extracted on a rectangular plate 20 cm by 200 cm. Determine the pressure on the plate in pascals.
17. A charge of 6.5 C flows through an element for 0.3 second. Determine the amount of current through the element.
18. The current through a certain element is measured to be 9.4 A. How much time will it take for 4 mC of charge to flow through the element?
19. If 42 J of energy is required to move 6 mC of charge though an element, what is the voltage across the element?
20. If 34 J of energy is required to move 3 mC of charge though an element, what is the voltage across the element?

CHAPTER 5 REVIEW EXERCISES

In problems 1–24, convert the quantity to the indicated unit. If necessary round the answer to two decimals.

1. 45 m to feet
2. 30 m to feet
3. 160 ft to meters
4. 120 ft to meters
5. 1.6 ft to centimeters
6. 1.9 ft to centimeters
7. 6.4 cm to inches
8. 3.8 cm to inches
9. 6.20 km to miles
10. 8.4 km to miles
11. 5.2 mi to kilometers
12. 6.8 mi to kilometers

13. 6.8 in. to centimeters

14. 7.5 in. to centimeters

15. 3 m to centimeters

16. 5.5 m to centimeters

17. 4 mm to centimeters

18. 3.5 mm to centimeters

19. 7 yd to feet

20. 5 mi to feet

21. 16 in. to feet

22. 8.5 in. to feet

23. 4 ft to centimeters

24. 15 m to feet

In exercises 25–32, change each of the given units.

25. 2.4 kW to watts

26. 325 mA to ampere

27. 48 V to kilovolt

28. 1400 kΩ to megaohm

29. 0.052 C to millicoulomb

30. 3.6 mA to microampere

31. 30,000 μA to ampere

32. 620 cm to meters

In exercises 33–38, give the symbol and the meaning of the given units.

33. millimeter

34. microampere

35. nanosecond

36. megawatt

37. kilovolt

38. picosecond

In exercises 39–46, write the abbreviation for each quantity.

39. 25 milligrams

40. 200 milliliters

41. 66 centimeters

42. 95 kiloliters

43. 41 kilograms

44. 22 microamps

45. 12 megawatts

46. 7 hectoliters

In exercises 47–52, change each of the following units.

47. 160 mA to ampere

48. 0.25 mA to microampere

49. 2.6 MΩ to kiloohm

50. 1600 Ω to kiloohm

51. 20 pF to nanofarad

52. 0.84 GHz to kilohertz

53. A plane climbs to an altitude of 3700 m. What is the plane's altitude in kilometers?

54. A human hair is 4.5×10^{-4} cm thick. What is the human hair thickness in millimeters?

55. A calculator is 1.8 mm thick. What is the calculated thickness in inches?

56. A car is clocked at a speed 15 ft/s. What is the car speed in meters per second?

57. A fence on a ranch measures a total of 2 miles long. Find the length of the fence in feet.

58. A tank contains 12 gal of fuel. How many liters of fuel are in the tank?

59. How many pounds does a 170 kg satellite weigh?

60. An iron bar weighs 3 lb. What is its weight in ounces?

61. A micro wheel weighs 0.065 oz. What is its weight in milligrams?

62. Mary drives 65 miles per hour. How many yards does Mary drive each hour?

63. Determine the force in newtons needed to hold a mass of 150 g in the air at sea level?

64. Determine the force in newtons needed to hold a mass of 50 g in the air at sea level?

65. A light uses 30 kJ of energy every minute. Find the power rating of the light.

66. Determine the power used by a motor that expends 70 kJ of energy every 40 seconds.

Chapter 6

Geometry

Learn as if you were going to live forever. Live as if you were going to die tomorrow.

■ Automatic Picture Transmission (APT) enabled meteorologists to obtain immediate local area cloud pattern photographs when the Nimbus satellite was within a 1700-mile range of a receiving station. The APT subsystem pioneered on Nimbus 1, provided direct readout of nighttime and daytime cloud coverage. It transmitted photographic data of synoptic meteorological conditions in areas 1200nmi square to over 300 ground stations in more than 43 countries.

INTRODUCTION

The study of measurements and properties of regions, figures, and solids formed by points, lines, and planes is called *geometry*. Point, line, and plane are the most popular undefined concepts in geometry; they are only described instead of defined. They are used for defining other terms in geometry. For example, geometry is used regularly in many jobs such as carpentry, machining, plumbing, auto-body repair, and drafting.

Fundamentals of Technical Mathematics. http://dx.doi.org/10.1016/B978-0-12-801987-0.00006-X

6.1 BASIC CONCEPTS IN GEOMETRY

A **point** is represented by a dot that has position and identifies a location, but it has no size and no dimension (no length, width, and height or thickness); it is usually named by a single capital letter.

Example 1

Point *A*:

A •

A **line** is a series of connected points that goes in two directions without ending (infinity); it has length only, and has no width or height or thickness. The name of a line usually can be a single lowercase letter such as *l* (if no points on the line are known) or two letters on the line with a line drawn over the letters such as, $\overline{AB}$ or $\overline{BA}$ (if any of two of its points are known such as *A* and *B*).

Example 2

Line *AB* drawn with arrows, or $\overline{AB}$:

Line *l*:

l

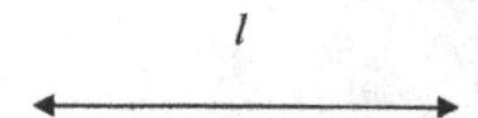

A **line segment** is a set of all points between two points (the end points) including the two points, which makes it a line that is a part of a line consisting of end points, and can be named as $\overline{AB}$ or $\overline{BA}$.

Example 3

Line segment *AB* ($\overline{AB}$) or line segment *BA* ($\overline{BA}$):

The end points of a line segment are the two points where the line segment begins and ends.

The length of a line segment is the distance between its end points.

Example 4

Line segment n:

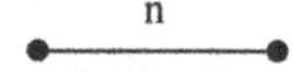

A **curved line** is a line with no part of it that is straight.

Example 5

Curve line *AB*:

A **ray** is a set of all points on one side of the end point, which makes it a part of a line with the end points, and can be named as $\overrightarrow{AB}$.

Example 6

Ray *AB* ($\overrightarrow{AB}$):

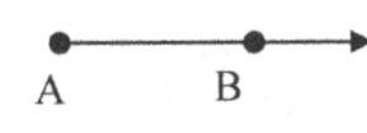

Ray *BA* ($\overrightarrow{BA}$):

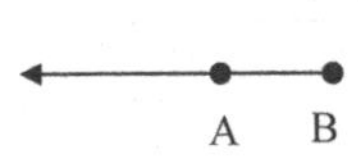

A **plane** is a set of points that forms a flat surface and goes without ending in all directions; it has length and width only, but has no thickness, no edges, and no boundaries.

Its name usually is presented in a capital letter located in any of its corners such as Q and usually pictured by a four-sided figure.

Example 7

Plane Q:

An **angle** is an intersection between two rays at the same end point. The end point is called the *vertex* of the angle, and the rays are called the *sides* of the angle. The symbol of an angle is usually presented as $\angle$ or $\measuredangle$. Angles also can be named with a number, a Greek alphabet letter, or a lower case letter.

Example 8

Vertex $= B$
Sides $= \overrightarrow{BA}$ and $\overrightarrow{BC}$
Angle $ABC(\angle ABC)$ or angle $CBA(\angle CBA)$

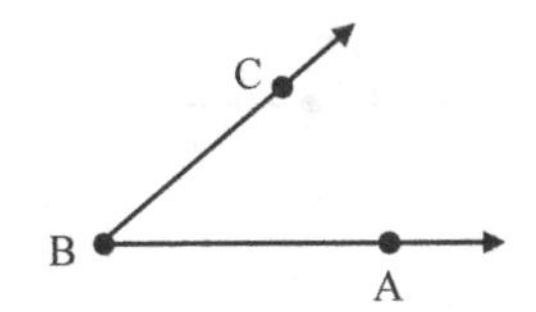

Example 9

Angle 25 ($\angle 25$)

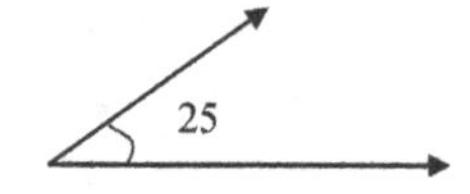

Angle m ($\angle m$)

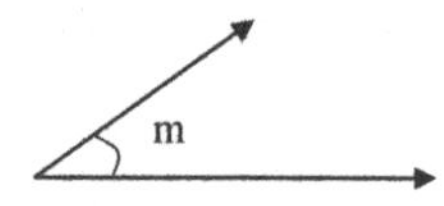

Angle alpha ($\angle \alpha$)

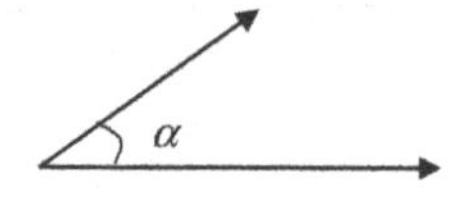

Example 10

Name the angle in four different ways.

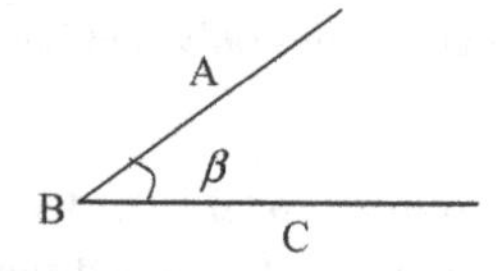

Solution

$\angle B$, $\angle \beta$, $\angle ABC$, and $\angle CBA$.

The **size** of an angle is found by the amount of rotation between the sides, and it is measured by a unit called a **degree** (°).

There are three common rotations:

1. 360 degrees (360°): is one complete rotation of a side about a vertex.

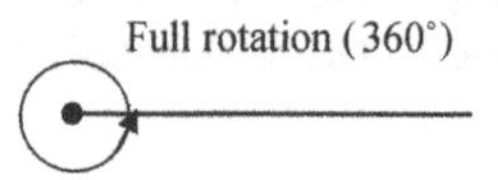

2. 180 degrees (180°) or straight line: is one-half rotation of a side about a vertex.

One-half rotation (180°)

3. 90 degrees (90°) or perpendicular (⊥): is one-quarter rotation of a side about a vertex.

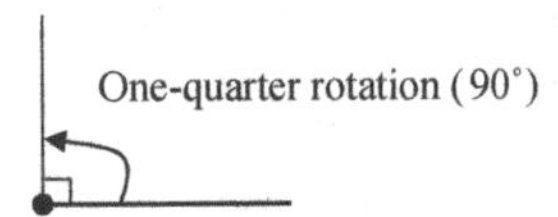

Perpendicular lines (⊥): are two lines that intersect to form a 90-degree angle.

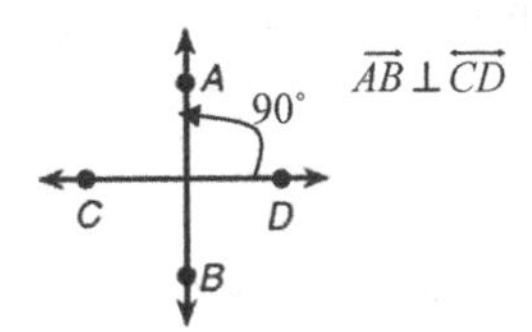

Parallel lines (||): are two lines that are in same plane and with the same distance apart at any point, that is, they never meet regardless of how far they are extended.

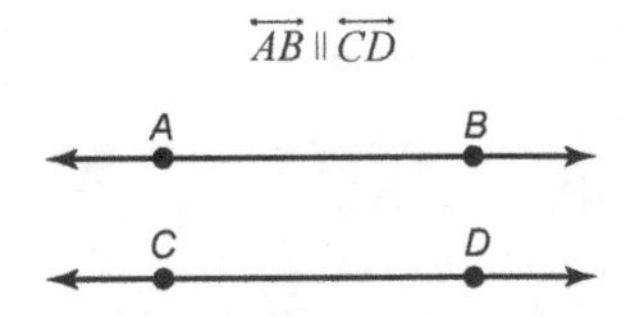

Example 11

Determine the perpendicular and parallel line segments in the figure.

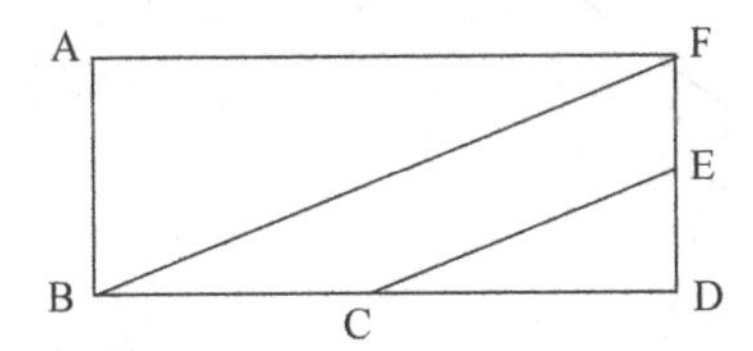

Solution

Perpendicular line segments are:

$AB \perp AF$

$BCD \perp DEF$

$AF \perp FED$

$AB \perp BCD$

Parallel line segments are:

$BE \parallel CE$

$AB \parallel FED$

$AF \parallel BCD$

A closed figure that is formed by three or more line segments is called a **polygon**. We usually identify polygons by their number of sides. There are six types of polygons:

1. Triangle: 3 sides

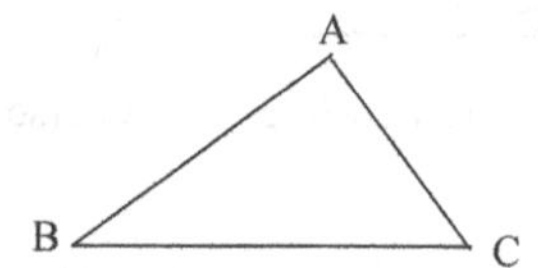

2. Quadrilateral: 4 sides

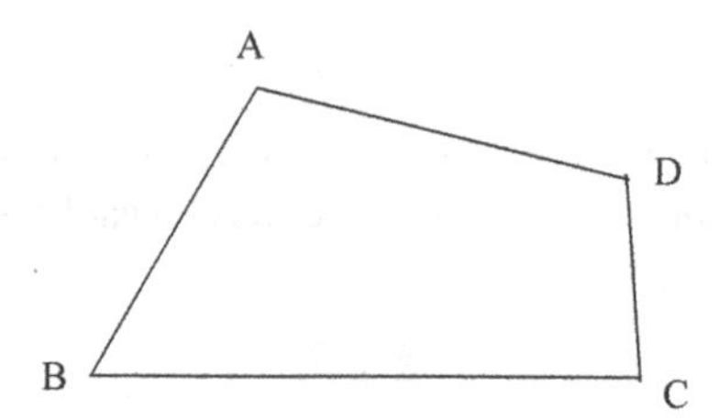

3. Pentagon: 5 sides

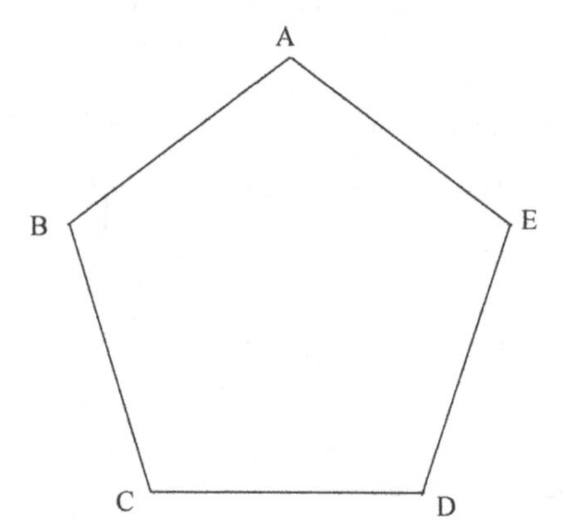

4. Hexagon: 6 sides

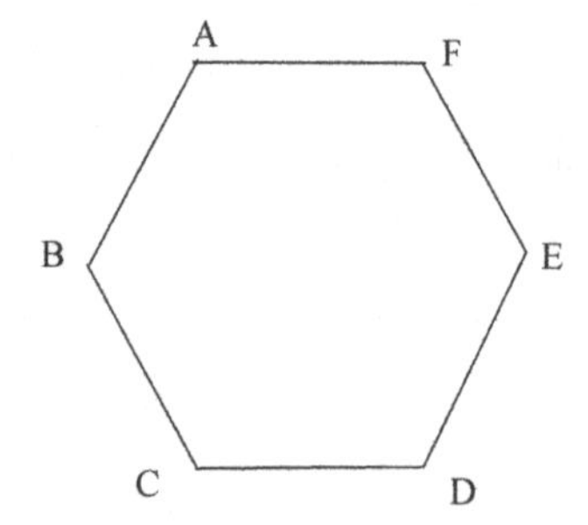

5. Octagon: 8 sides

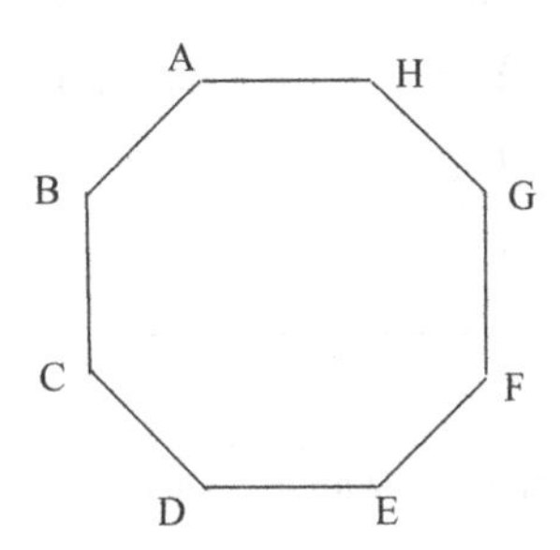

6. Decagon: 10 sides

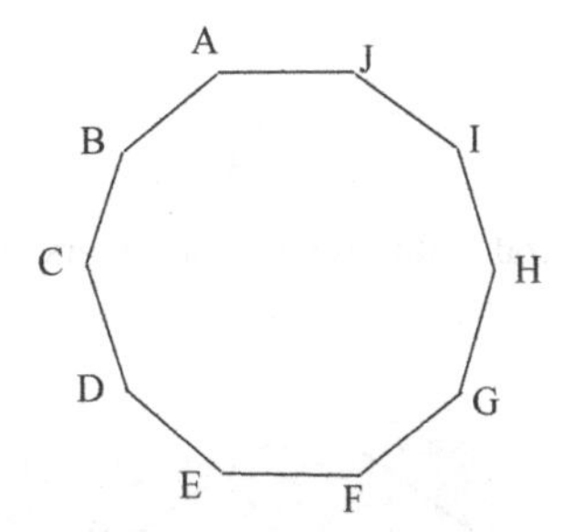

Quadrilaterals are divided into subclasses based on their sides and angles, and the common ones are:

1. Parallelogram: is a quadrilateral with both pairs of opposite sides parallel.

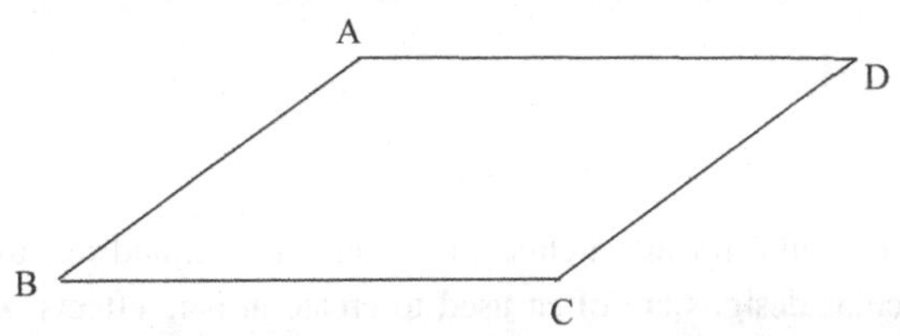

2. Trapezoid: is a quadrilateral with only one pair of opposite sides parallel.

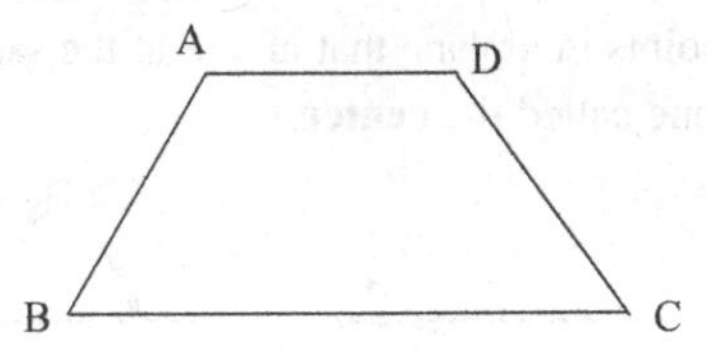

3. Square: is a parallelogram with all sides equal in length and all angles equal to 90°.

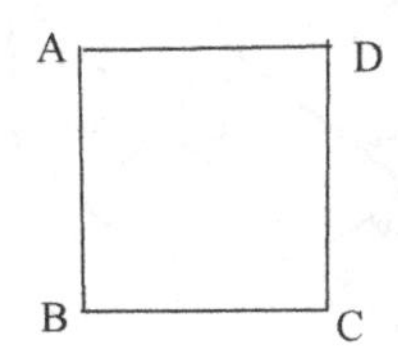

4. Rectangle: is a parallelogram with all angles equal to 90°.

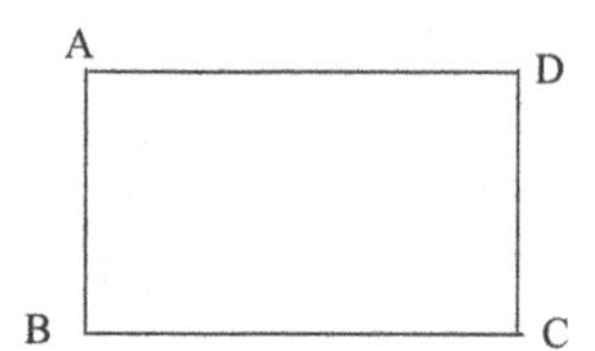

5. Rhombus: is a parallelogram with all sides equal in length.

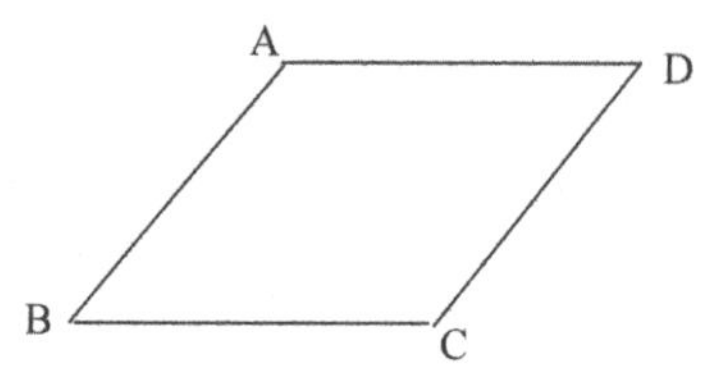

6. Kite: is a quadrilateral with exactly two pairs of distinct congruent constructive sides.

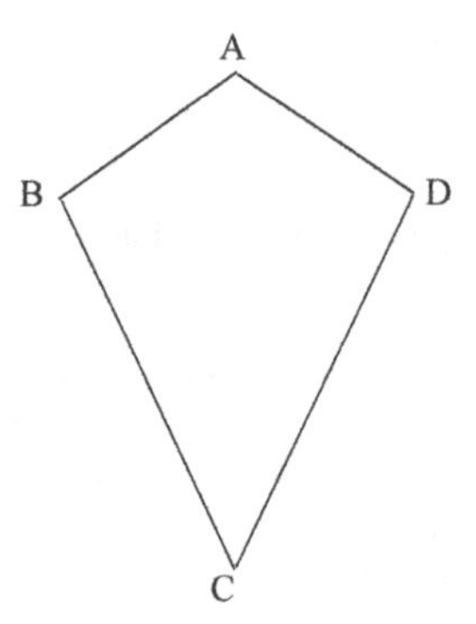

Circles are essential for art, architecture, construction, and manufacturing. Indeed, circular designs are often used to create artistic effects. Also, machines operate by the use of combinations of gears and pulleys.

A **circle** is a set of points in a plane that all are at the same distance from a fixed point in the plane called the **center**.

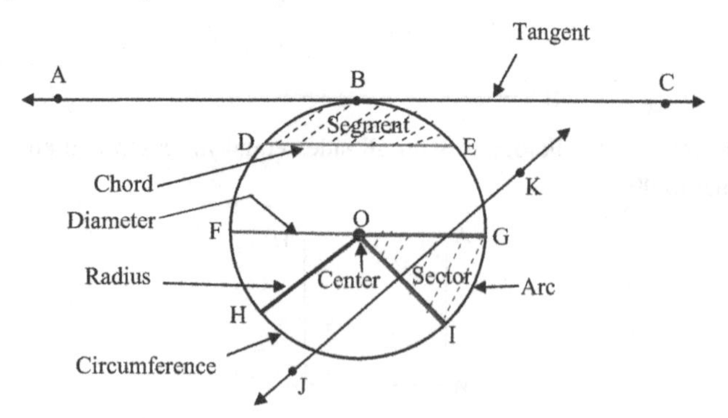

There are terms commonly used to describe the properties of circles, including the following:

A **circumference** is the length of the curved line that forms the circle. See the above figure.

A **diameter** is a chord that passes through the center of a circle. In the above figure, *FG is* a diameter.

A **chord** is a straight line segment that joins two points on the circle. In the above figure, *DE* is a chord. Diameter *FOG* is also a chord of the circle.

A **radius** is a straight line segment that connects the center of a circle with any point on the circle. A radius is equal to one-half the diameter of a circle. In the above figure, *HO* is a radius. Also, *FO*, *OG*, and *OI* are radii of the circle.

An **arc** is that part of a circle between any two points on the circle. In the above figure, an arc is *IG* ($\overset{\frown}{IG}$).

A **tangent** is a straight line that touches the circle at only one point. The point on the circle touched by the tangent is called the **point of tangency**. In the above figure, $\overleftrightarrow{AC}$ is a tangent and point *B* is the point of tangency.

A **secant** is a straight line passing through a circle and intersecting the circle at two points. In the above figure, $\overleftrightarrow{JK}$ *is* a secant.

A **sector** is a figure formed by two radii and the arc intercepted by the radii. In the above figure, the shaded portion *GOI is* a sector.

A **segment** is a figure formed by an arc and the chord joining the end points of the arc. In the above figure, the shaded portion *BDE* is a segment.

6.1 EXERCISES

1. Define geometry.
2. Write each statement using symbols.
   - **a.** Segment *AB* is parallel to segment *CD*.
   - **b.** Line *MN* is perpendicular to line *JK*.
   - **c.** Segment *TJ* is parallel to line *FM*.
3. Refer to the following figure, identify all the line segments.

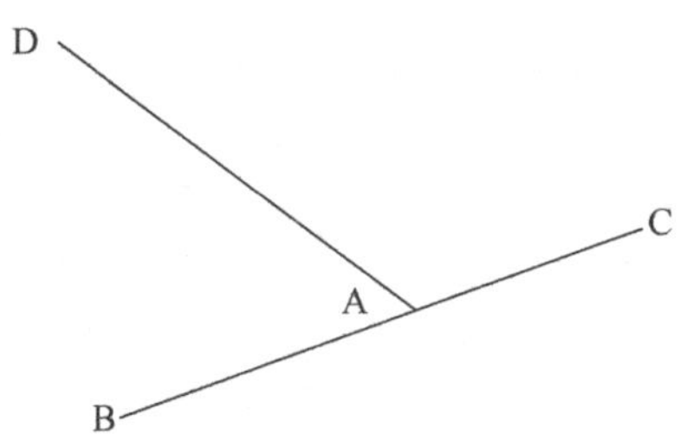

4. Refer to the following figure, identify all the line segments.

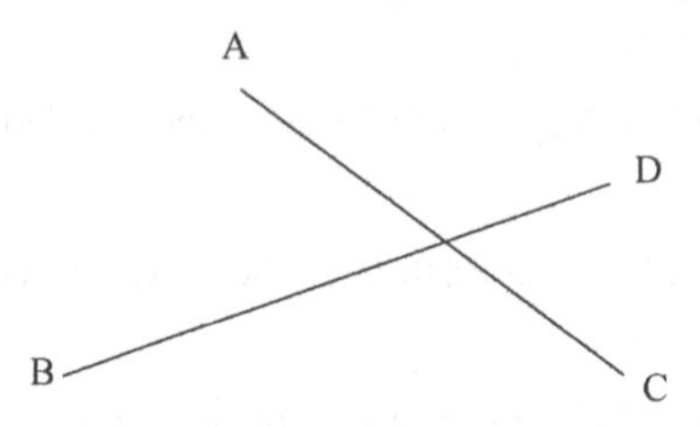

5. Refer to the following figure, identify all the line segments.

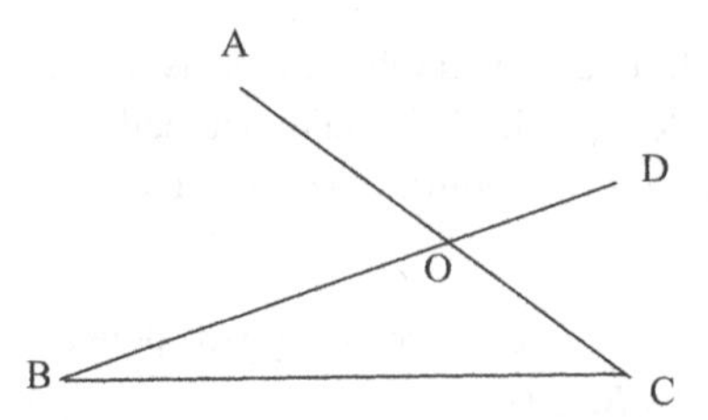

6. Refer to the following figure, identify all the line segments.

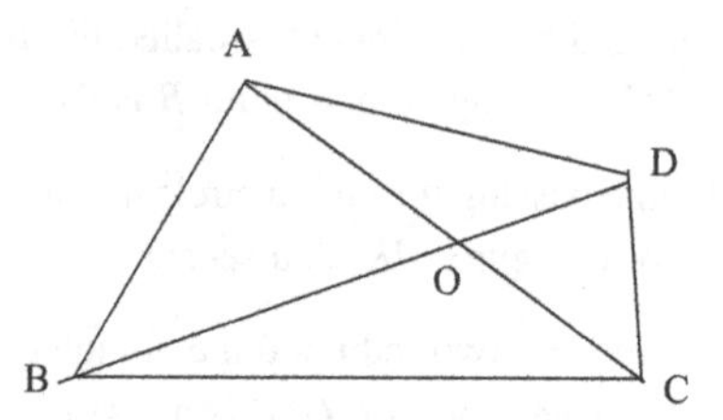

7. Name all of the angles of the polygon.

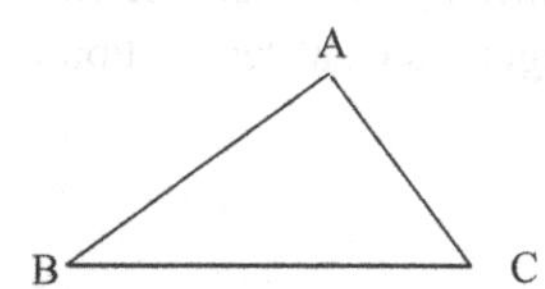

8. Name all of the angles of the kite.

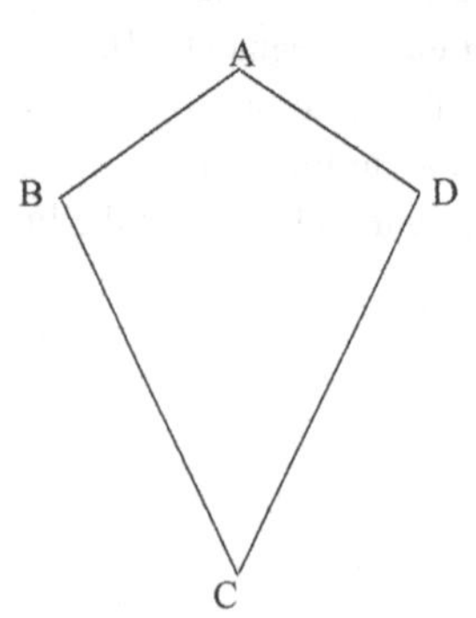

9. How many sides do the following have?
 a. Triangle
 b. Quadrilateral
 c. Pentagon

10. How many sides do the following have?
 a. Hexagon
 b. Octagon
 c. Decagon

11. Identify the line segments that are parallel or perpendicular.

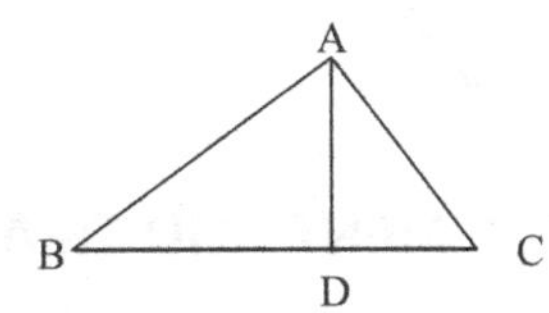

12. Identify the line segments that are parallel or perpendicular.

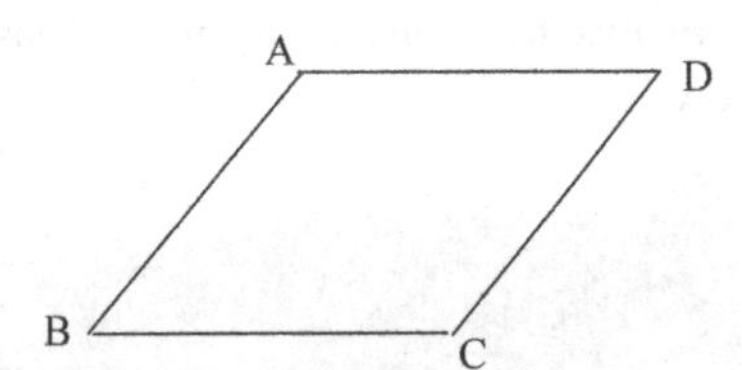

13. Refer to the following figure. Write the word that identifies each of the following.
 a. AD
 b. BC
 c. Point O
 d. EO

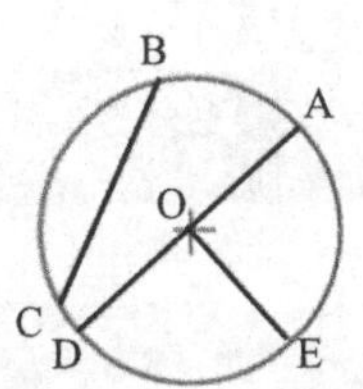

14. Refer to the following figure. Write the word that identifies each of the following.
 a. $\overset{\frown}{DE}$
 b. $\overleftrightarrow{BC}$
 c. Point A
 d. $\overleftrightarrow{FG}$
 e. $\overline{DE}$

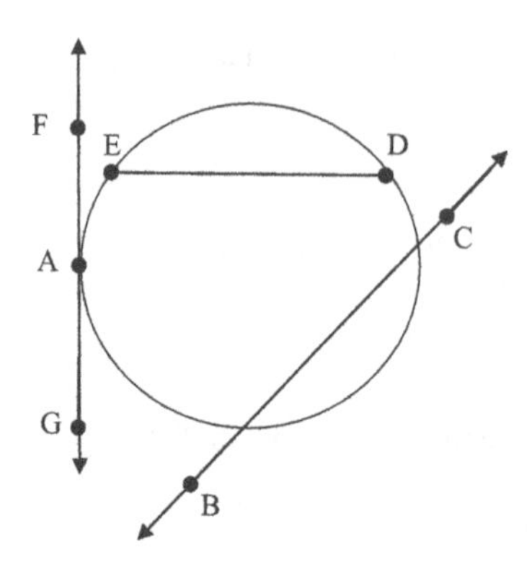

6.2 ANGLE MEASUREMENT AND TRIANGLES

Angles are measured as part of a circle. The standard unit for measuring an angle is degrees. The circle is divided into 360 parts. Each part is one degree, written as $1°$. A **protractor** is an instrument for measuring angles as in the figures given below.

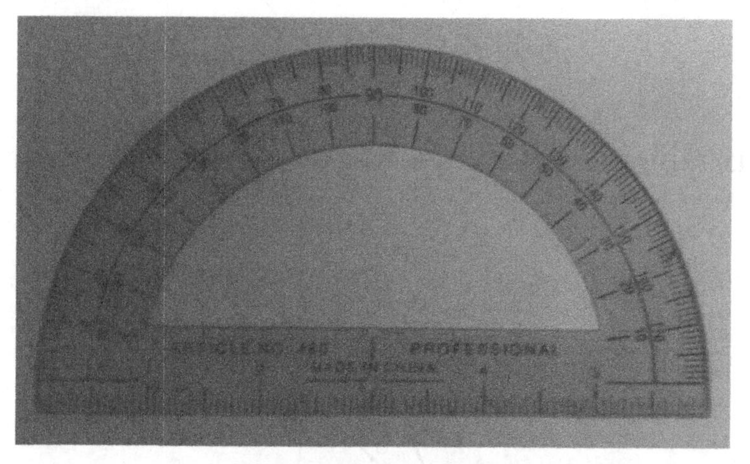

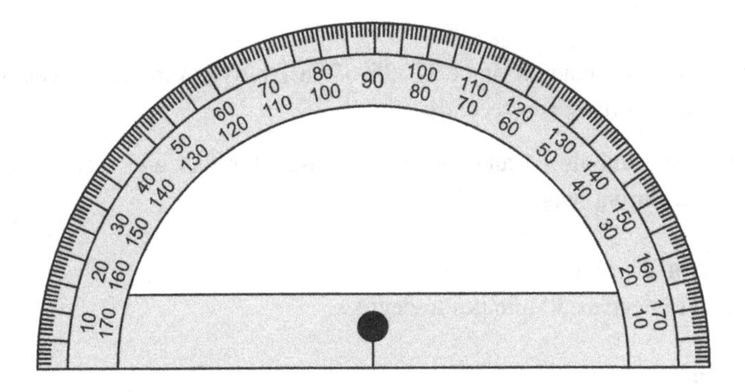

Example 1

What are the steps to measure the angle $\angle ACB$ in the below figure?

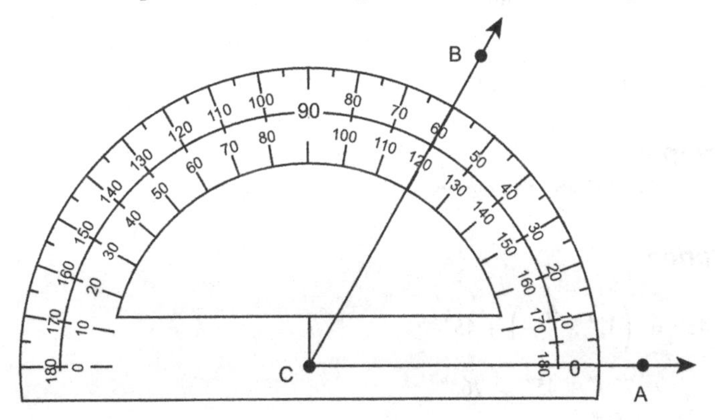

Solution

1. Place the midpoint mark of the protractor over the vertex C of the angle $\angle ACB$.
2. Align the line $\overleftrightarrow{CA}$ with the 0° mark on the protractor,
3. The angle $\angle ACB$ is the number where line $\overleftrightarrow{CA}$ crosses the scale of the protractor that is 60°.

One circle is divided into 360 degrees (°). A degree is divided into 60 equal parts called a **minutes** (′). A minute is divided into 60 equal parts called a **second** (″). Angles are represented as decimal parts of a degree that can be expressed as degrees, minutes, and seconds. The relationship between degrees, minutes, and seconds is shown in Table 6.1.

Table 6.1 Units of Angular Measure in Degrees, Minutes, and Seconds

1 circle = 360°	$1^\circ = \frac{1}{360}$ of a circle
$1^\circ = 60'$	$1'' = \frac{1}{60^\circ}$
$1' = 60''$	$1'' = \frac{1}{60'}$

For example, we read the angle $37° 25' 47''$ as, thirty-seven degrees twenty-five minutes forty-seven seconds.

Note: The symbols ($'$) and ($''$) are also used for feet and inches when measuring length.

Example 2

Change 13 degrees 30 minutes to degrees.

Solution

$$30 \text{ min} = 30\cancel{\text{min}} \times \frac{1 \text{ degree}}{60\cancel{\text{min}}} = 0.5 \text{ degree}$$

Thus, $13° 30' = 13° + 0.5° = 13.5°$.

Example 3

Change $57° 12' 45''$ to degrees.

Solution

$$12' 45'' = \left(12' \times \frac{60''}{1'}\right) + 45'' = 720'' + 45'' = 765''$$

$$1° = 60' \times \frac{60''}{1'} = 3600''$$

$$765'' \times \frac{1°}{3600''} = \frac{765°}{3600} = 0.21°$$

Thus, $57° 12' 45'' = 57° + 0.21° = 57.21°$.

Example 4

Express $45.35°$ as degrees and minutes.

Solution

Multiply the decimal part of the degrees by 60 minutes to obtain minutes.

$$0.35° = 0.35° \times \frac{60'}{1°} = 21'$$

Combine degrees and minutes.

$45.35° = 45° 21'$

Example 5

Express $38.3765°$ as degrees, minutes, and seconds.

Solution

Multiply the decimal part of the degrees by 60 minutes to obtain minutes.

$$0.3765^\circ \times \frac{60'}{1^\circ} = 22.59'$$

Multiply the decimal part of the minutes by 60 seconds to obtain seconds.

$$0.59' \times \frac{60''}{1'} = 35.4'' = 35''(\text{rounded to the nearest whole second}).$$

Combine degrees, minutes, and seconds.

$38^\circ + 22' + 35'' = 38^\circ\ 22'\ 35''$

Thus, $38.3765^\circ = 38^\circ\ 22'\ 35''$.

In machining operations, dimensions at times are computed to seconds in order to ensure accurate and the proper functioning of parts.

Example 6

Find the measure of angle $\angle\theta = \angle BAC$ in the below illustration.

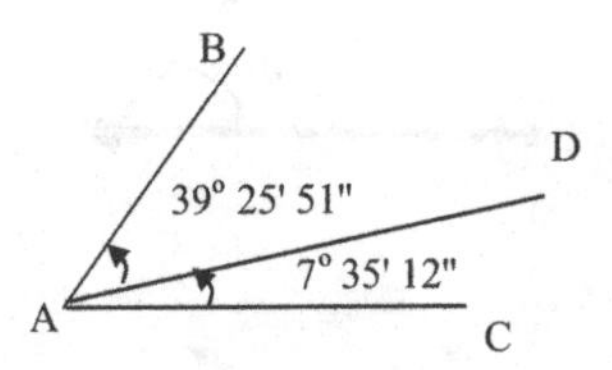

Solution

$$\begin{array}{rr} & 39^\circ 25' 51'' \\ + & 7^\circ 35' 12'' \\ \hline & 46^\circ 60' 63'' \end{array}$$

So we can say by using Table 6.1 that $46^\circ\ 60'\ 63'' = 47^\circ\ 1'\ 3''$.

Thus, the sum of the two angles is $47^\circ\ 1'\ 3''$.

6.2 EXERCISES

1. Write the value of angle A in the figure given below.

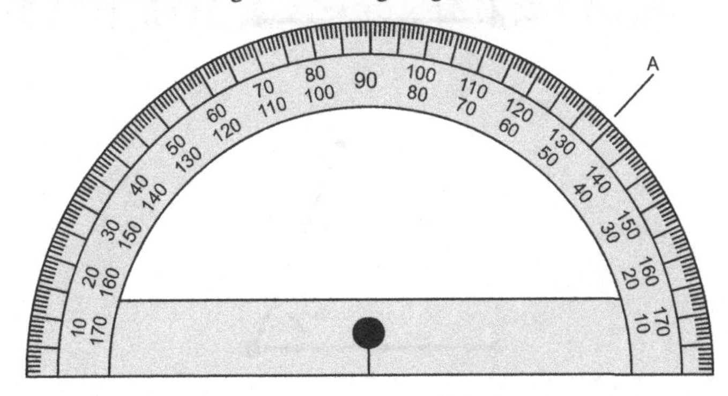

2. Write the value of angle B in the figure given below.

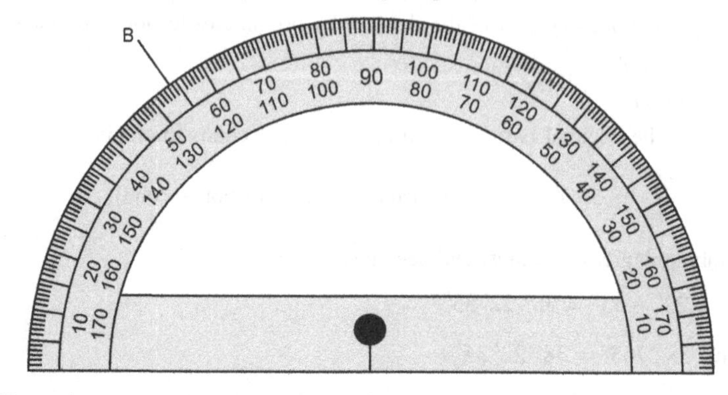

Use a protractor to measure the angles in exercises 3–6.

3.

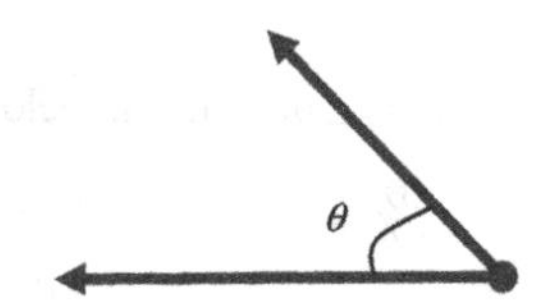

4.

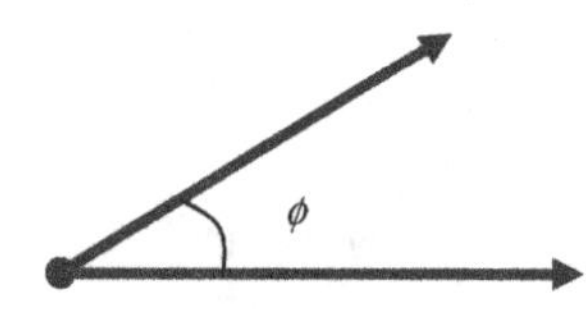

5.

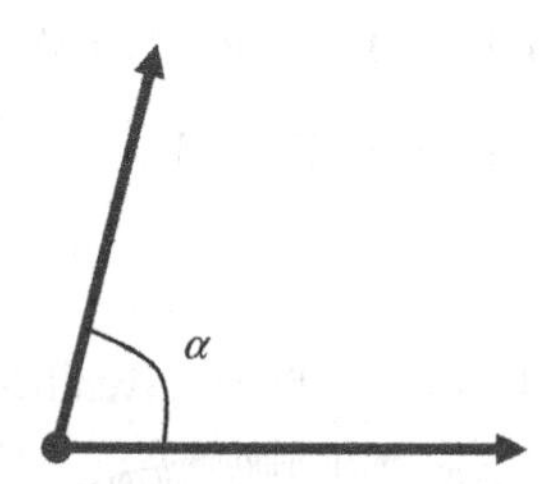

6.

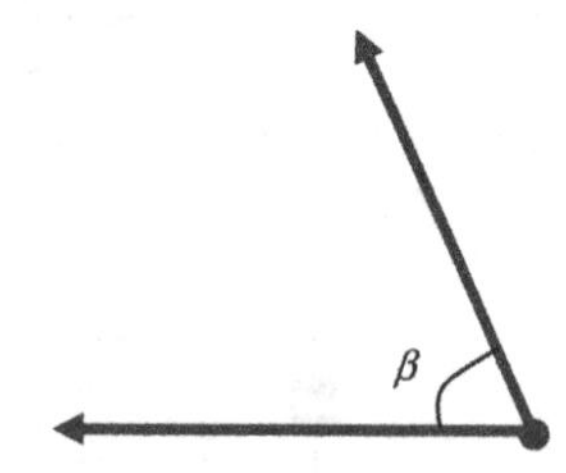

Change the following degrees and minutes to degrees. Round the answers to two decimal places where needed in exercises 7–12.

7. 25° 34′
8. 48° 52′
9. 15° 40′
10. 74° 15′
11. 7° 12′
12. 20° 43′

Change the following degrees, minutes, and seconds to degrees. Round the answers to three decimal places where needed in exercises 13–18.

13. 12° 18′ 30″
14. 20° 59′ 24″
15. 83° 59′ 17″
16. 115° 57′ 7″
17. 146° 44′ 6″
18. 107° 0′ 1″

Change the following degrees to degrees and minutes. Round the answers to the nearest minute where needed in exercises 19–24.

19. 75.6°
20. 12.06°
21. 114.92°
22. 51.12°
23. 39.13°
24. 0.514°

Change the following degrees to degrees, minutes, and seconds. Round the answers to the nearest second where needed in exercises 25–30.

25. 89.7°
26. 102.18°
27. 23.34°
28. 241.03°
29. 68.275°
30. 18.5384°

31. Determine the value of the angle θ in the below illustration.

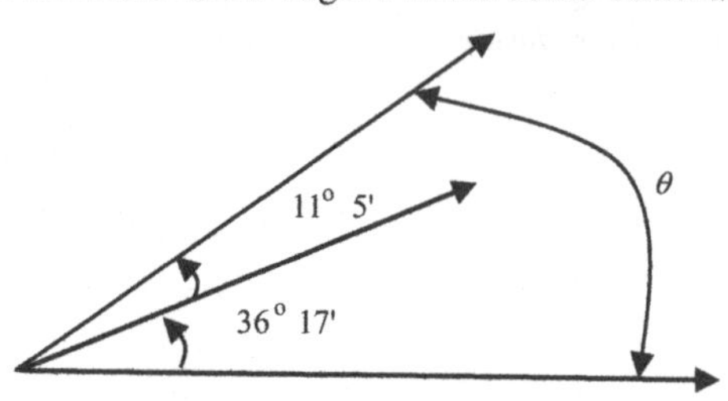

32. Determine the value of the angle θ in the below illustration.

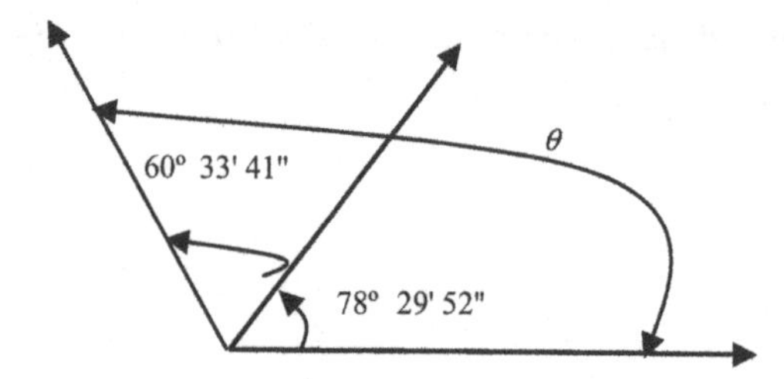

6.3 PERIMETER AND CIRCUMFERENCE IN GEOMETRY

Length, area, and volume (capacity) of an object are essential measurements in engineering and technical shop problems. Indeed, perimeters play a very important part of overall design in the manufacture of metal stampings.

The **perimeter** of a figure is the total length of the sides of the figure.

The perimeter (P) of a *polygon* with n sides is given by

$$P_{polygon} = s_1 + s_3 + \cdots + s_n$$

where s_1, s_3,..., s_n are the individual lengths of each of the sides of the polygon.

Example 1

Calculate the perimeter of the machine part shown.

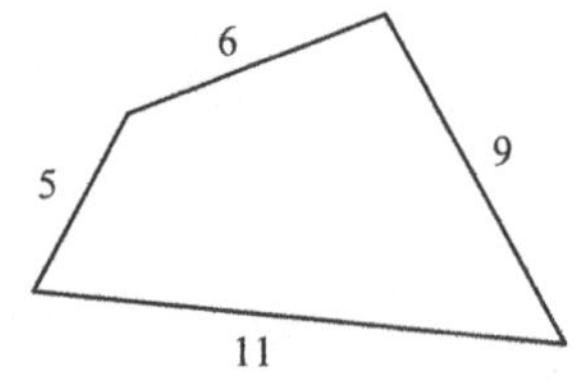

Solution

Perimeter $= P = 5 + 6 + 9 + 11 = 31$.

An *equilateral polygon* is a polygon with its sides in equal lengths. Therefore, we can present its perimeter as

$$P_{\substack{equilateral\\polygon}} = ns$$

where n is the number of sides of the polygon, and s is the length of any one side.

For example, a *square* is an equilateral polygon, thus its perimeter is given by

$$P_{square} = 4\,s$$

s

s s

s

Also, the perimeter for a rhombus is given by

$$P_{\text{rhombus}} = 4\,s$$

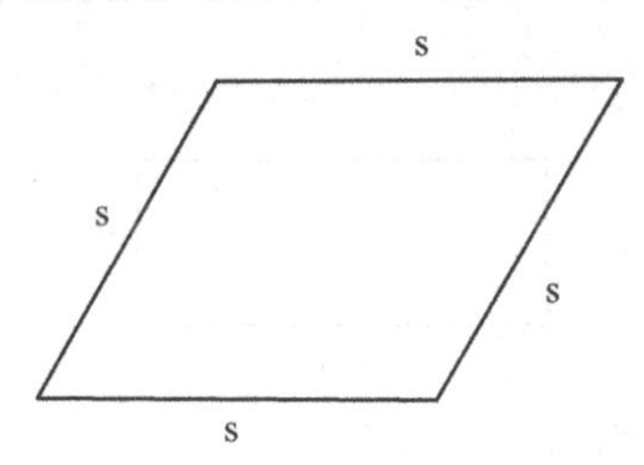

Example 2

Find the perimeter of a square whose sides measure as shown in the following figure.

2 ft.

2 ft. 2 ft.

2 ft.

Solution

Perimeter $= P = 4 \times 2 = 8$ ft.

The perimeter for a *parallelogram* (its opposite sides are equal), is given by

$$P_{parallelogram} = 2a + 2b = 2(a + b)$$

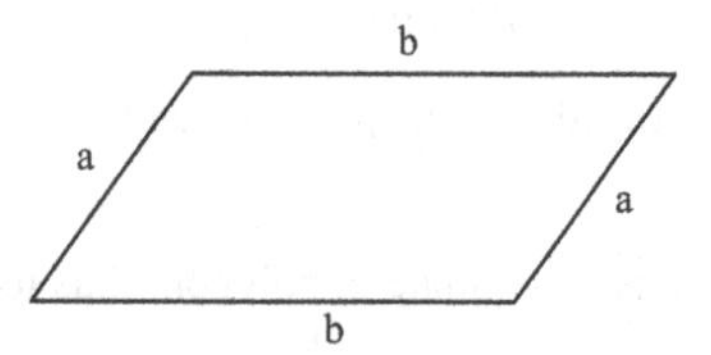

The perimeter for a *rectangle* (its opposite sides are equal), is given by

$$P_{rectangle} = 2a + 2b = 2(a + b)$$

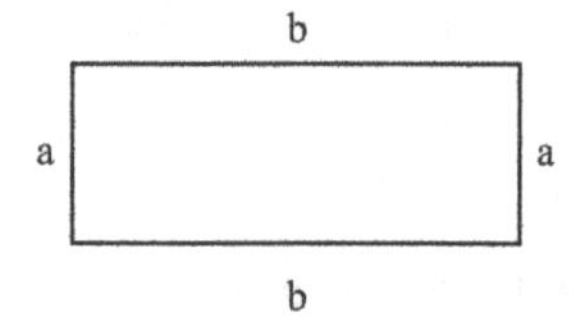

Example 3

Find the perimeter of a rectangle whose sides measure as in the following figure.

5 in.

2 in. 2 in.

5 in.

Solution

Perimeter $= P = (2 \times 5) + (2 \times 2) = 14$ in.

The perimeter, P, of a triangle is given by the

$$P_{triangle} = a + b + c$$

where a, b, and c are the side lengths of the triangle.

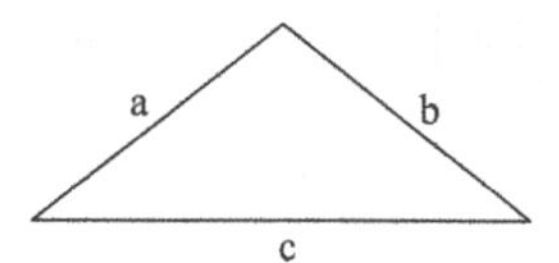

The perimeter for an *isosceles triangle* (its two sides are equal) is given by

$$P_{\substack{isosceles\\triangle}} = a + 2b$$

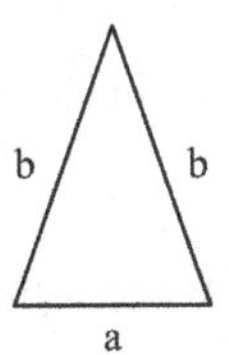

The perimeter for an *equilateral triangle* (all sides are equal) is given by

$$P_{\substack{equilateral\\triangle}} = 3a$$

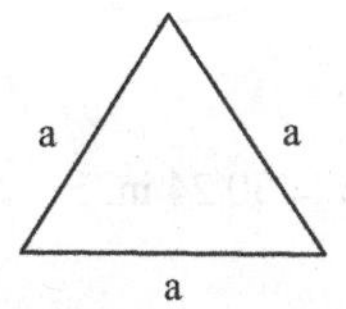

Example 4

Find the perimeter of a triangle whose sides measure as in the following figure.

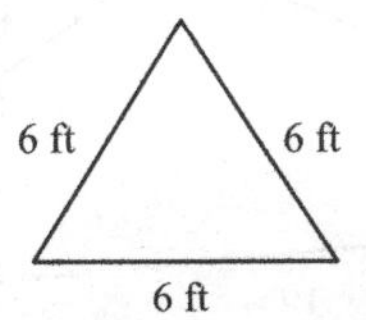

Solution

Perimeter $= P = 3 \times 6 = 18$ ft.

The perimeter of a circle (or distance around a circle) is called the *circumference* of the circle, that is, the boundary of the circle.

$$Circumference = C = \pi \times d$$

where d is length of the diameter of the circle, and the number π (pi) $= 3.14$ or $3\frac{1}{7}$ or $\frac{22}{7}$.

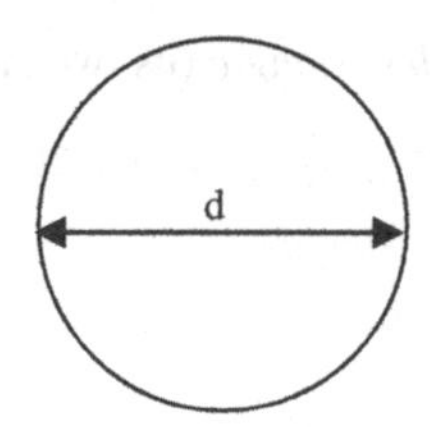

$d = 2r$, where r is the radius of the circle.

Example 5

Find the circumference of a fly wheel with a diameter of 16 in. as in the following figure.

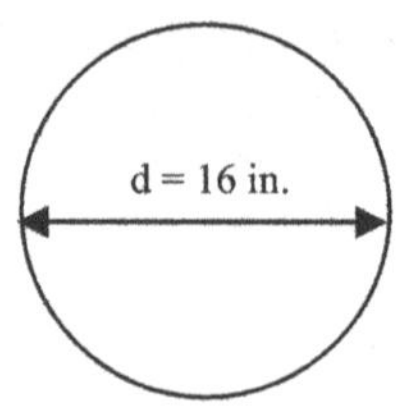

Solution

Circumference $= C = \pi \times 16 = 50.24$ in.

6.3 EXERCISES

1. Calculate the perimeter of the machine part shown.

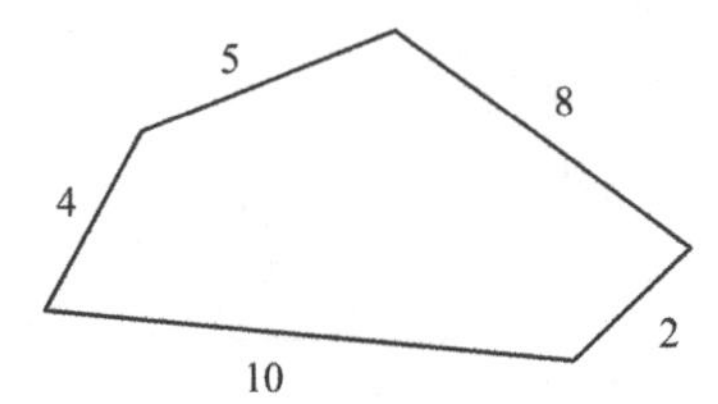

2. Calculate the perimeter of the machine part shown.

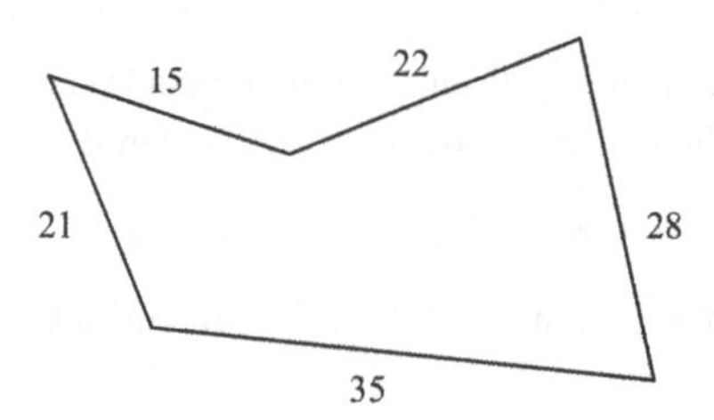

3. Calculate the perimeter of the polygon as shown.

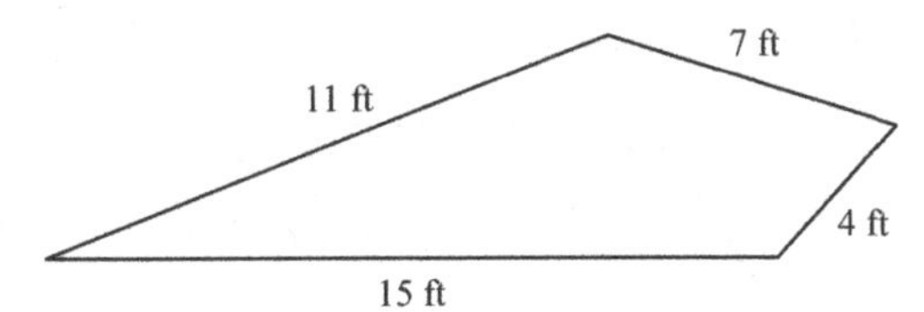

4. Calculate the perimeter of the polygon as shown.

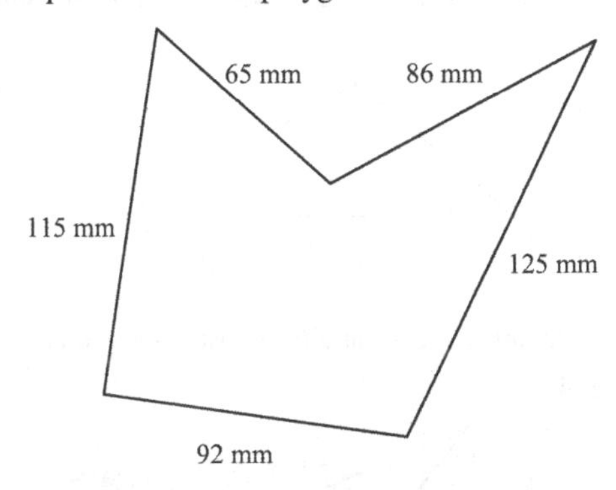

5. Find the perimeter of a rectangle whose sides measure as in the following figure.

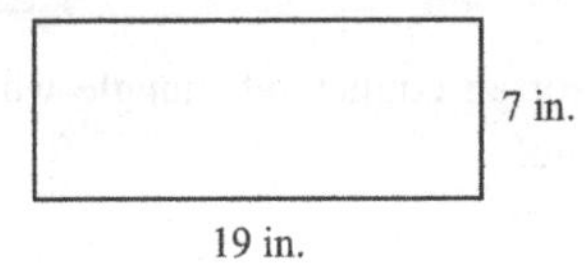

6. Find the perimeter for a rhombus whose sides measure as in the following figure.

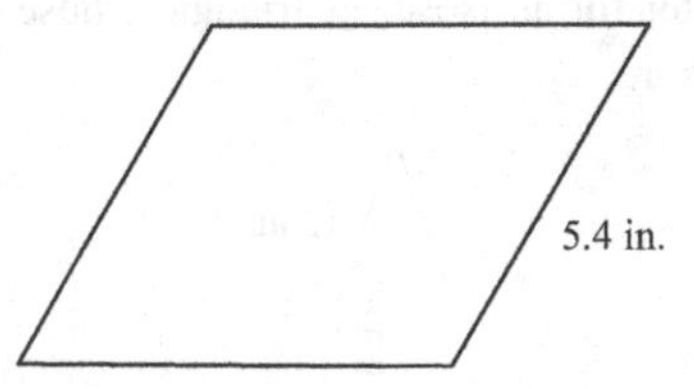

7. Find the perimeter of a square whose sides measure as in the following figure.

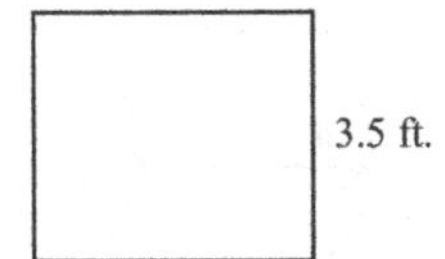

8. Find the perimeter for a parallelogram whose sides measure as in the following figure.

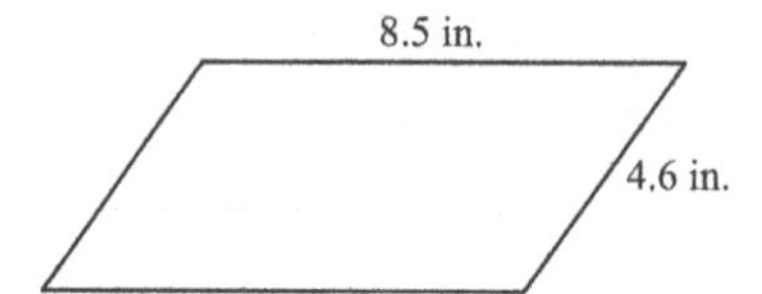

9. Find the perimeter of a triangle whose sides measure as in the following figure.

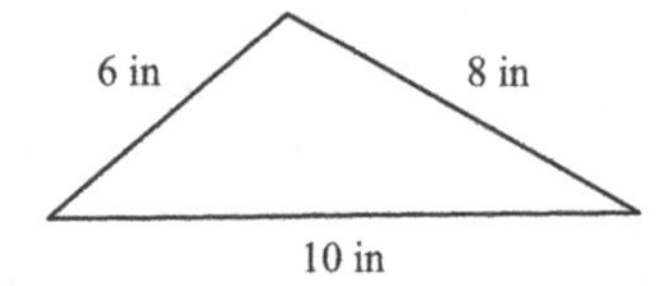

10. Find the perimeter of a triangle whose sides measure as in the following figure.

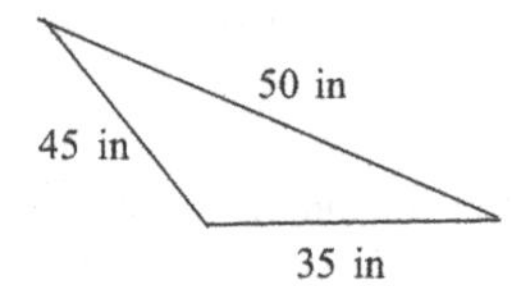

11. Find the perimeter for an equilateral triangle whose sides measure as in the following figure.

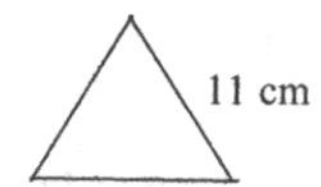

12. Find the perimeter for an isosceles triangle whose sides measure as in the following figure.

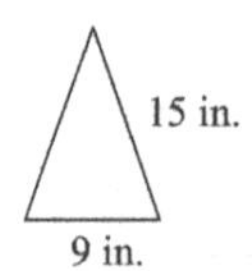

13. Find the circumference of a fly wheel with a diameter of 20 in. as in the following figure.

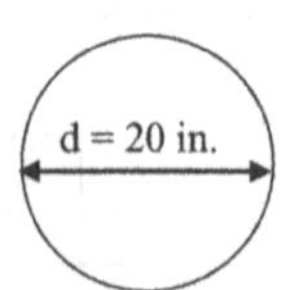

14. Find the circumference of a 30-millimeter radius shaft as in the following figure.

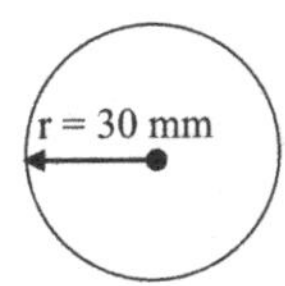

15. A land is covered with grass and other low plants suitable for grazing animals, especially cattle or sheep. The owner of the land wants to build a new fence for a pasture that is a rectangle 155 yd long and 64 yd wide. Find the perimeter of the pasture.

16. A round porthole window has a diameter of 24 in. Find the circumference of the circle.

6.4 AREA IN GEOMETRY

Many occupations require computations of areas in estimating job material quantities and costs. The *area* of a region is the number of units needed to completely cover the region. Area is given in units of length squared, such as m^2, ft^2, and $in.^2$.

The *area of a square* is equal to the length of any side of the square squared, that is,

$$A_{square} = s^2$$

where s is the length of any side of the square.

The *area of a rectangle* is equal to the product of length and width.

$$A_{rectangle} = l \times w$$

where l is length, and w is the width.

The area of the yellow square is 16 square units. That means, 16 square units are needed to cover the surface enclosed by the square.

Example 1

The area of the below square is 4 square units. That means, 4 square units are needed to cover the surface enclosed by the square.

Example 2

A square platform has a 12-ft length for a side. Find the area of the platform.

Solution

$$A_{square} = s^2$$
$$A_{square} = (12)^2 = 144 \text{ ft}^2$$

Example 3

A rectangle platform is 10 ft in length and 6 ft wide. Find the area of the platform.

Solution

$$A_{rectangle} = l \times w$$
$$A = 10 \times 6 = 60 \text{ ft}^2$$

The *area of a parallelogram* is equal to the product of the base and height.

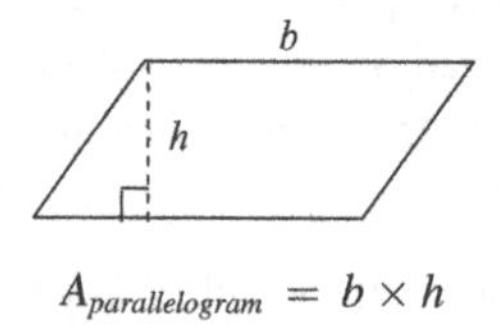

$$A_{parallelogram} = b \times h$$

where b is the base and h is the height.

Example 4

A parallelogram platform has a 7-mm base and 4-mm height. What is the area of the platform?

Solution

$$A_{parallelogram} = b \times h$$
$$A = 7 \times 4 = 28 \text{ mm}^2.$$

The *area of a trapezoid* is equal to one-half the product of the height and the sum of the bases.

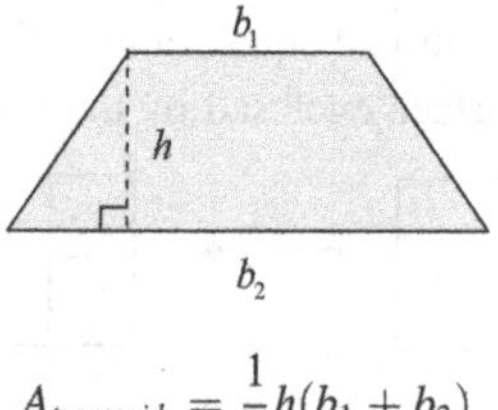

$$A_{trapezoid} = \frac{1}{2}h(b_1 + b_2)$$

where b_1 and b_2 are the bases, and h is the height.

Example 5

A trapezoid platform is bases 15 and 32 cm and height 12 cm. What is the area of the platform?

Solution

$$A_{trapezoid} = \frac{1}{2}h(b_1 + b_2)$$

$$A = \frac{1}{2}(12)(15 + 32)$$

$$A = 6(47) = 282 \text{ cm}^2$$

The *area of a triangle* is equal to one-half the product of the base and height.

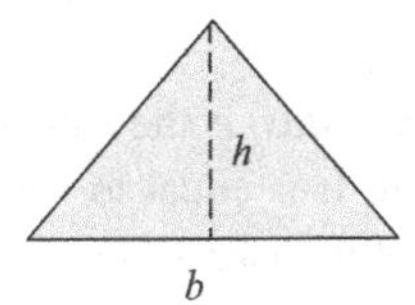

$$A_{triangle} = \frac{1}{2}bh$$

Example 6

Find the area of the triangle as in the figure.

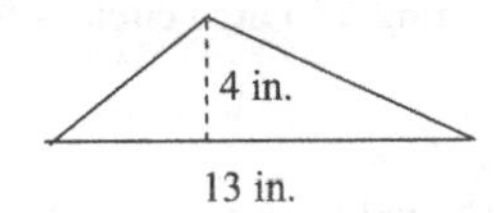

Solution

$$A_{triangle} = \frac{1}{2}bh$$

$$A = \frac{1}{2}(13)(4) = 26 \text{ in.}^2$$

The *area of a circle* is equal to one-half the product of the base and height.

$$A_{circle} = \pi r^2 \text{ or } A_{circle} = \pi\left(\frac{d^2}{4}\right)$$

where $r = \frac{d}{2}$, r is the radius, and d is the diameter.

Example 7

What is the area of a cross-section of 0.143 in. wire?

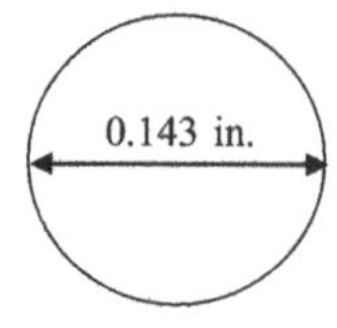

Solution

$$A_{circle} = \pi\left(\frac{d^2}{4}\right)$$

$$A = \pi\left(\frac{(0.143)^2}{4}\right) = 0.0161 \text{ in.}^2$$

The a*rea of a circular ring (annulus)* is the area between a large circle and a smaller circle that is within it. It is equal to subtraction of the area of the smaller circle from the area of the larger circle.

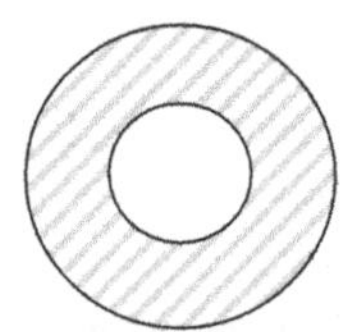

Area of circular ring = Large circle – Small circle

Example 8

Find the area of the ring (shaded portion) as in the figure.

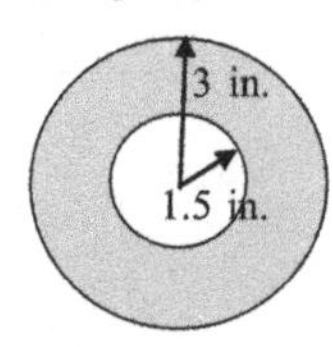

Solution

$A_{circle} = \pi r^2$

Area of large circle $= \pi(3)^2 = 28.27$ in.^{2}

Area of small circle $= \pi(1.5)^2 = 7.07$ in.^{2}

Area of ring $= 28.27 - 7.07 = 21.2$ in.^{2}

The *area of a sector* is the product of the area of the circle and the fraction of the area of the circle (central angle to 360°).

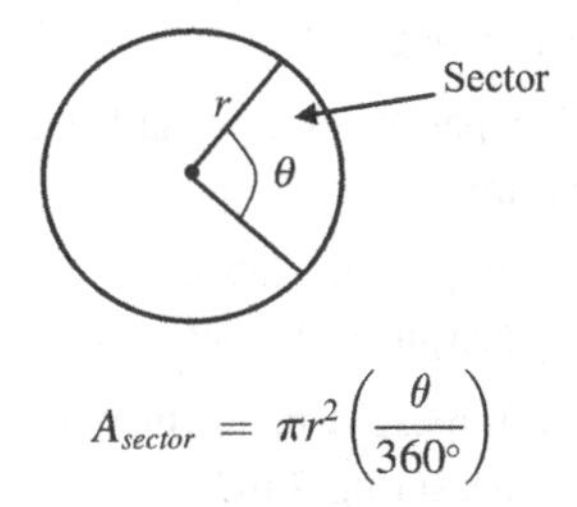

$$A_{sector} = \pi r^2 \left(\frac{\theta}{360°}\right)$$

Example 9

Find the area of a sector of 60° in a circle with a radius of 8 in.

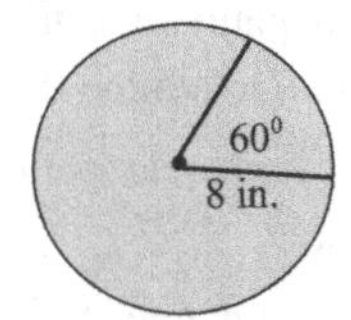

Solution

$$A_{sector} = \pi r^2 \left(\frac{\theta}{360°}\right)$$

$$A = \pi(8)^2\left(\frac{60}{360}\right) = 33.5 \text{ in.}^2$$

6.4 EXERCISES

1. Find the area of a square with sides 7 in. in length.
2. Find the area of a square with sides 16 in. in length.
3. A rectangular platform is 25 ft long and 12 ft wide. Find the area of the platform.
4. Find the area of a rectangle that has a length of 11 cm and a width of 6 cm.
5. How many square feet are there in 1 square yard?
6. Determine the area, in square meters, of a rectangular strip of sheet stock 16 cm wide and 2.75 m long.
7. Determine the area, in square meters, of a rectangular strip of sheet stock 35 cm wide and 5.25 m long.

8. A rectangle X-ray film measures 25 cm long and 46 cm. Find its area.
9. What is the area of a parallelogram platform whose base measures 14 cm and height measures 8 cm?
10. What is the area of a parallelogram platform whose base measures 9 cm and height measure 4 cm?
11. What is the area of a trapezoid platform with bases measuring 22 m and 37 m, and height measuring 18 m?
12. What is the area of a trapezoid platform with bases measuring 13 m and 19 m, and height measuring 7 m?
13. A contractor is building a patio in a triangle shape with a base of 20 in. and a height of 24 in. Determine the area.
14. A contractor is building a patio in a triangle shape with a base of 20 in. and a height of 82 in. Determine the area.
15. A contractor is pouring a concrete patio in a circular area with a 24-in. diameter. Find the area.
16. A contractor is pouring a concrete patio in a circular area with a 12-in. radius. Find the area.
17. Find the area of the ring in the cross-section of a circular concrete pipe with an outer diameter of 9 in. and an inner diameter of 5 in.
18. Calculate the cross-sectional area of the bushing as shown.

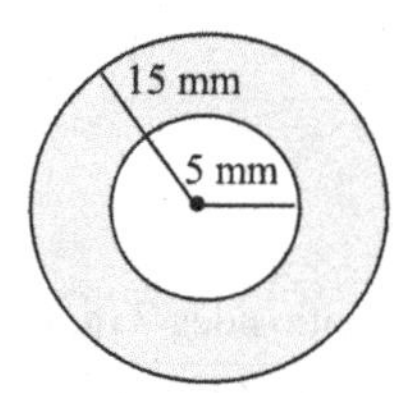

19. What is the area of the sector shown in the figure?

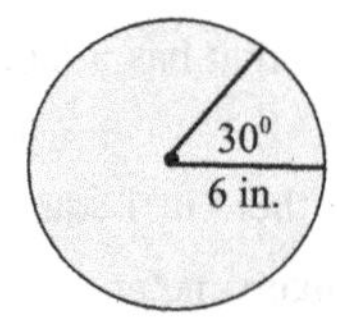

20. What is the area of the sector shown in the figure?

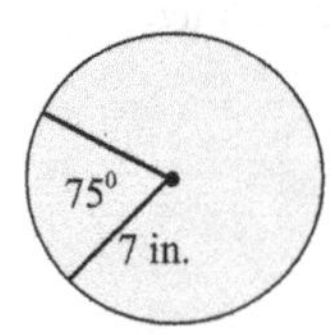

6.5 VOLUME IN GEOMETRY

The computation of volumes is required in various occupations such as by heating and air-conditioning technicians who calculate the volume of air in a building to find the heating and cooling system requirements. Also, the displacement of an automobile engine is based on the calculation of volume of its cylinders. *Volume* gives information about the amount of three-dimensional (3D) space an object occupies. Volume is measured in cubic units, for example, ft^3, $in.^3$, m^3.

The *volume of a rectangular solid (box)* is equal to the product of length, width, and height.

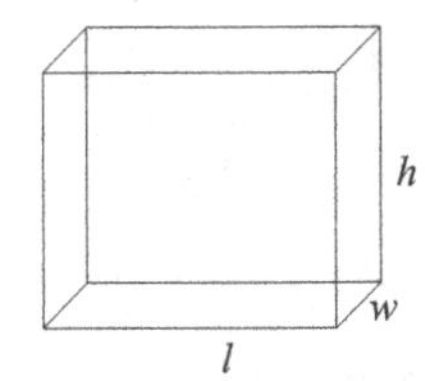

$$V_{rectangular} = l \times w \times h$$

where l is the length, w is the width, h is the height.

Example 1

What is the volume of the rectangular solid in the following figure?

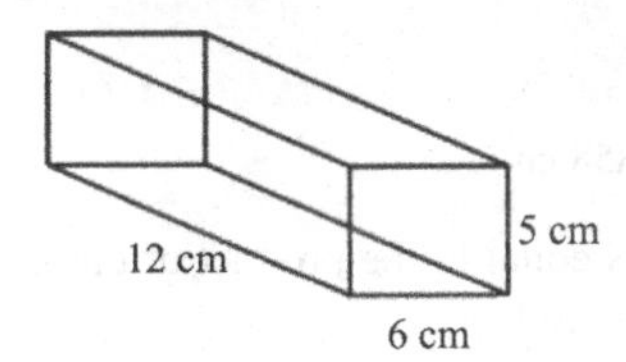

Solution

$V = l \times w \times h$

$V = 12 \times 6 \times 5 = 360 \text{ cm}^3$

The *volume of a cube* is equal to the product of the length of any edge by itself twice.

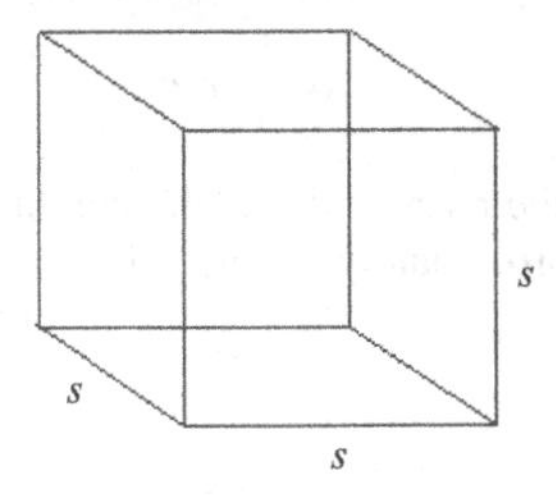

$$V_{cube} = s^3$$

where s is the length, width, and height of the cube.

Example 2

Find the volume of a cube if the length of one side is 4 cm.

Solution

$V_{cube} = s^3$

$V = (4)^3 = 64\ cm^3.$

The *volume of a cylinder* is equal to the product of the area of one of the bases multiplied by the height of the object.

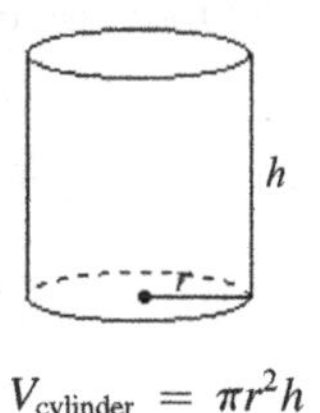

$$V_{cylinder} = \pi r^2 h$$

where r is the radius, and h is the height.
π is a number that approximately equals 3.14.

Example 3

Find the volume of a cylindrical canister with radius 5 cm and height 9 cm.

Solution

$V_{cylinder} = \pi r^2 h$

$V = \pi(5)^2(9) = 706.858\ cm^3.$

The *volume of a cone* is equal to the product of one-third of the area of the base by the height of the object.

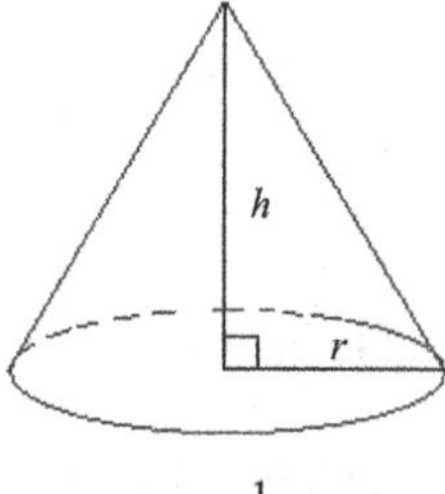

$$V_{cone} = \frac{1}{3}\pi r^2 h$$

where r is the radius of the base, and h is the height.
π is a number that is approximately equals 3.14.

Example 4

Find the volume of a cone with radius 4.50 in. and height 2.35 in.

Solution

$$V_{\text{cone}} = \frac{1}{3}\pi r^2 h$$

$$V = \frac{1}{3}\pi(4.5)^2(2.35) = 49.834 \text{ in}^3$$

The *volume of a pyramid is* equal to the product of one-third of the area of the base by the height of the object.

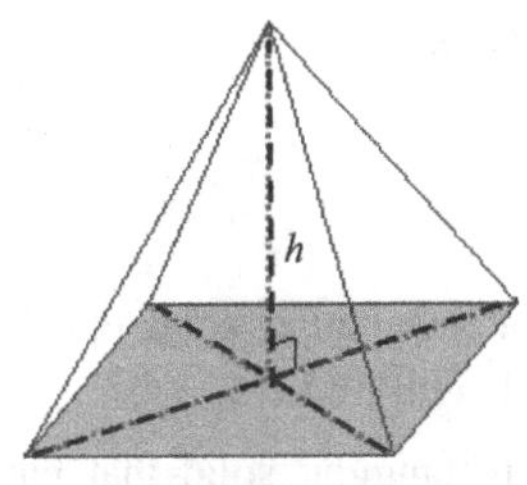

$$V_{pyramid} = \frac{1}{3}Bh$$

where B is the area of the base, and h is the height.

Example 5

A pyramid has a square base of side 8 cm and a height of 14 cm. Find its volume.

Solution

$$V_{pyramid} = \frac{1}{3}Bh$$

$$V = \frac{1}{3}(8)(14) = 37.333 \text{ cm}^3.$$

The *volume of a sphere is* equal to the product of four-thirds of pi and the cube of the radius.

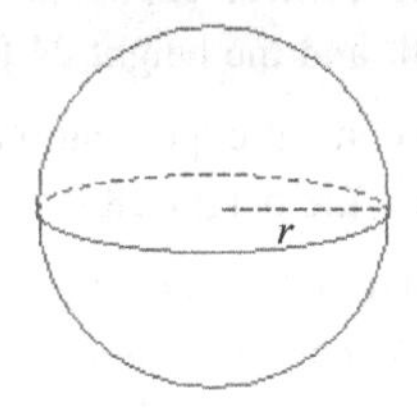

The volume of a sphere is given by the formula:

$$V_{sphere} = \frac{4}{3}\pi r^3$$

Example 6

Find the volume of a sphere of radius 3.80 cm.

Solution

$$V_{sphere} = \frac{4}{3}\pi r^3$$

$$V = \frac{4}{3}\pi(3.80)^3 = 229.847 \text{ cm}^3.$$

6.5 EXERCISES

1. Find the volume of a rectangular solid that has a length of 8 cm, a width of 6 cm, and a length of 5 cm.
2. Find the volume of rectangular solid that has a length of 14 cm, a width of 8 cm, and a length of 7 cm.
3. What is the volume of a rectangular room with the dimensions 14 ft 8 in. by 12 ft 4 in. by 10 ft 2 in.?
4. What is the volume of a rectangular room with the dimensions 18 ft 4 in. by 16 ft 2 in. by 12 ft 6 in.?
5. Find the volume of a crate measuring 3.60 ft by 5.60 ft by 7.80 ft.
6. Find the volume of a crate measuring 5.20 ft by 7.30 ft by 8.50 ft.
7. Find the volume of a cube if the length of one side is 3 cm.
8. Find the volume of a cube if the length of one side is 5 cm.
9. Find the volume of a container in the shape of a cylinder with radius 10 in. and height 12 in.
10. Find the volume of a container in the shape of a cylinder with diameter 18 in. and height 7 in.
11. Find the volume of the conical cupola atop a round tower with a radius of the base of 6 ft, and the height 24 ft.
12. Find the volume of the conical cupola atop a round tower with diameter of the base 16 ft, and the height 28 ft.
13. Find the volume of a pyramid with a square base whose sides measure 4 in. each and whose height is 6 in.

14. Find the volume of a pyramid with a square base whose sides measure 5 m each and whose height is 9 m.
15. Find the volume of a sphere with a radius 4.8 m.
16. Find the volume of a sphere with a diameter 12 m.

CHAPTER 6 REVIEW EXERCISES

1. Identify the figure.

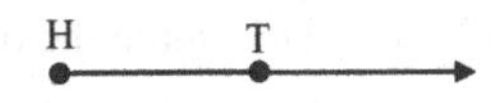

2. Identify the figure.

3. Determine the value of the angle θ in the below illustration.

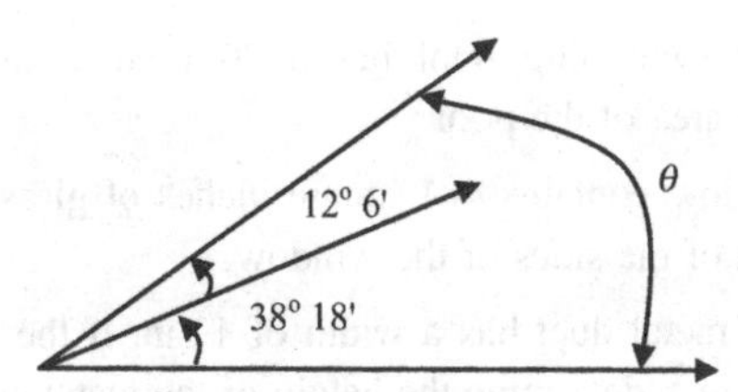

4. Determine the value of the angle θ in the below illustration.

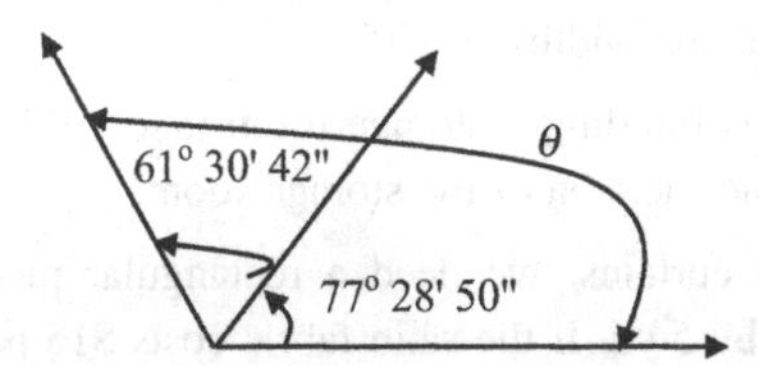

5. Compute the perimeter of the following figure.

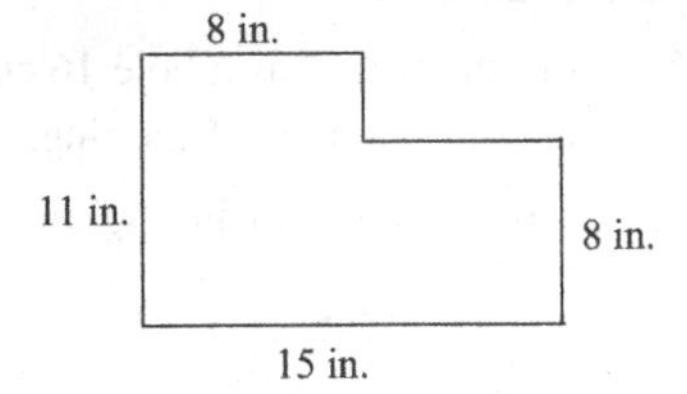

6. Compute the perimeter of the following figure

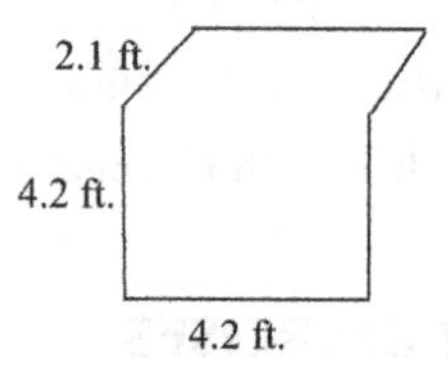

7. A wheel rolls along level ground; it makes one complete revolution as it travels a distance of 87.46 in. Determine the diameter of the wheel.
8. A parallelogram-shaped lot is shown in the figure. How many feet of fence will be needed to fence the lot?

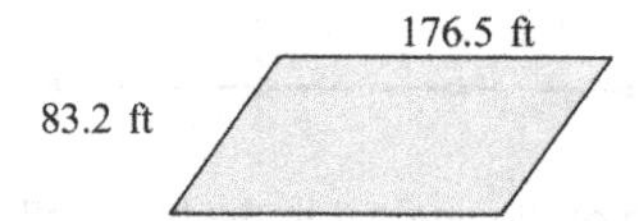

9. A rectangular swimming pool has a 20 ft width and 46 ft length. Determine the area of the pool.
10. A square window contains 481 square inches of glass. Determine the length of each of the sides of the window.
11. A rectangular metal duct has a width of 12 in. If the area of a cross-section is 108 in.^{2}, determine the height of the metal and the perimeter of a cross-section.
12. A rectangular piece of sheet metal has an area of 1932 in.^{2}. Its length is 84 in. What is the width?
13. A homeowner is building a storage room that will be 24 in. long and 18 in. wide. Find the area of the storage room.
14. To make satin curtains, we need a rectangular piece of fabric that measures 7 yd by 5 yd. If the satin fabric costs $15 per square yard, what is the cost of the fabric?
15. A kitchen measures 12 ft by 17 ft. If a square foot of tile costs $4.50, how much can you spend on tile?
16. A flat steel ring 58 cm in diameter has a hole 16 cm in diameter in its center. Determine the area of one face of the ring.
17. Find the area of the sector in the following figure.

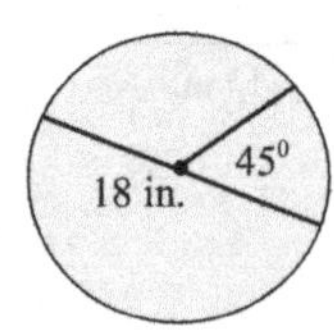

18. The water from a certain sprinkler nozzle will reach the shaded area as in the figure. Determine how many square yards the water will reach.

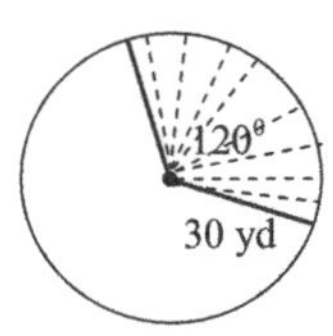

19. Determine the volume of a rectangular solid that has a length of 7 m, a width of 5 m, and a height of 2 m.

20. Determine the volume of a rectangular solid that has a length of 12 m, a width of 9 m, and a height of 3 m.

21. How many cubic feet of space are there in a storage area 11 ft long, 9 ft wide, and 10 ft high?

22. How many cubic feet of concrete are needed for a retaining wall 70 ft long, 7.7 ft high, and 2.65 ft wide?

23. Find the volume of a cube if the length of one side is 2 cm.

24. Find the volume of a cube if the length of one side is 7 cm.

25. Find the volume of a cylindrical grain storage container with a radius of 22 ft and height of 76 ft.

26. Find the volume of a lawn roller in the shape of a cylinder with a diameter of 40 cm and height of 98 cm.

27. Find the volume of a fire sprinkler that sprays water in the shape of a cone with a diameter of 26 ft and height of 12 ft.

28. Find the volume of a high pressure atomizing oil burner that propels a spray of oil in the shape of a cone with a diameter of 6.04 in. and a height of 1.60 in.

29. Find the volume of a pyramid with a base of 26 square feet and a height of 8 ft.

30. Find the base area of a pyramid with a volume of 180 cm^3 and a height of 12 cm.

31. Find the volume of a sphere with a radius of 4.2 m.

32. Find the volume of a stainless steel ball-bearing that contains balls that are each 1.60 cm in diameter.

Chapter 7

Trigonometry

Forget about all the reasons why something may not work. You only need to find one good reason why it will.

Dr Robert Anthony

■ Heinrich Hertz (1857—1894), a German physicist, he made possible the development of radio, television, and radar by proving that electricity can be transmitted in electromagnetic waves. He explained and expanded the electromagnetic theory of light that had been put forth by Maxwell. He was the first person who successfully demonstrated the presence of electromagnetic waves, by building an apparatus that produced and detected the VHF/UHF radio waves. His undertakings earned him the honor of having his surname assigned to the international unit of frequency (one cycle per second).

INTRODUCTION

Many applications in engineering, science, and technology require the use of *trigonometry*. For many years scientists used trigonometry in the computation of astronomy and geography. Trigonometry technology has many applications in engineering, mathematics, physics, and chemistry, and is useful in designing machines, electric and electronic circuits, and in architecture.

Fundamentals of Technical Mathematics. http://dx.doi.org/10.1016/B978-0-12-801987-0.00007-1

Trigonometry is a branch of mathematics that is primarily concerned with the relationships between angles and sides of geometry.

7.1 UNITS OF ANGLES

Angles are often measures in degree (°) unit. We use the symbol "∠" to represent "angle" or common Greek letters to denote an angle such as α (alpha), β (beta), and θ (theta). Units can be measured in radians, seconds, or minutes.

180 degrees (180°) = π radian

1 degree (1°) = 60 minutes = 60′

1 minute = 60 seconds

1 degree (1°) = 3600 seconds

1 circle = 2π radian = 360°

Example 1

(a) Convert 320 minutes to degrees

(b) Convert 12,000 seconds to degrees

(c) Convert 270° to radians

(d) Convert $\frac{3\pi}{5}$ to degrees

Solution

(a) $320\,\cancel{\text{minutes}} \times \frac{1^\circ}{60\,\cancel{\text{minutes}}} = 5.33^\circ$

(b) $12,000\,\cancel{\text{seconds}} \times \frac{1^\circ}{3600\,\cancel{\text{seconds}}} = 3.33^\circ$

(c) $270^\circ \times \frac{\pi \text{ radian}}{180^\circ} = \frac{3\pi}{2}$ rad

(d) $\frac{3\cancel{\pi}}{5} \times \frac{180^\circ}{\cancel{\pi}} = 108^\circ$

7.1 EXERCISES

1. Convert 75° to minutes
2. Convert 540 minutes to degrees
3. Convert 15,000 seconds to minutes
4. Convert 90° to radians
5. Convert $\frac{2\pi}{5}$ to degrees

6. Convert $\frac{\pi}{4}$ to degrees
7. Convert 2 circles to degrees
8. Convert 2 circles to minutes
9. Convert 2 circles to seconds
10. Convert 1° to radians
11. Convert 30° to radians
12. Convert 60 minutes to radians
13. An airplane traveled for three hours. How many degrees did it move?
14. If a circle is 360°, how many radians does it have?

7.2 TYPES OF ANGLES AND RIGHT TRIANGLE

Angles are used daily. Angles can virtually measure anything, as we will see using geometry and trigonometry. In this section we focus on the main common types of angles as listed below:

Right angle is an angle equal to 90°:

B

Angle = 90°

C A

Acute angle is any angle less than 90°:

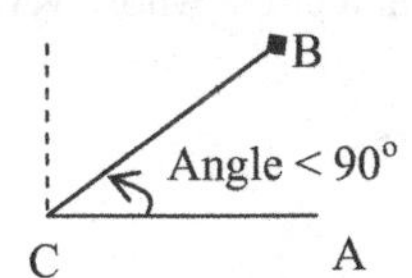

Obtuse angle is any angle greater than 90°:

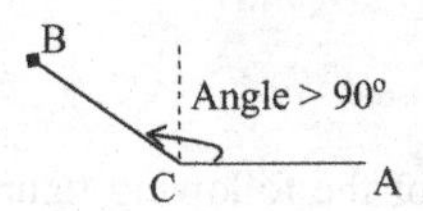

Straight angle is an angle measured equal to 180°:

Angle = 180°

B A

Zero angle is an angle measured equal to 0°:

0° angle

Complementary angles are angles whose measures have a sum equal to 90°:

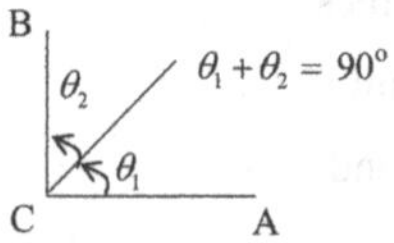

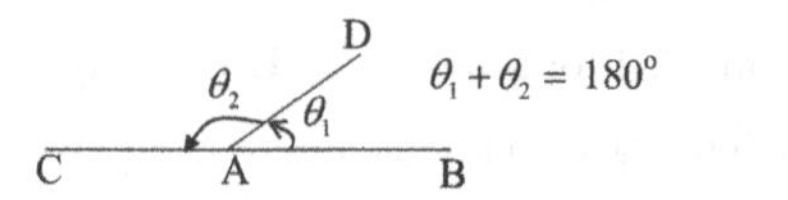

Supplementary angles are angles whose measures have a sum equal to 180°.

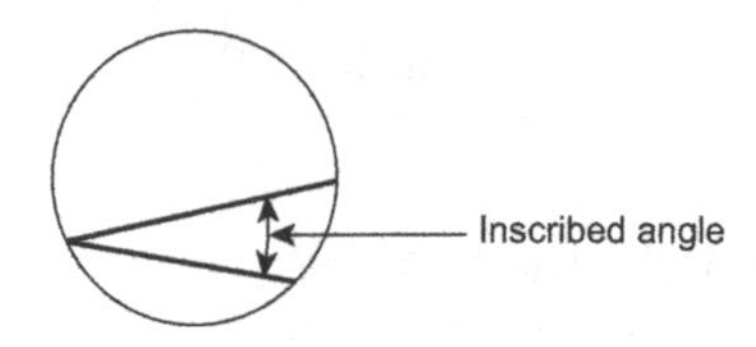

Central angle is an angle whose vertex is located at the center of the circle and whose sides are radii. In this figure, $\angle CAB$ is a central angle.

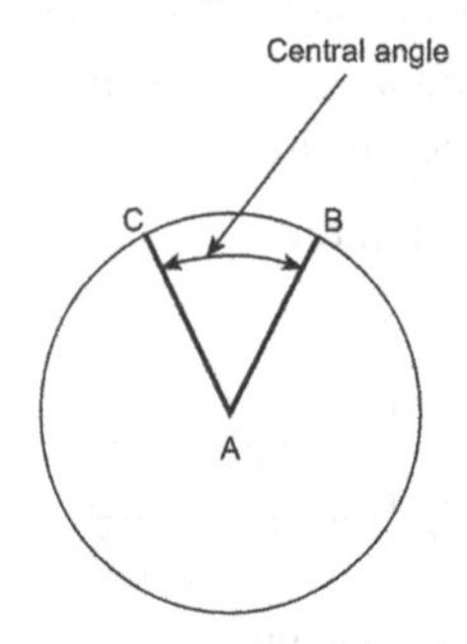

Inscribed angle is an angle in a circle whose vertex is located on the circle and whose sides are chords:

Example 1

Identify the type of angles of the following figures:

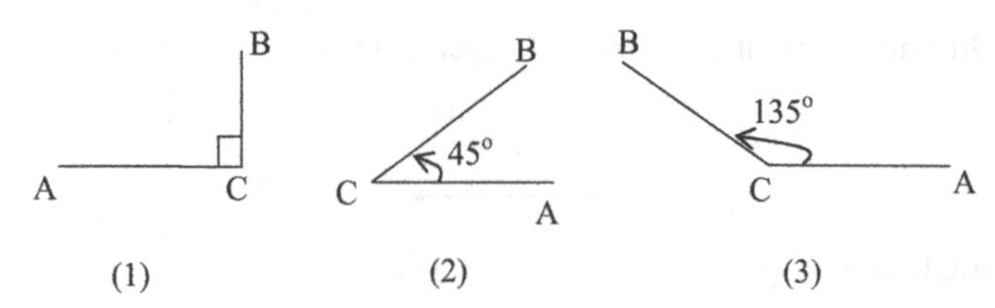

Solution

(1) is right angle because the angle is 90°

(2) is an acute angle because the angle 45° is less than 90°

(3) is an obtuse angle because the angle 135° is greater than 90°

Example 2

Determine the central and inscribed angles.

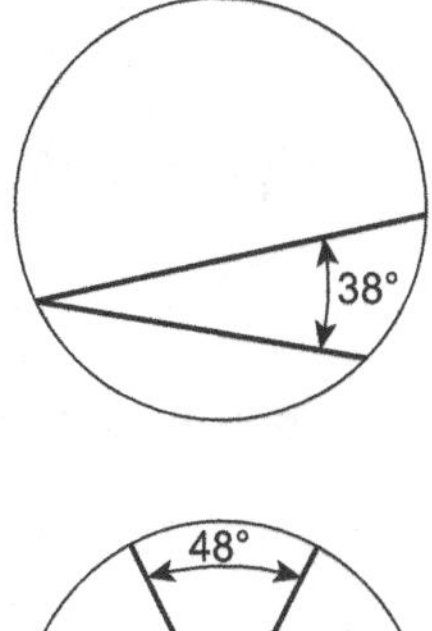

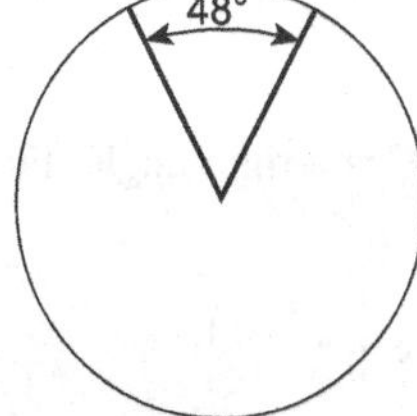

Solution

The central angle is 48° and inscribed angle is 38°.

7.2.1 Right triangle

A *right triangle* is any triangle with a 90° angle.

B

Hypotenuse side = c

α

Opposite side of θ = a

θ

C

A

Adjacent side of θ = b

Pythagorean theorem: For any right angle, the square of the length of the hypotenuse is equal to the sum of the squares of the lengths of the other two sides, the opposite and adjacent. (Hypotenuse side)^2 = (Opposite side)^2 + (Adjacent side)^2, that is,

$c^2 = a^2 + b^2$, therefore, $c = \sqrt{a^2 + b^2}$.

We use the capital letters A, B, and C on each vertex as the names of angles. So $\angle BAC = \angle A = \theta$, $\angle ABC = \angle B = \alpha$, and $\angle ACB = \angle C = 90°$. We use lower case letters to represent the sides of the triangle.

Example 1

In the triangle ABC, angle C is a right angle. Find the value of the hypotenuse x.

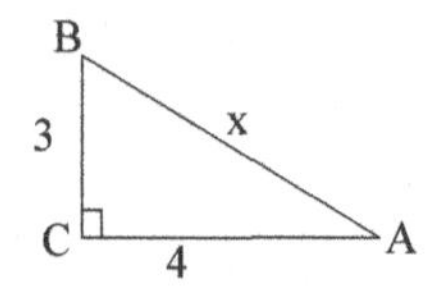

Solution

By using the Pythagorean theorem, $x = \sqrt{(3)^2 + (4)^2} = \sqrt{9 + 16} = \sqrt{25} = 25.$

Example 2

In the triangle ABC, angle C is a right angle. Find the value of c.

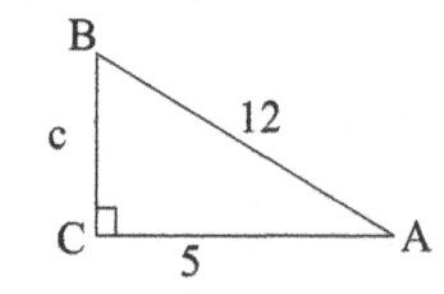

Solution

By using the Pythagorean theorem,

$$12 = \sqrt{(5)^2 + (c)^2}$$

$$(12)^2 = 25 + c^2$$

$$c = \sqrt{144 - 25}$$

$$c = 10.91$$

7.2 EXERCISES

In exercise 1–10, identify the type of angles:

1. 89°

2. 90°

3. 40°

4. 100°

5. 91°

6. 0°

7. 180°

8. 15°

9. 45°

10. 135°

11. Find the complementary angle of the following angles:

(a) 65°

(b) 75°

(c) 45°

(d) 85°

12. Find the supplementary angle of the following angles:

(a) 125°

(b) 145°

(c) 115°

(d) 165°

For exercises 13 and 14 use the following figure.

13. Find c.

b = 2

c

a = 3

14. Prove $b = 2$.

Find the length of the hypotenuse in exercises 15–18.

15.

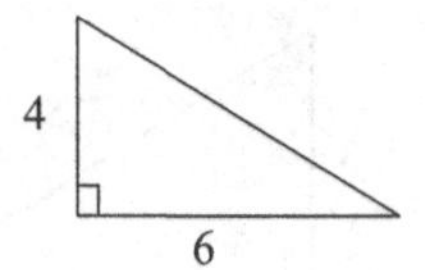

16.

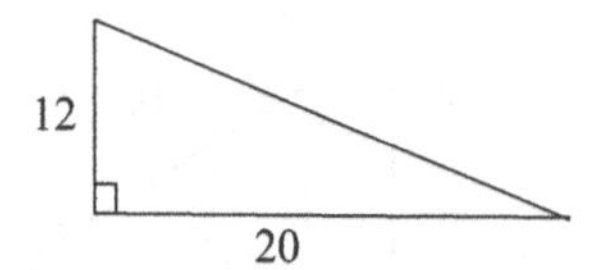

17.

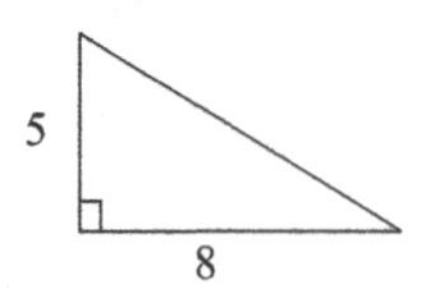

18.

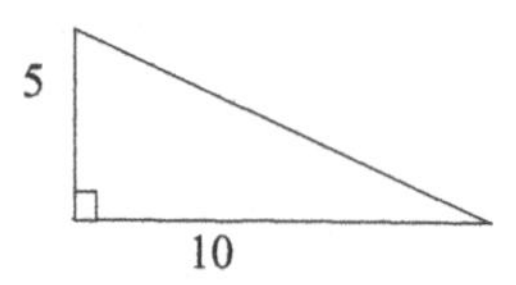

Find the length of the leg in exercises 19–22.

19.

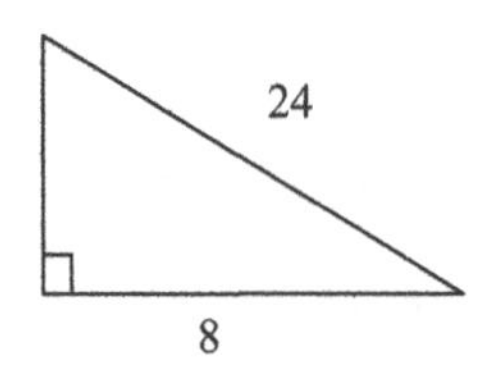

20.

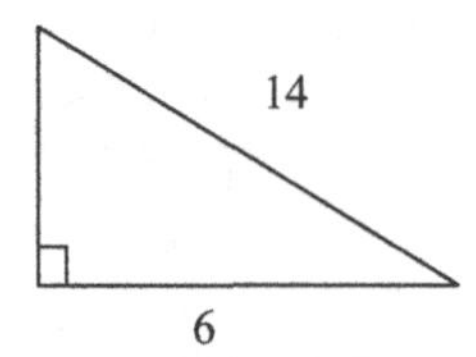

21.

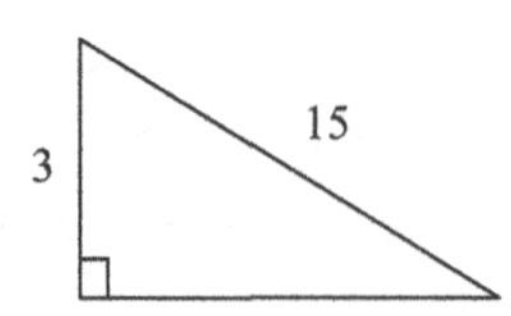

22.

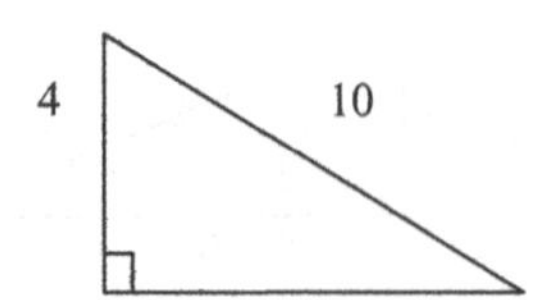

23. The Pythagorean theorem can be used to find the missing side on any triangle. True or false?

24. Determine the length of *AB* in right triangle *ABC*.

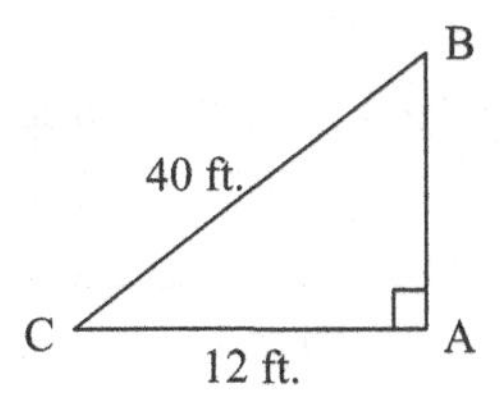

25. Find the length (*d*) of the diagonal brace in a metal rectangular frame with length 18 in. and width 12 in.

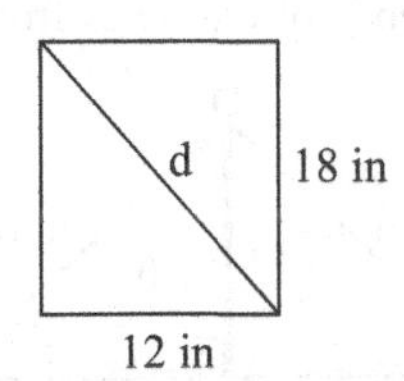

26. Find the length (*d*) of the diagonal brace for a rectangular doorframe that is 9 ft high and 4 ft wide.

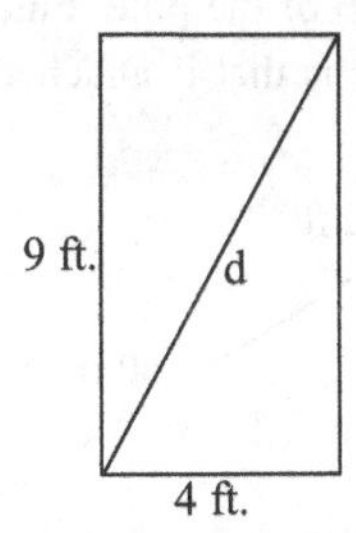

27. A concrete slab forms a triangle if two sides measure 12 cm and 7 cm, and the altitude to the base is 5 cm. Find the length (*AC*) of the base of the triangle.

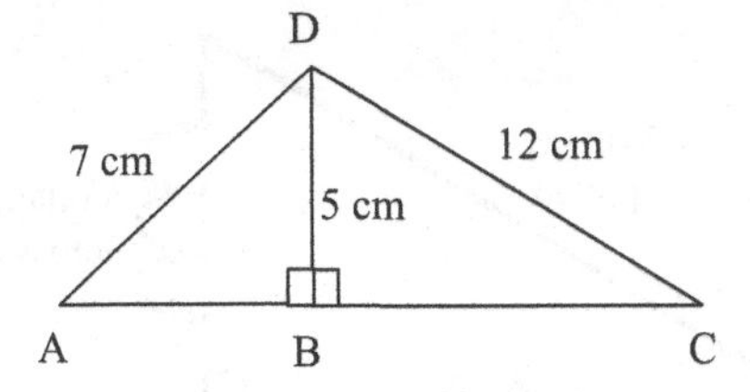

28. A steel plate forms a triangle if two of the sides measure 16 cm and 10 cm, and the altitude to the base is 6 cm. Find the length (AC) of the base of the triangle.

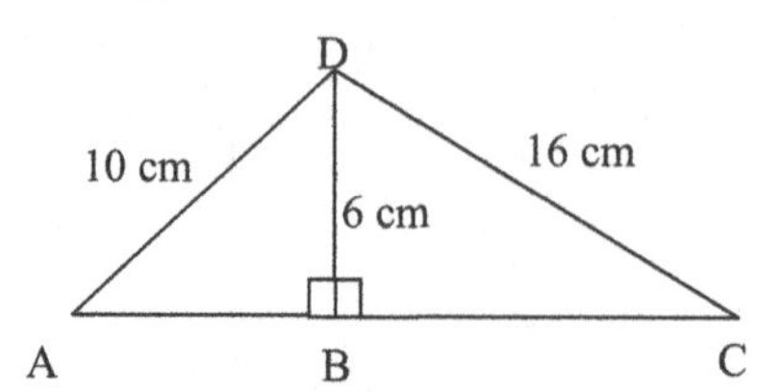

29. Suppose that a pole is on level ground and perpendicular to the ground with two brace wires on its sides. Each wire is attached at the top of the pole and is 30 ft long and the pole is 25 ft high. How far (AC) are the grounded ends of the wires from each other?

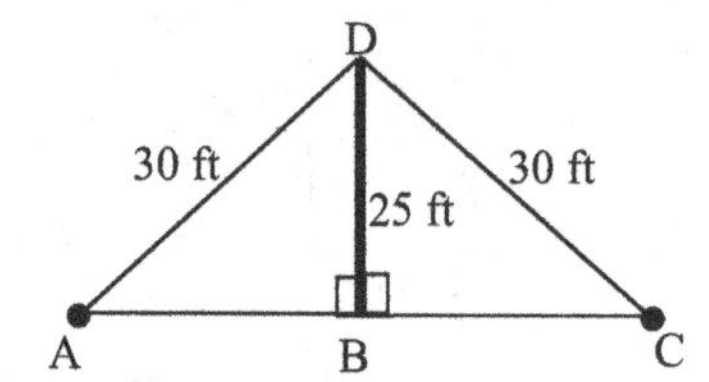

30. A guy-wire 60 ft long is attached to a pole 50 ft high. The wire is attached 6 ft from the top of the pole. Find how far (AB) from the base of the pole the wire is that is attached to the ground.

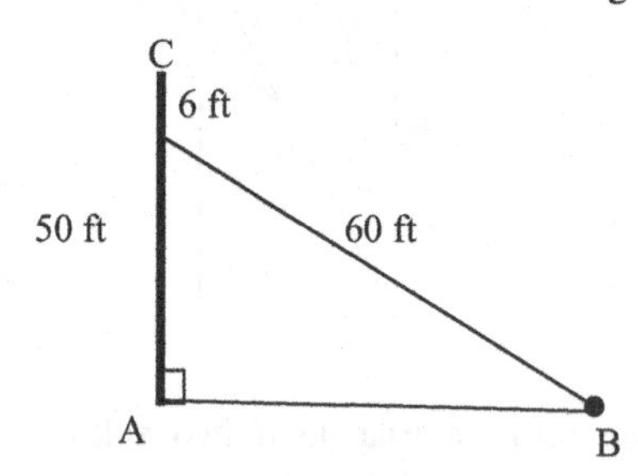

31. A circuit has voltage across the resistance of 86 volts and voltage across the coil of 92 volts. Find the applied voltage.

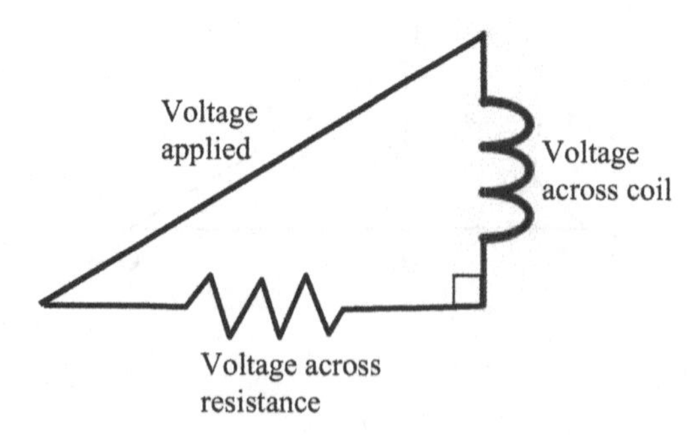

32. A circuit has voltage across the coil of 345 volts and the voltage applied is 620 volts. Find the voltage across the resistance.

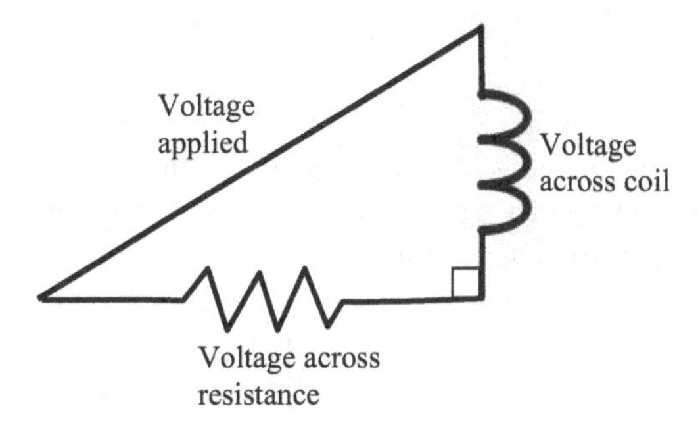

7.3 TRIGONOMETRIC FUNCTIONS

There are six trigonometric functions in reference to any right triangle: sine (abbreviated as sin), cosine (abbreviated as cos), and tangent (abbreviated as tan). Their reciprocals are cosecant (abbreviated as csc), secant (abbreviated as sec), and cotangent (abbreviated as cot). Based on the following figure, we can define the trigonometric functions for the angle θ as:

$$\sin\theta = \frac{\text{length of } \textit{opposite} \text{ side of } \theta}{\text{length of hypotenuse}} = \frac{a}{c},$$

$$\cos\theta = \frac{\text{length of adjacent side of } \theta}{\text{length of hypotenuse}} = \frac{b}{c},$$

$$\tan\theta = \frac{\text{length of } \textit{opposite} \text{ side of } \theta}{\text{length of adjacent side of } \theta} = \frac{a}{b},$$

$$\csc\theta = \frac{\text{length of hypotenuse}}{\text{length of } \textit{opposite} \text{ side of } \theta} = \frac{c}{a},$$

$$\sec\theta = \frac{\text{length of hypotenuse}}{\text{length of adjacent side of } \theta} = \frac{c}{b},$$

$$\cot\theta = \frac{\text{length of adjacent side of } \theta}{\text{length of } \textit{opposite} \text{ side of } \theta} = \frac{b}{a}$$

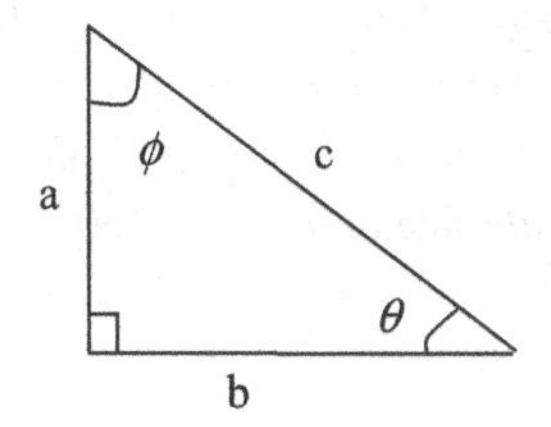

It is worth knowing that the sum of the acute angles of any triangle is equal to 90°. So,

$$\phi + \theta = 90^\circ$$

Note that $\tan\theta = \dfrac{\sin\theta}{\cos\theta}$.

The 90-45-45 and 90-60-30 degree triangles as in the following figures are special triangles with very common angles used in trigonometric functions. These values are worth memorizing.

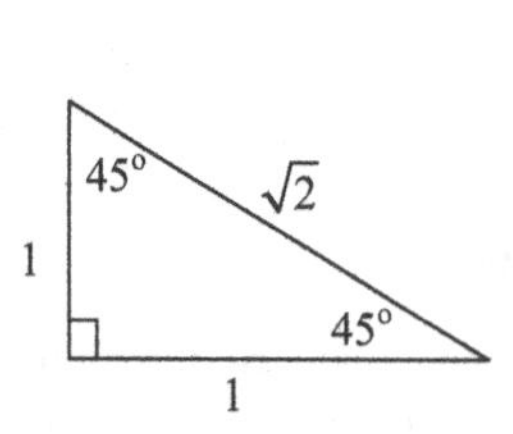

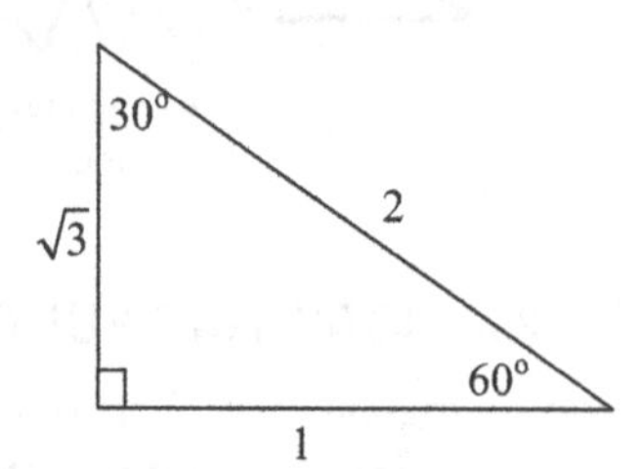

By using the figure we know the triangle functions such as $\sin 45^\circ = \dfrac{1}{\sqrt{2}}$ and $\cos 30^\circ = \dfrac{\sqrt{3}}{2}$.

Example 1

Find the six trigonometric ratios for angle θ in the figure.

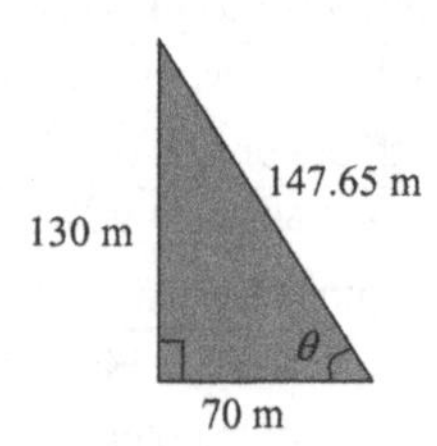

Solution

$$\sin\theta = \frac{\text{length of } \textit{opposite} \text{ side of } \theta}{\text{length of hypotenuse}} = \frac{130}{147.65} = 0.88,$$

$$\cos\theta = \frac{\text{length of adjacent side of } \theta}{\text{length of hypotenuse}} = \frac{70}{147.65} = 0.47,$$

$$\tan\theta = \frac{\text{length of } \textit{opposite} \text{ side of } \theta}{\text{length of adjacent side of } \theta} = \frac{130}{70} = \frac{13}{7} = 1.86,$$

$$\csc\theta = \frac{\text{length of hypotenuse}}{\text{length of } \textit{opposite} \text{ side of } \theta} = \frac{147.65}{130} = 1.14,$$

$$\sec\theta = \frac{\text{length of hypotenuse}}{\text{length of adjacent side of } \theta} = \frac{147.65}{70} = 2.11,$$

$$\cot\theta = \frac{\text{length of adjacent side of } \theta}{\text{length of } \textit{opposite} \text{ side of } \theta} = \frac{70}{130} = \frac{7}{13} = 0.54$$

Example 2

Find the tangent of angle θ for the triangle in the figure.

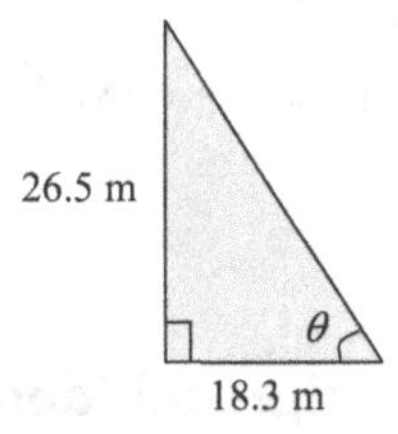

Solution

We know the opposite and adjacent sides of angle θ. So we use the tangent ratio: $\tan\theta = \frac{\text{length of } \textit{opposite} \text{ side of } \theta}{\text{length of adjacent side of } \theta} = \frac{26.5}{18.3} = 1.45$

Example 3

Based on the right triangle in the figure,

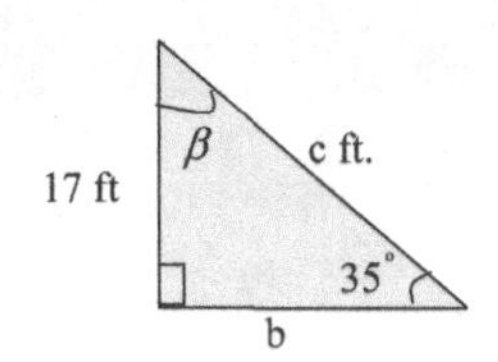

(a) find angle β

(b) the length of side c

(c) the length of side b

Solution

(a) We are given in the figure the measure of one acute angle and the length of one side:

$\beta = 90° - 35° = 55°$.

(b) We then used the sine ratio to find c.

$$\sin 35° = \frac{\text{length of } \textit{opposite} \text{ side of } \theta}{\text{length of hypotenuse}} = \frac{17}{c}$$

But,
Sin 35° = 0.57, so

$$0.57 = \frac{17}{c} \Rightarrow c = \frac{17}{0.57} = 29.82 \text{ ft.}$$

(c) By using the Pythagorean theorem,

$$b = \sqrt{(29.82)^2 - (17)^2} = \sqrt{889.2324 - 289} = \sqrt{600.2324}$$
$$= 24.50 \text{ ft.}$$

7.3.1 Trigonometry reciprocal identities

The reciprocal identities of trigonometric functions for an angle θ are:

$$\sin\theta = \frac{1}{\csc\theta}, \quad \csc\theta = \frac{1}{\sin\theta}$$
$$\cos\theta = \frac{1}{\sec\theta}, \quad \sec\theta = \frac{1}{\cos\theta}$$
$$\tan\theta = \frac{1}{\cot\theta}, \quad \cot\theta = \frac{1}{\tan\theta}$$

Example 1

Find the reciprocal identities of trigonometric functions for:

(a) $\sin\theta = 0.53$

(b) $\tan\phi = 0.35$

(c) $\cos\beta = 0.47$

Solution

(a) $\csc\theta = \frac{1}{\sin\theta} = \frac{1}{0.53} = 1.87$

(b) $\cot\phi = \frac{1}{\tan\phi} = \frac{1}{0.35} = 2.86$

(c) $\sec\beta = \frac{1}{\cos\beta} = \frac{1}{0.47} = 2.13$

7.3.2 Trigonometry complementary (cofunction identities) functions

By looking at the following figure, we see that $\alpha + \beta = 90°$, which means that the angle β is the complementary angle for angle α.

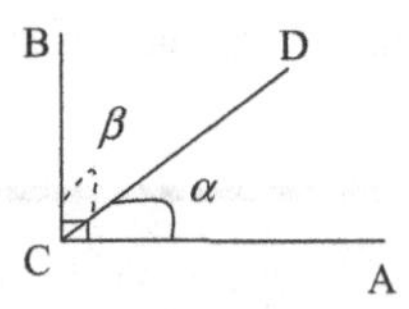

The cofunction identities (complementary) of trigonometric functions for an acute angle θ are:

$$\sin\theta = \cos(90-\theta),\quad \cos\theta = \sin(90-\theta)$$
$$\tan\theta = \cot(90-\theta),\quad \cot\theta = \tan(90-\theta)$$
$$\sec\theta = \csc(90-\theta),\quad \csc\theta = \sec(90-\theta)$$

It is necessarily to know that you should not keep the same function when you do the complementary functions!!

Example 1

Find the complementary functions for the following trigonometric functions:

(a) cos 30°

(b) sec 65°

(c) cot 20°

Solution

We can use the above rules to get the complementary functions as:

(a) $\cos 30° = \sin 60°$

(b) $\sec 65° = \cos 25°$

(c) $\cot 20° = \tan 70°$

7.3.3 Trigonometry supplementary functions

The following figure shows that $\alpha + \beta = 180°$, which means that the angle β is the supplementary angle for angle α.

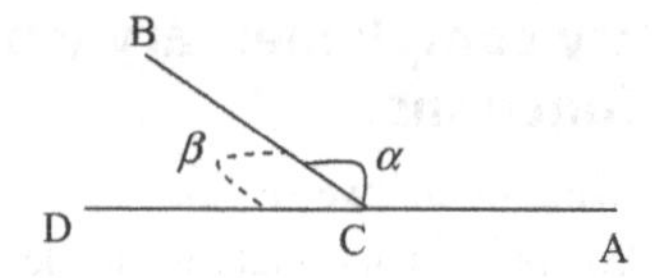

The supplementary functions of trigonometric functions for an angle θ between 0° and 180° are:

$$\sin\theta = \cos(180-\theta), \quad \cos\theta = \sin(180-\theta)$$
$$\tan\theta = \cot(180-\theta), \quad \cot\theta = \tan(180-\theta)$$
$$\sec\theta = \csc(180-\theta), \quad \csc\theta = \sec(180-\theta)$$

It is necessary to know that you should not keep the same function when you do the supplementary functions!!

Example 1

Find the supplementary functions for the following trigonometric functions:

(a) sin 135°

(b) tan 115°

(c) sec 140°

Solution

By using the supplementary functions rules, we find that

(a) sin 135° = cos 45°

(b) tan 115° = cot 65°

(c) sec 140° = cos 40°

7.3.4 Inverse trigonometric functions

Inverse functions are called *arcfunction*, for example, arcsine, arccosine, and arctangent.

We can measure the angle value by knowing any two sides of a right triangle using inverse functions, for example, the inverse function of sin is $\sin^{-1}$, therefore, if $\sin\theta = \frac{a}{c}$, then we can find the angle θ by using the inverse function of the sin for the ratio of the $\frac{a}{c}$, which is $\theta = \sin^{-1}\left(\frac{a}{c}\right)$.

Example 1

Write the following angles of the trigonometric functions in the form of their inverse functions:

(a) $\cos\alpha = \frac{m}{n}$

(b) $\tan\beta = \frac{k}{g}$

Solution

The angles α and β can be found as:

(a) $\alpha = \cos^{-1}\left(\frac{m}{n}\right)$

(b) $\beta = \tan^{-1}\left(\frac{k}{g}\right)$

Example 2

Find the angle α for the shown triangle in the figure.

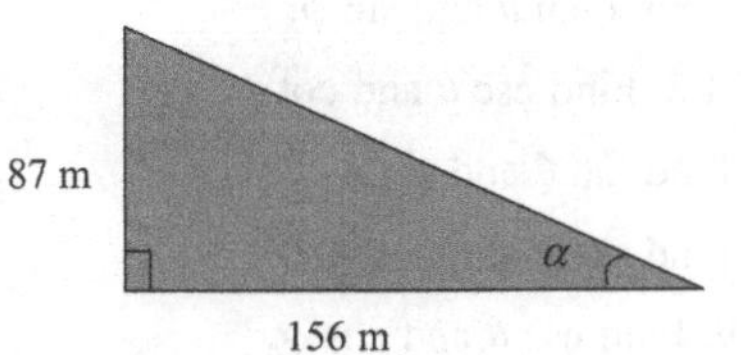

Solution

If $\tan\alpha = \frac{87}{156}$, then

$$\alpha = \tan^{-1}\left(\frac{87}{156}\right) = 29.15°$$

7.3 EXERCISES

1. Find the six trigonometric ratios for the angle θ in the figure.

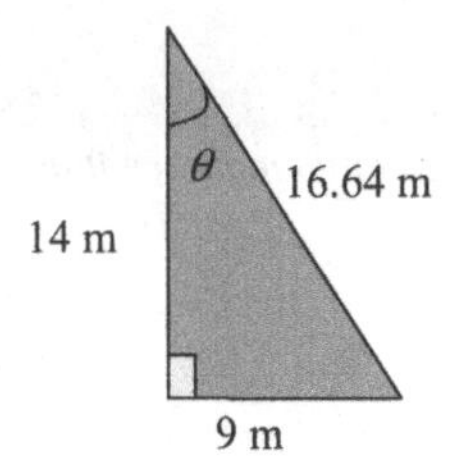

2. Find the six trigonometric ratios for the angle θ in the figure.

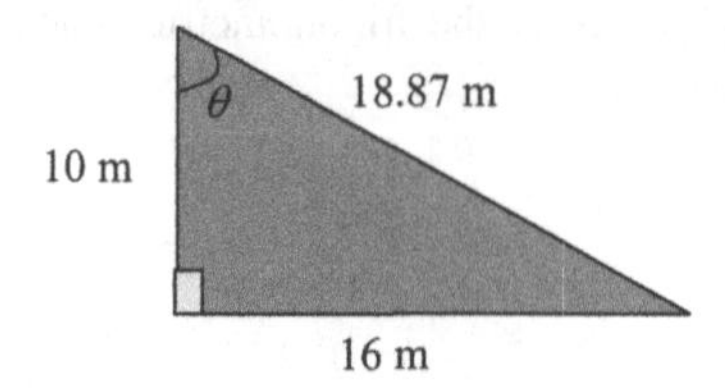

In exercises 3–10, find the indicated trigonometric ratio for the figure.

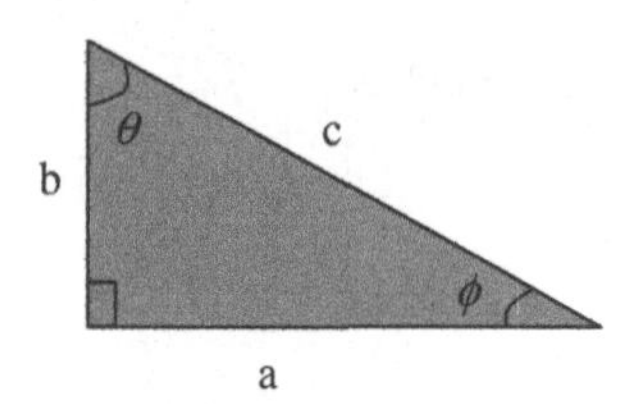

3. $b = 214$, $c = 294$. Find $\sin\theta$ and $\csc\phi$.
4. $a = 0.068$, $c = 0.082$. Find $\sin\phi$ and $\sec\theta$.
5. $b = 2.4$, $a = 3$. Find $\tan\theta$ and $\sin\phi$.
6. $c = 8.64$, $b = 4.2$. Find $\csc\theta$ and $\cot\phi$.
7. $a = 6$, $b = 8$. Find $\tan\theta$ and $\sin\phi$.
8. $a = 1$, $b = 1$. Find $\cot\theta$ and $\cos\phi$.
9. $b = 10$, $c = 19$. Find $\csc\theta$ and $\cos\phi$.
10. $a = 9$, $c = 28$. Find $\cos\theta$ and $\cot\phi$.

In exercises 11–16, find the complementary functions for the following trigonometric functions.

11. $\sin 60°$
12. $\tan 25°$
13. $\sec 30°$
14. $\csc 40°$
15. $\cot 10°$
16. $\cos 80°$

In exercises 17–22, find the supplementary functions for the following trigonometric functions.

17. $\sin 125°$
18. $\cos 105°$

19. tan 170°

20. csc 130°

21. cot 145°

22. sec 160°

In exercises 23–26, find the reciprocal identities of trigonometric functions for:

23. $\csc \theta = 1.15$

24. $\sec \phi = 1.67$

25. $\cot \beta = 0.76$

26. $\tan \alpha = 0.35$

7.4 UNIT CIRCLE AND OBLIQUE TRIANGLES

Positive angle and negative angle can be defined based on the quadrant locations. The below figure shows the unit circle.

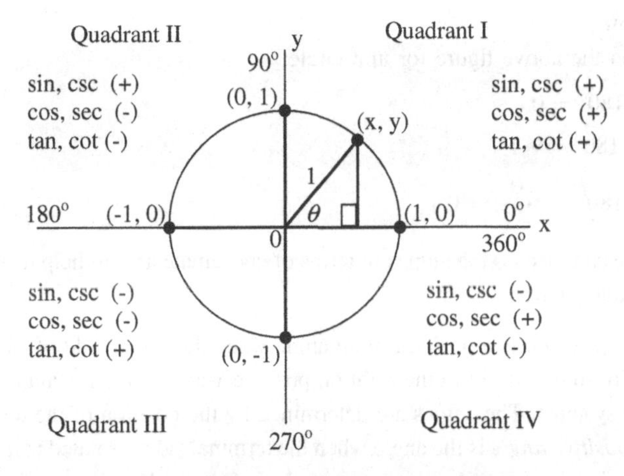

For the unit circle $x = \cos \theta$, $y = \sin \theta$, $\frac{y}{x} = \tan \theta$, and the radius is equal to 1. The sign of each trigonometric function can be different based on their region location in the circle. In Quadrant I, all the trigonometric functions are positive (+); in Quadrant II, only sin and csc are positive (+); in Quadrant III, only tan and cot are positive (+); and in Quadrant IV, only cos and sec are positive (+).

The trigonometric function values for angle 0° and 360° are the same.

Based on the above figure, we can find the following:

When $\theta = 0°$: $\sin 0° = 0$, $\cos 0° = 1$, $\tan 0° = 0$, $\csc 0° = \dfrac{1}{\sin\theta} = \dfrac{1}{0} = \infty$, $\sec 0° = \dfrac{1}{\cos\theta} = \dfrac{1}{1} = 1$, and $\cot 0° = \dfrac{1}{\tan\theta} = \dfrac{1}{0} = \infty$.

When $\theta = 90°$: $\sin 90° = 1$, $\cos 90° = 0$, $\tan 90° = \dfrac{1}{0} = \infty$, $\csc 90° = \dfrac{1}{\sin\theta} = \dfrac{1}{1} = 1$, $\sec 90° = \dfrac{1}{\cos\theta} = \dfrac{1}{0} = \infty$, and $\cot 90° = \dfrac{1}{\tan\theta} = \dfrac{1}{\infty} = 0$.

Example 1

Using the unit circle diagram (not calculator), find the values of the following trigonometric functions:

1. $\sin 180°$
2. $\cos 180°$
3. $\tan 180°$

Solution

Based on the above figure for unit circle:

(a) $\sin 180° = 0$

(b) $\cos 180° = -1$

(c) $\tan 180° = \dfrac{0}{-1} = 0$

Now, we can discuss the angles in terms of coordinate axis to help us studying oblique triangles.

Standard position is a position of an angle when the initial side of an angle extends from the origin to the right on positive x-axis in the rectangular coordinate systems. The angles are determined by the position of the terminal side. A *positive angle* is the angle when the terminal side is rotated in a counterclockwise direction. A *negative angle* is the angle when the terminal side is rotated in a clockwise direction. As seen from the figure below:

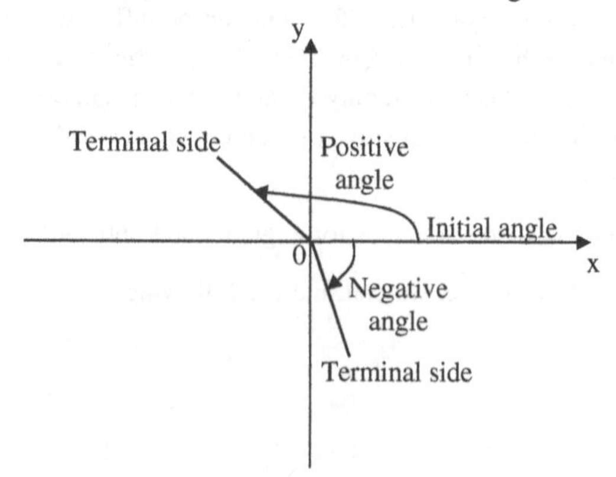

Now, we can define the trigonometric functions based on the following figure.

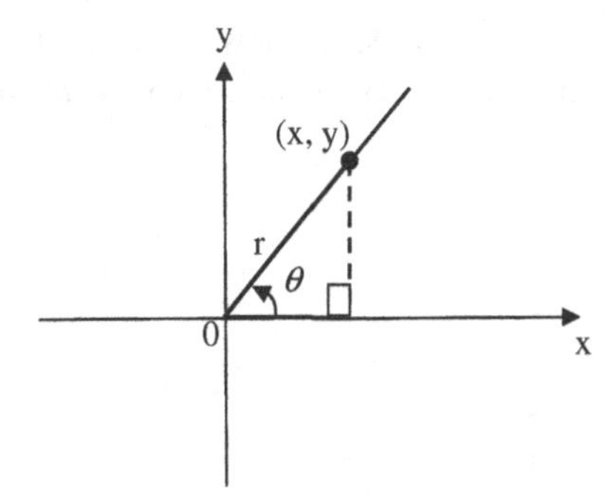

We know that by applying the Pythagorean theorem, $r = \sqrt{x^2 + y^2}$,

$$\sin\theta = \frac{y}{r}$$

$$\cos\theta = \frac{x}{r}$$

$$\tan\theta = \frac{y}{x}$$

$$\csc\theta = \frac{r}{y}$$

$$\sec\theta = \frac{r}{x}$$

$$\cot\theta = \frac{x}{y}$$

Example 2

Find the values of the trigonometric functions of the angle α with a terminal side passing through the point (3, 4) as in the figure below.

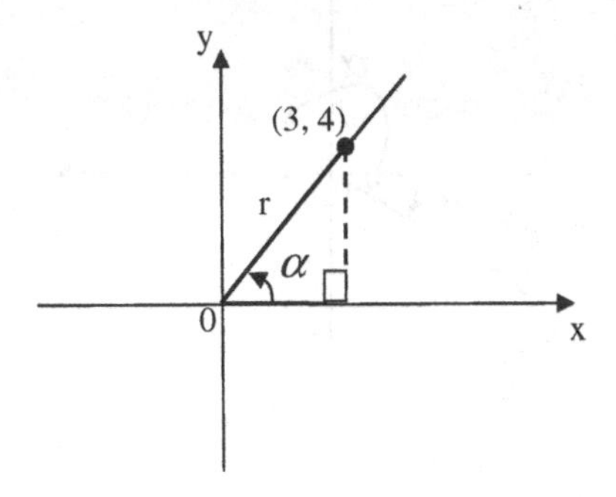

Solution

By applying the Pythagorean theorem,

$r = \sqrt{x^2 + y^2} = \sqrt{3^2 + 4^2} = 5$. So now we know that $x = 3$, $y = 4$, $r = 5$, so we can find the values of the trigonometric functions of the angle α:

$$\sin \alpha = \frac{4}{5}, \quad \cos \alpha = \frac{3}{5}$$

$$\tan \alpha = \frac{4}{3}, \quad \cot \alpha = \frac{3}{4}$$

$$\sec \alpha = \frac{5}{3}, \quad \csc \alpha = \frac{5}{4}$$

If the terminal side is not in the first quadrant, then we use a *reference angle*. A *reference angle* β for a given angle α is the positive acute angle formed by the x-axis and the terminal side of the α as illustrated in the figures below.

If the α is in the second quadrant, then the reference angle β can be found by $\beta = 180° - \alpha$.

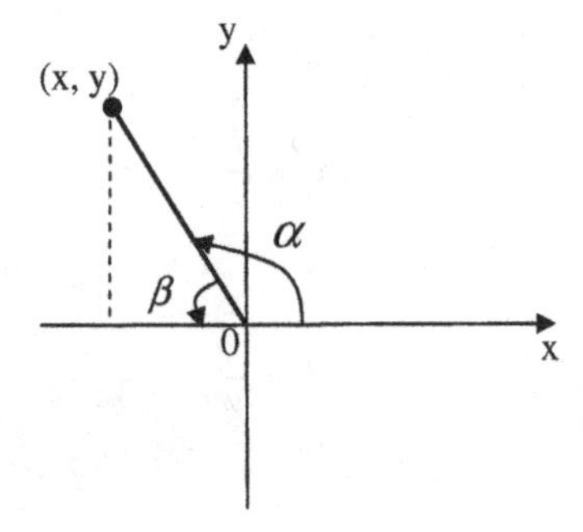

If the α is in the third quadrant, then the reference angle β can be found by $\beta = \alpha - 180°$.

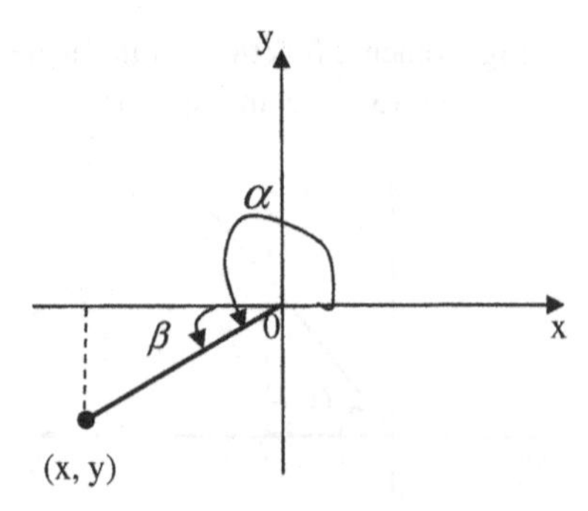

If the α is in the fourth quadrant, then the reference angle β can be found by $\beta = 360° - \alpha$.

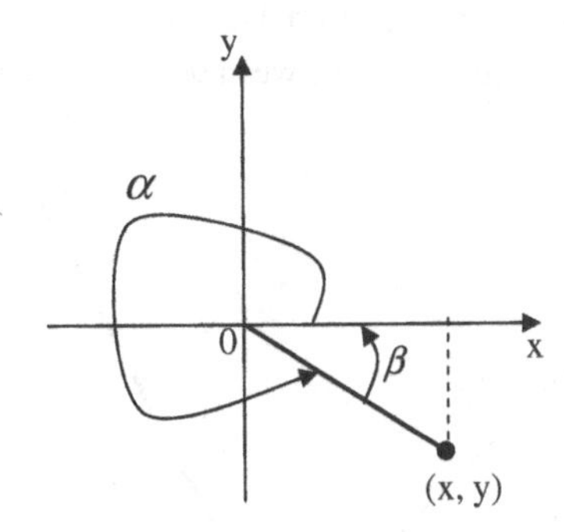

Example 3

The reference angle for the following angles:

(a) 220°

(b) 165°

(c) 335°

Solution

(a) 220° is an angle in the third quadrant, therefore, its reference angle is $\beta = 220° - 180° = 40°$.

(b) 165° is an angle in the second quadrant, therefore, its reference angle is $\beta = 180° - 165° = 15°$.

(c) 335° is an angle in the fourth quadrant, therefore, its reference angle is $\beta = 360° - 335° = 25°$.

Example 4

Find the sin 130°, cos 130°, and tan 130° for the figure.

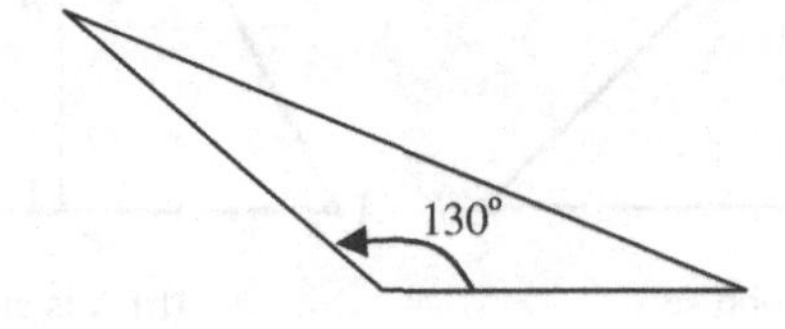

Solution

The angle is in the second quadrant, therefore,

$$\sin 130° = \sin(180° - 130°) = \sin 50° = 0.766$$
$$\cos 130° = -\cos(180° - 130°) = -\cos 50° = -0.643$$
$$\tan 130° = -\tan(180° - 130°) = -\tan 50° = -1.192$$

An *oblique triangle* is a triangle that does not have a right angle. This includes both acute and obtuse triangles. An *acute triangle* is a triangle with three acute angles (angles less than 90 degrees). An *obtuse triangle* is a triangle with one obtuse angle (angle between 90° and 180°) and two acute angles. For example,

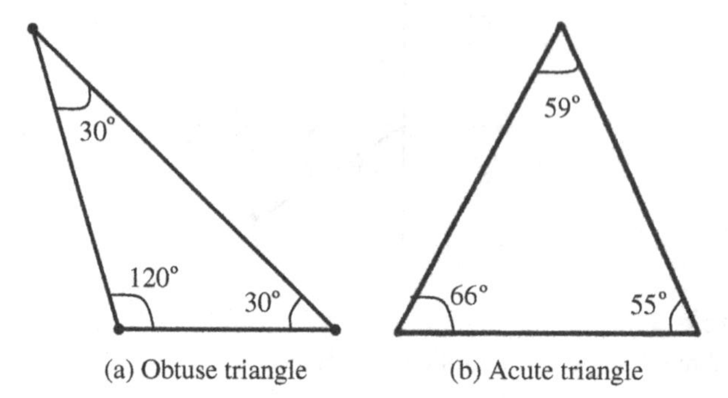

(a) Obtuse triangle (b) Acute triangle

There are two common methods used to solve oblique triangles: the law of sines and the law of cosines.

7.4.1 The law of sines

If ABC is a triangle with sides a, b, and c, then

$$\frac{\sin A}{a} = \frac{\sin B}{b} = \frac{\sin C}{c}.$$

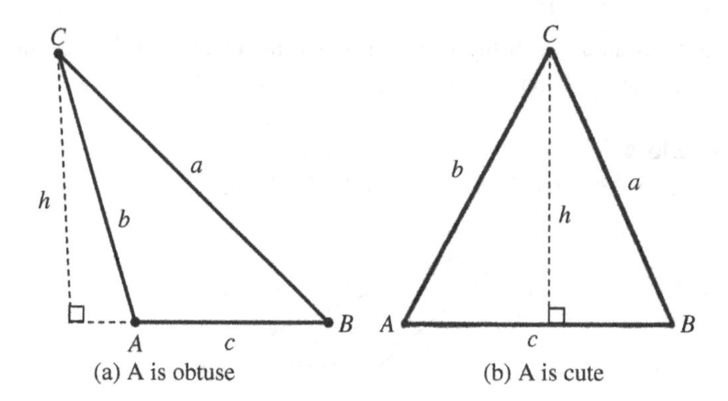

(a) A is obtuse (b) A is cute

Example 1

Find the lengths b and c, and the angle θ for the figure.

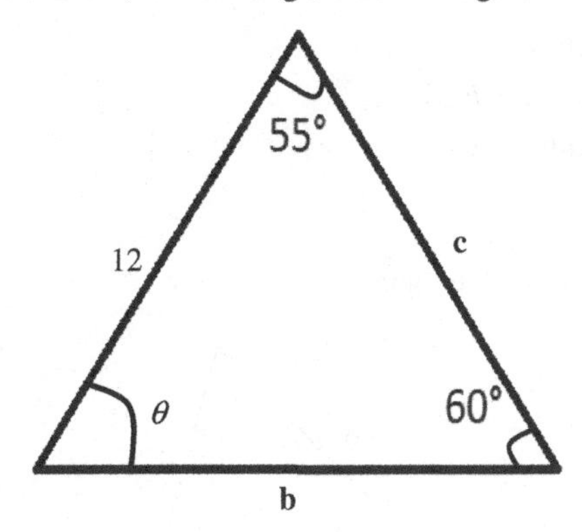

Solution

The third angle θ of the triangle is

$$\theta = 180° - (60° + 55°) = 180° - 115° = 65°$$

Now, by using the sines law, we have

$$\frac{\sin 60°}{12} = \frac{\sin 55°}{b}$$

$$b = \frac{\sin 55°}{0.0722} = 11.35$$

and

$$\frac{\sin 60°}{12} = \frac{\sin 65°}{c}$$

$$c = \frac{\sin 65°}{0.0722} = 12.55.$$

7.4.2 The law of cosines

The law of cosines is another relationship between the sides and the angles in any triangle. When we are given two sides of a triangle and the angle between the two sides, or we are given three sides of a triangle, it is not possible to solve the triangle with the law of sines. But we can use the law of cosines.

If ABC is a triangle with sides a, b, and c, then the following equations are valid:

$$a^2 = b^2 + c^2 - 2bc \cos A$$
$$b^2 = a^2 + c^2 - 2ac \cos B$$
$$c^2 = a^2 + b^2 - 2ab \cos C$$

Example 1

Find the length (d) for the figure.

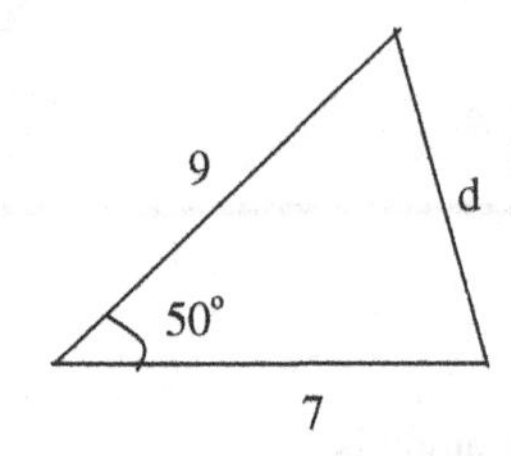

Solution

$$a^2 = b^2 + c^2 - 2bc \cos A$$
$$d^2 = 7^2 + 9^2 - 2(7)(9)\cos 50°$$
$$d^2 = 49 + 81 - 126(0.643)$$
$$d^2 = 130 - 81.02 = 48.98$$
$$d = \sqrt{48.98} \cong 7$$

7.4 EXERCISES

1. By using the unit circle diagram (not calculator), find the values of the following trigonometric functions:
   - **(a)** csc 180°
   - **(b)** sec 180°
   - **(c)** cot 180°
   - **(d)** sin 270°
2. Find the values of the following trigonometric functions:
   - **(a)** csc 115°
   - **(b)** sec 130°
   - **(c)** cot 90°
   - **(d)** sin 147°

In exercises 3 and 4, find the exact value of the trigonometric functions of the given angle α.

3.

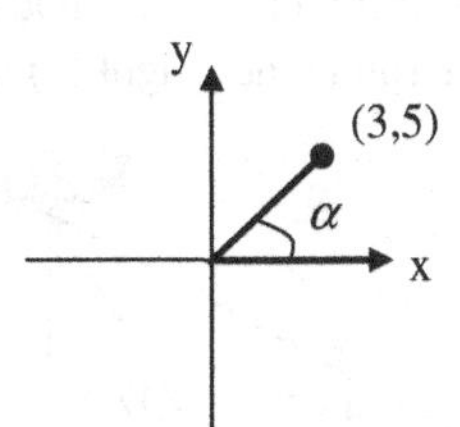

4.

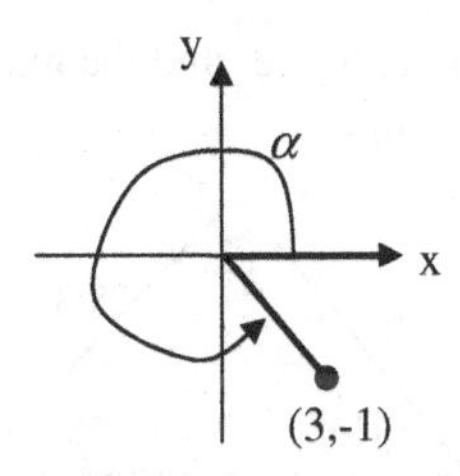

5. Given a triangle as in the figure, find the remaining side and angles.

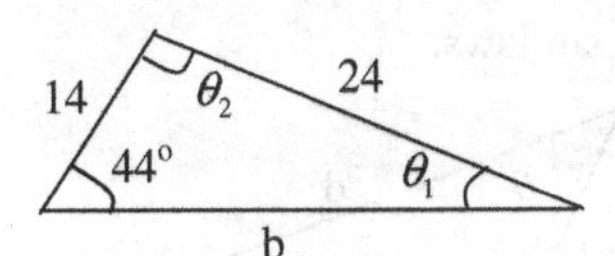

6. Given a triangle as in the figure, find the remaining sides and angle.

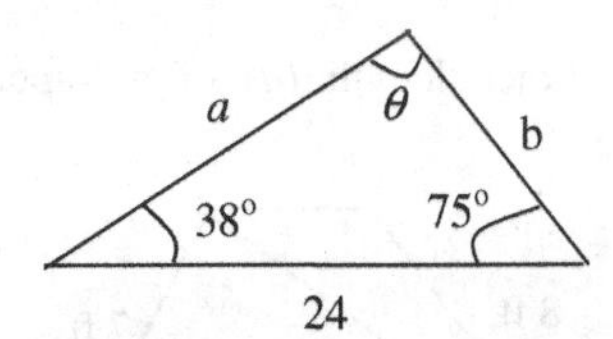

7. Determine the lengths of the two steel supports a_1 and a_2 as shown in the figure.

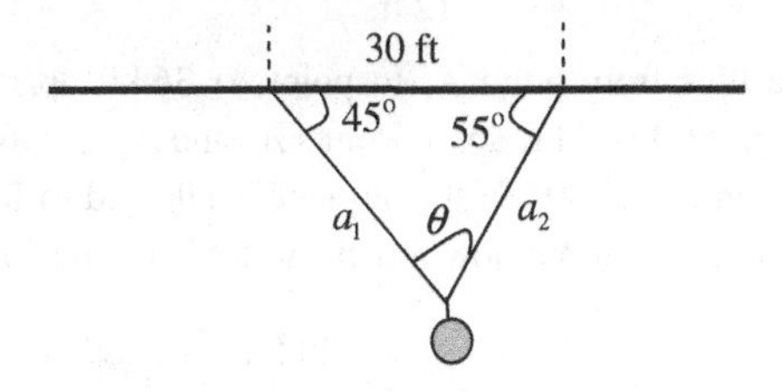

8. An airplane is sighted from points A_1 and A_2 on ground level. Points A_1 and A_2 are 467 ft apart. At point A_1, the angle of elevation to the airplane is 43.2° and for point A_2 the angle of elevation to the airplane is 47.5°. Determine the height (h) of the airplane.

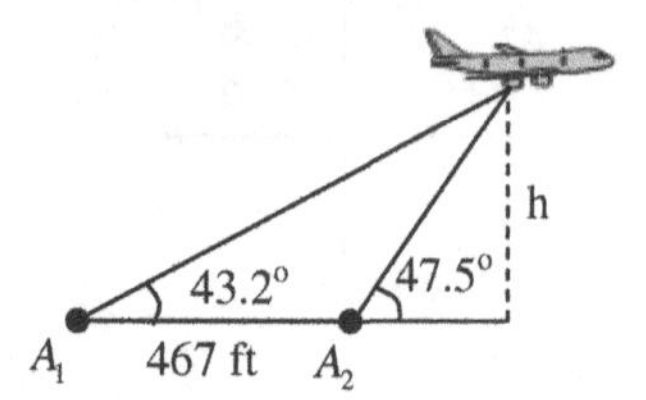

9. Given a triangle as in the figure, find the remaining side (d) and angles. **Hint**: use cos and sin laws.

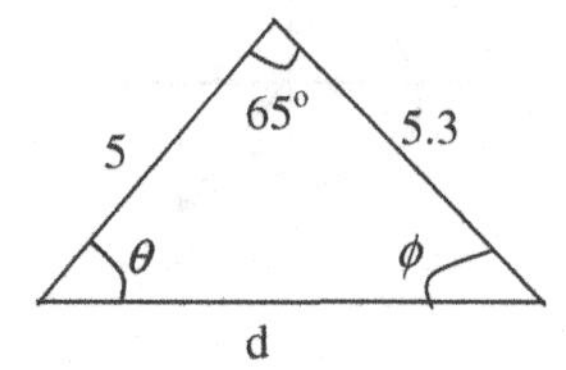

10. Given a triangle as in the figure, find the remaining side (d) and angles. **Hint**: use cos and sin laws.

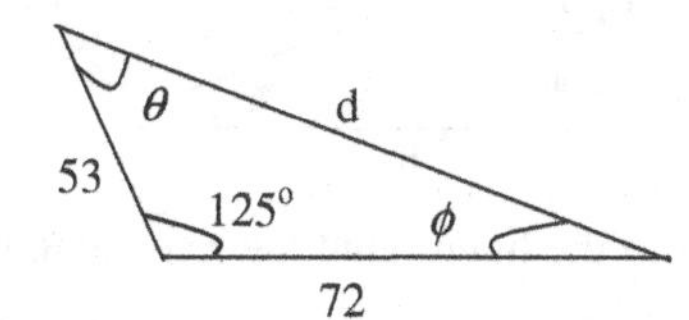

11. Find the diagonal brace length (d) of a trapezoid metal frame as shown in the figure.

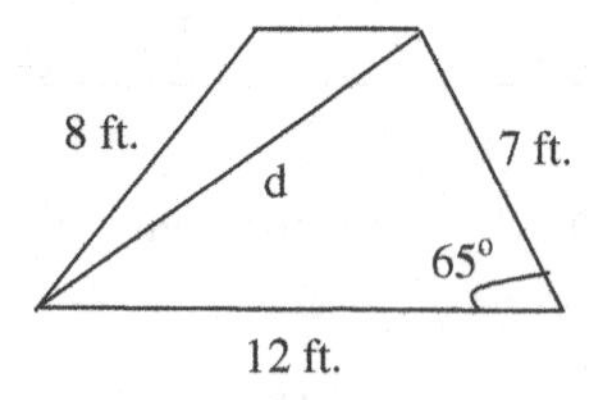

12. An airplane flies from point A_1 to point A_2 36 kft apart, and from to point A_2 to point A_3 29 kft apart. Points A_1 and A_2 are 467 ft apart. At point A_2 the angle is 150°. If the airplane is allowed to fly directly from point A_1 to point A_3, how much shorter is the new allowed route?

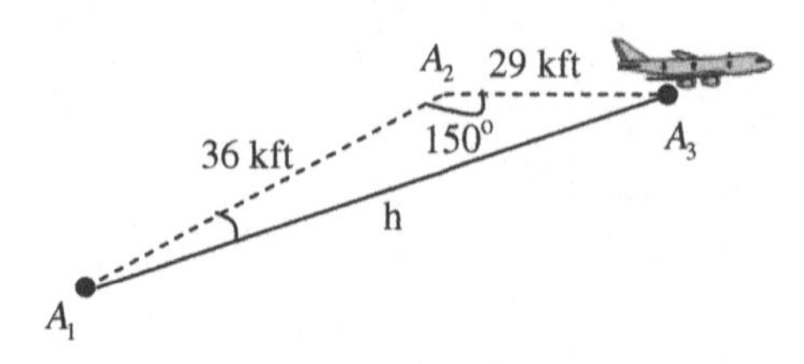

CHAPTER 7 REVIEW EXERCISES

Express the given angle measurements in radians (leave the answers in terms of π) for exercises 1–8.

1. 30°

2. 72°

3. 12°

4. 75°

5. 240°

6. 253°

7. 135°

8. 300°

Express the given angle measurements in degrees for exercises 9–14.

9. $\frac{3\pi}{4}$

10. $\frac{7\pi}{9}$

11. $\frac{2\pi}{9}$

12. $\frac{3\pi}{2}$

13. $\frac{6\pi}{5}$

14. $\frac{4\pi}{5}$

15. Find the length of side *b* of the triangle in the figure.

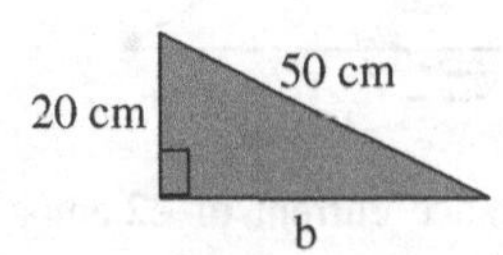

16. Find the length of side *a* of the triangle in the figure.

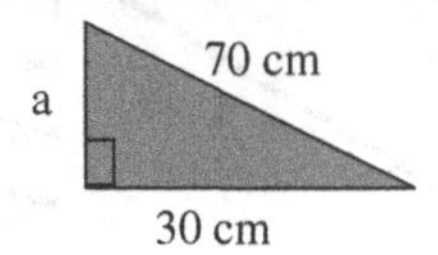

17. Find the length of side c of the triangle in the figure.

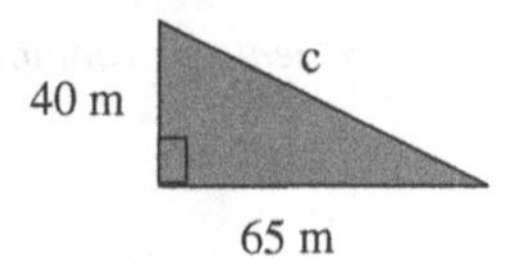

18. Find the length of side c of the triangle in the figure.

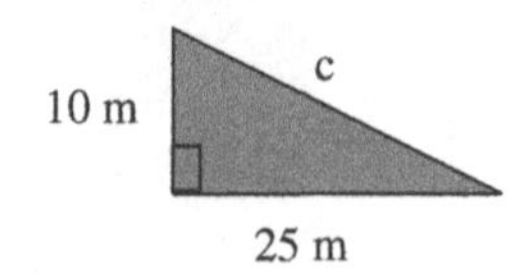

19. John resets his trip odometer at the bottom of a hill and then proceeds to drive up the hill until he comes to a sign that says you are 2 mi high. John's trip odometer reading says John has driven 2.6 mi. Determine how far the horizontal distance (b) is from John's initial starting point.

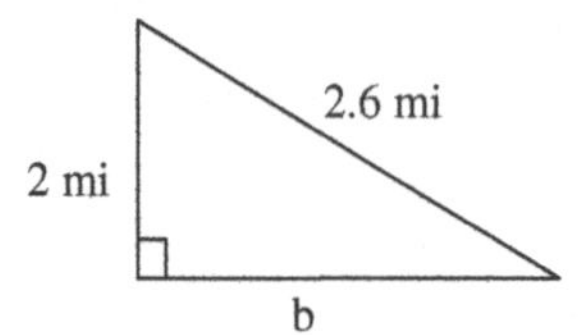

Answer = using Pythagorean theorem

20. A boat travels 6 mi due south and then travels due east for 12 mi as in the figure. Determine how far the boat is from its starting point.

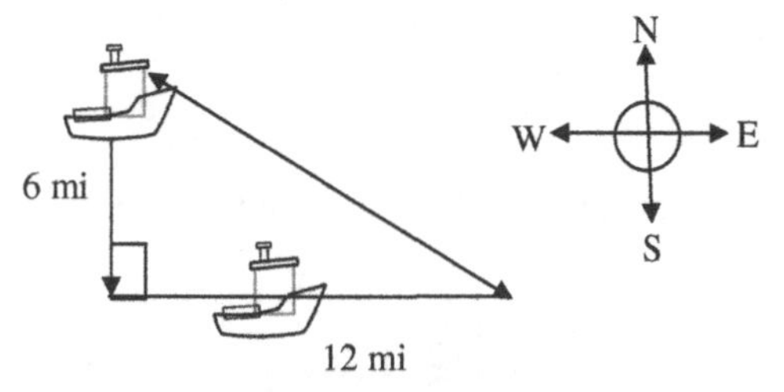

21. A circuit has a resistance current of 22 amps and the total current is 46 amps. Find the coil current.

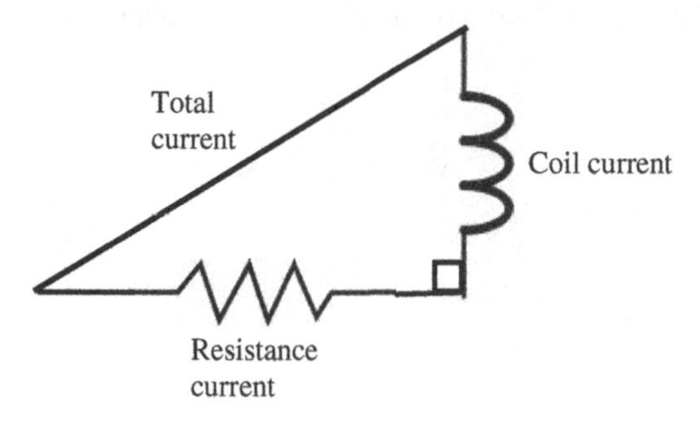

22. A circuit has a resistance current of 60.4 amps and the coil current is 75.6 amps. Find the total current.

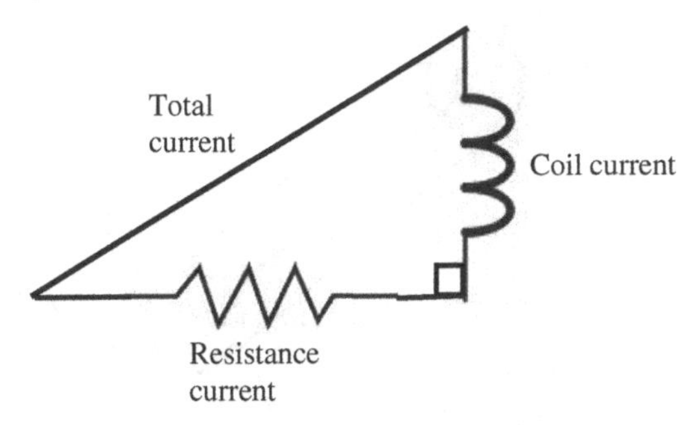

23. A circuit has impedance 170 ohms and resistance 116 ohms. Find the reactance of the circuit.

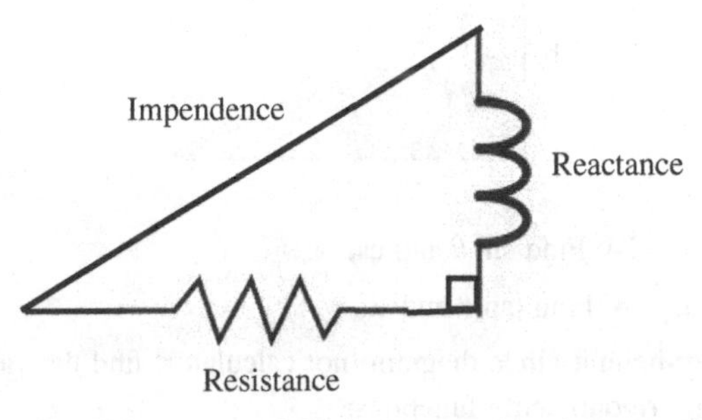

24. A circuit has reactance of 30.2 ohms and resistance of 56.5 ohms. Find the impedance of the circuit.

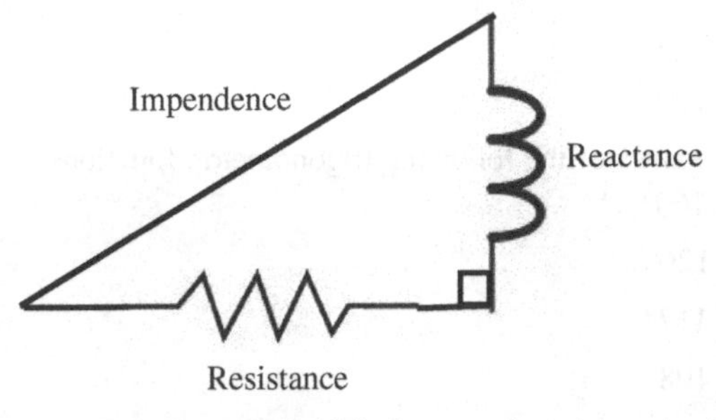

25. A circuit has reactance of 40.3 ohms and resistance of 73.4 ohms. Find the impedance of the circuit.

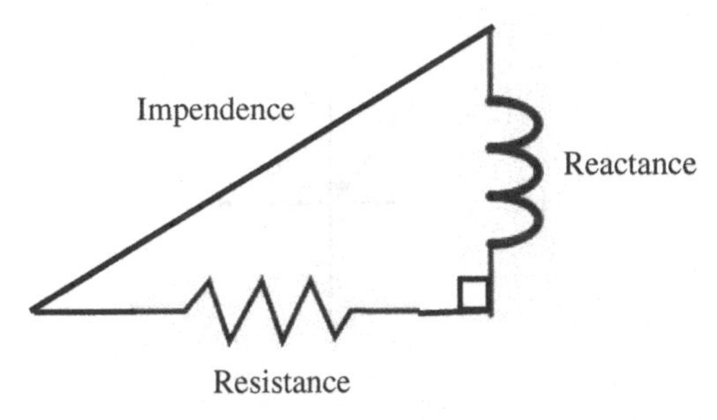

26. Find the distance (d) between the centers of two pulleys if one is placed 11 in. to the left and 7 in. above the other.

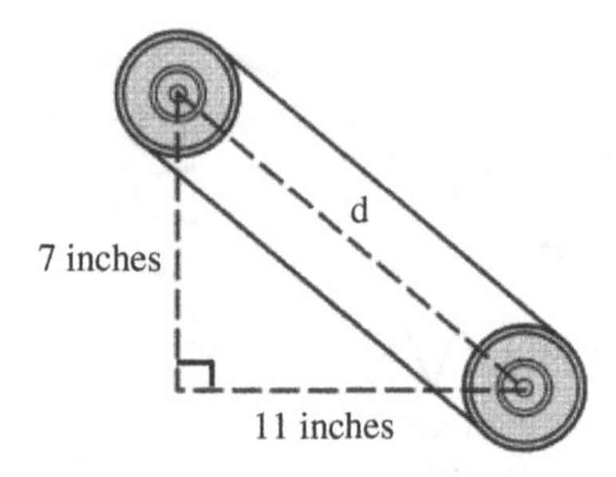

In exercises 27 and 28, find the indicated trigonometric ratio for the figure.

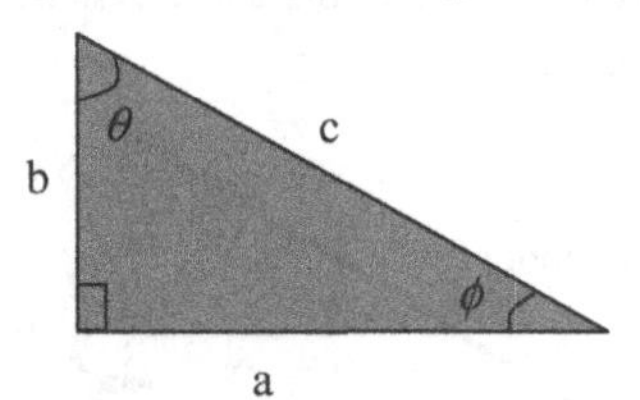

27. $b = 21$, $c = 24$. Find $\sin\theta$ and $\csc\phi$.

28. $c = 6.8$, $a = 4$. Find $\tan\theta$ and $\sin\phi$.

29. By using the unit circle diagram (not calculator) find the values of the following trigonometric functions:

(a) $\cos 270°$

(b) $\tan 270°$

(c) $\csc 270°$

(d) $\sec 270°$

30. Find the values of the following trigonometric functions:

(a) $\cos 160°$

(b) $\tan 126°$

(c) $\csc 117°$

(d) $\sec 108°$

In exercises 31 and 32, find the exact value of the trigonometric functions of the given angle α.

31.

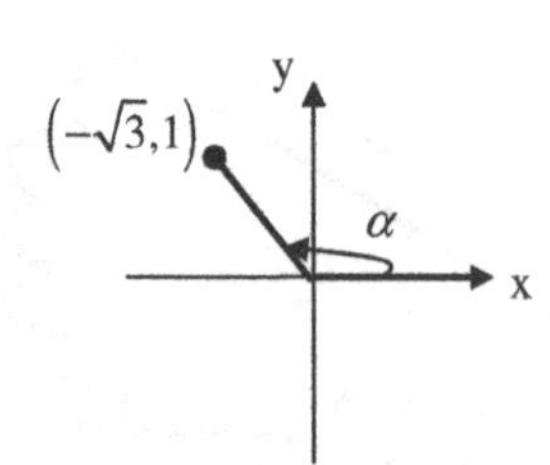

32.

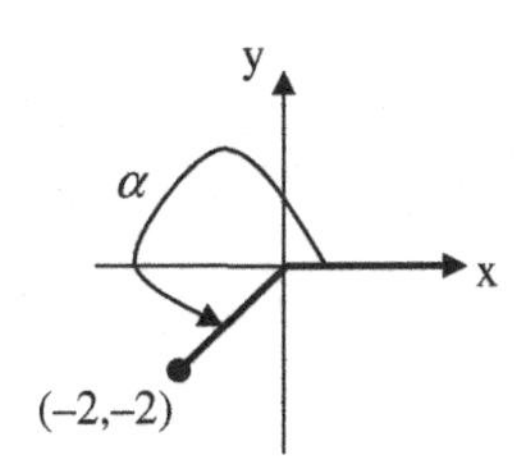

33. Determine the length (L) of supports of the shelf shown as in the figure.

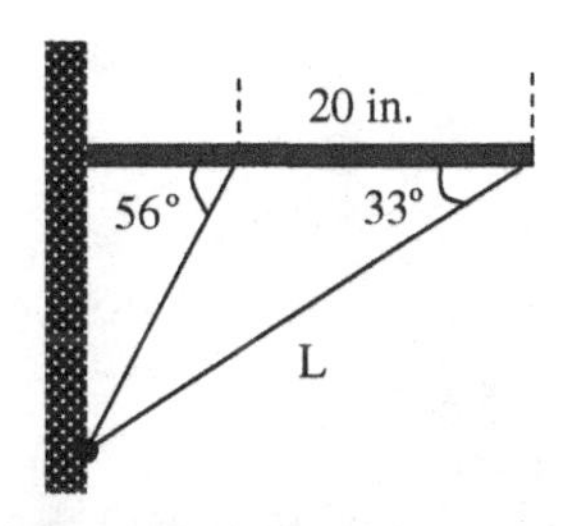

34. Determine the length (L) of the brace required to support the street-light shown as in the figure.

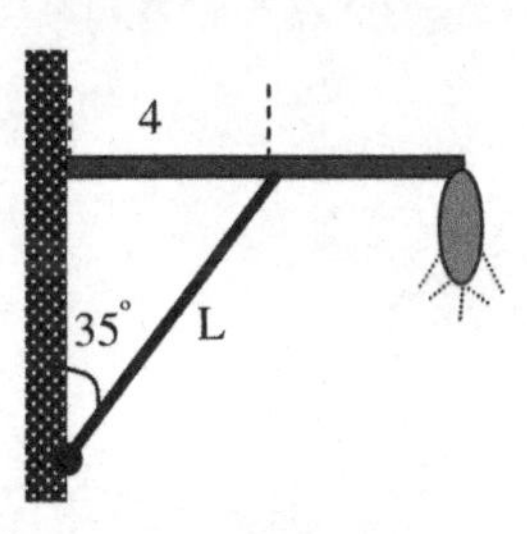

Chapter 8

Matrices, Determinants, and Vectors

In three words I can sum up everything I've learned about life: it goes on.

Robert Frost

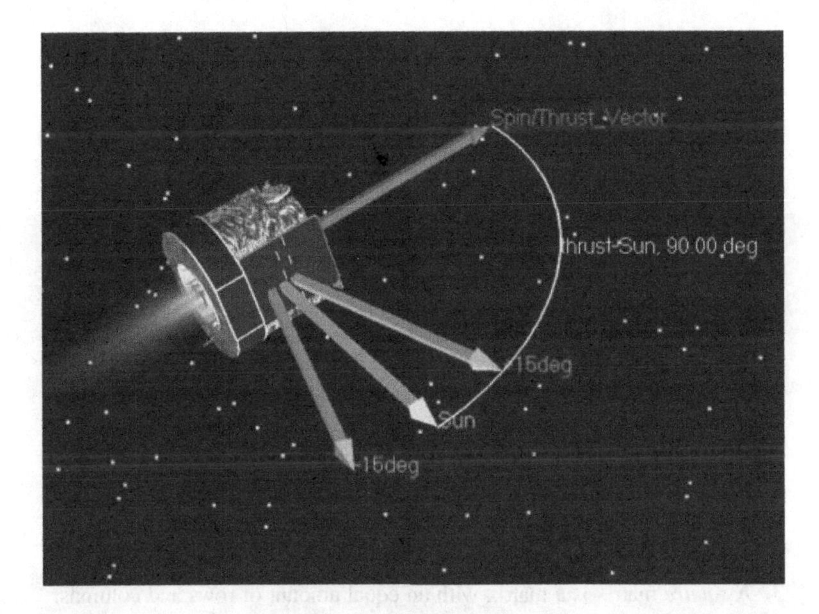

The capability of a satellite tool kit to do utilization as a systems analysis companion by incorporating vectors, axes, and angles, and vector- and axis-based viewing represents a means to more quickly understand design conditions, allowing the system engineer to get to the next iteration more quickly, understand results of heritage system tools more quickly and thoroughly, and resolve results in seconds.

Analyze system design constraints based on satellite body axes orientation at specific mission times against vectors (i.e., Sun, Moon, ground station…) to define launch window.

Fundamentals of Technical Mathematics. http://dx.doi.org/10.1016/B978-0-12-801987-0.00008-3

INTRODUCTION

In this chapter we focus on the basic calculations and rules of matrices, determinants, and vectors.

8.1 MATRICES

Matrices are used to organize data in several areas such as management, science, engineering, and technology.

Definition of matrix

A *matrix* is an array of numbers, enclosed by brackets.

The numbers are called *elements* or *entries* of the matrix. The elements are arranged in rows (horizontal) or columns (vertical), which determine the size (dimension or order) of the matrix.

Size of a matrix = number of rows × number of columns.

It can be read as the size of a matrix and is equal to number of rows "by" number of columns.

There are several popular types of matrices:

1. A *column* matrix (column vector) is a matrix that contains only one column.
2. A *row* matrix (row vector) is a matrix that contains only one row.
3. A *square* matrix is a matrix with an equal amount of rows and columns.
4. A *null* (zero) matrix is a matrix in which all elements are zero.
5. A *diagonal* matrix is a matrix in which all of the elements not on the diagonal of a square matrix are 0.
6. A *unit* (identity) matrix is a diagonal matrix in which the elements on the main diagonal are 1.

Location of an element in a matrix:

Let $A = \begin{bmatrix} a_{11} & a_{12} & a_{13} & a_{14} \\ a_{21} & a_{22} & a_{23} & a_{24} \\ a_{31} & a_{32} & a_{33} & a_{34} \\ a_{41} & a_{42} & a_{43} & a_{44} \end{bmatrix}$ is a square matrix with size 4×4

a_{11} is element a at row 1 and column 1.
a_{12} is element a at row 1 and column 2.
a_{32} is element a at row 3 and column 2.

Example 1

Determine the size of each matrix and name the type of matrix.

1. $A = \begin{bmatrix} -1 & 3 & 6 \end{bmatrix}$

2. $B = \begin{bmatrix} 0 \\ 2 \\ 4 \end{bmatrix}$

3. $C = \begin{bmatrix} 1 & 0 & -1 \\ 3 & 2 & 5 \\ 6 & 4 & 1 \end{bmatrix}$

4. $O = \begin{bmatrix} 0 & 0 & 0 \\ 0 & 0 & 0 \end{bmatrix}$

5. $D = \begin{bmatrix} 15 & 0 & 0 \\ 0 & 12 & 0 \\ 0 & 0 & 14 \end{bmatrix}$

6. $E = \begin{bmatrix} 1 & 0 & 0 \\ 0 & 1 & 0 \\ 0 & 0 & 1 \end{bmatrix}$

Solution

(a) Size of matrix $A = 1 \times 3$; row matrix
(b) Size of matrix $B = 3 \times 1$; column matrix
(c) Size of matrix $C = 3 \times 3$; square matrix
(d) Size of matrix $O = 2 \times 3$; zero matrix
(e) Size of matrix $D = 3 \times 3$; diagonal matrix
(f) Size of matrix $D = 2 \times 3$; unit matrix

Example 2

Let $\begin{bmatrix} p & m \\ a & v \\ u & t \end{bmatrix}$ find the following:

(a) The element in row 1 column 1
(b) The element in row 2 column 1

(c) The element in row 3 column 2
(d) The element in row 2 column 2

Solution

(a) The element in row 1 column 1 is p
(b) The element in row 2 column 1 is a
(c) The element in row 3 column 2 is t
(d) The element in row 2 column 2 is v

8.1.1 **Operations with matrices and their properties**

The study of matrices properties is based on the following:

8.1.1.1 Equality of matrices

Two matrices are equal if they have the same size and their corresponding elements are equal.

Example 1

Let $A = \begin{bmatrix} 1 & -4 \\ 5 & 3 \end{bmatrix}$ and $B = \begin{bmatrix} 2x & w \\ z-2 & k+1 \end{bmatrix}$

If matrices A and B are equal, find the value of x, w, z, and k.

Solution

$1 = 2x \longrightarrow x = 1/2$
$w = -4$
$z - 2 = 5 \longrightarrow z = 7$
$3 = k + 1 \longrightarrow k = 2$

8.1.1.2 Addition of matrices

Two matrices with the same size can be summed by sums of the corresponding elements of the two matrices.

8.1.1.3 Subtraction of matrices

Two matrices with the same size can be subtracted by taking the difference of the corresponding elements of the two matrices.

Example 1

Let $A = \begin{bmatrix} 1 & -2 \\ 4 & 3 \end{bmatrix}$ and $B = \begin{bmatrix} 0 & 3 \\ 5 & 6 \end{bmatrix}$, find

(a) $A + B$
(b) $A - B$

Solution

(a) $A + B = \begin{bmatrix} 1+0 & -2+3 \\ 4+5 & 3+6 \end{bmatrix} = \begin{bmatrix} 1 & 1 \\ 9 & 9 \end{bmatrix}$

(b) $A - B = \begin{bmatrix} 1-0 & -2-3 \\ 4-5 & 3-6 \end{bmatrix} = \begin{bmatrix} 1 & -5 \\ -1 & -3 \end{bmatrix}$

We can use MATLAB to calculate addition and subtraction of matrices:

```
>> A=[1 -2;4 3];
>> B=[0 3;5 6];
>> A+B
ans =
     1     1
     9     9
>> A-B
ans =
     1    -5
    -1    -3
```

We can use Maple to calculate addition and subtraction of matrices:

```
> A := Matrix([[1,-2], [4,3]]) :
> B := Matrix([[0,3], [5,6]]) :
> Matrix(A + B)
```

$$\begin{bmatrix} 1 & 1 \\ 9 & 9 \end{bmatrix}$$

```
> Matrix(A - B)
```

$$\begin{bmatrix} 1 & -5 \\ -1 & -3 \end{bmatrix}$$

8.1.1.4 Additive identity matrix (zero matrix)

Let A and 0 be matrices with the same size, then $A + 0 = A$, where is 0 called zero matrix.

Example 1

Let $A = \begin{bmatrix} 7 & 3 \\ 1 & 4 \end{bmatrix}$ and $0 = \begin{bmatrix} 0 & 0 \\ 0 & 0 \end{bmatrix}$, then $A + 0 = \begin{bmatrix} 7 & 3 \\ 1 & 4 \end{bmatrix} + \begin{bmatrix} 0 & 0 \\ 0 & 0 \end{bmatrix} = \begin{bmatrix} 7+0 & 3+0 \\ 1+0 & 4+0 \end{bmatrix} = \begin{bmatrix} 7 & 3 \\ 1 & 4 \end{bmatrix}$

8.1.1.5 Additive inverse (negative) matrix

Let A and $-A$ be matrices with the same size, then $A + (-A) = 0$, where $-A$ is called the additive inverse of matrix A.

Example 1

If $A = \begin{bmatrix} 5 & 2 \\ 4 & 1 \end{bmatrix}$ and $-A = \begin{bmatrix} -5 & -2 \\ -4 & -1 \end{bmatrix}$, then $A + (-A) = \begin{bmatrix} 5 & 2 \\ 4 & 1 \end{bmatrix} + \begin{bmatrix} -5 & -2 \\ -4 & -1 \end{bmatrix} = \begin{bmatrix} 5+(-5) & 2+(-2) \\ 4+(-4) & 1+(-1) \end{bmatrix} = \begin{bmatrix} 0 & 0 \\ 0 & 0 \end{bmatrix} = 0$

8.1.1.6 Commutative property of matrix addition

Let A and B be matrices with the same size, then $A + B = B + A$.

Example 1

Let $A = \begin{bmatrix} 6 & 5 \\ 2 & 1 \end{bmatrix}$ and $B = \begin{bmatrix} 3 & 2 \\ 1 & 5 \end{bmatrix}$, then $A + B = \begin{bmatrix} 6 & 5 \\ 2 & 1 \end{bmatrix} + \begin{bmatrix} 3 & 2 \\ 1 & 5 \end{bmatrix} = \begin{bmatrix} 9 & 7 \\ 3 & 6 \end{bmatrix}$ and $B + A = \begin{bmatrix} 3 & 2 \\ 1 & 5 \end{bmatrix} + \begin{bmatrix} 6 & 5 \\ 2 & 1 \end{bmatrix} = \begin{bmatrix} 9 & 7 \\ 3 & 6 \end{bmatrix}$, therefore, $A + B = B + A$.

8.1.1.7 Associative property of matrix addition

Let A, B, and C be matrices with the same size, then $(A + B) + C = A + (B + C)$.

Example 1

Let $A = \begin{bmatrix} 1 & 3 \\ 7 & 8 \end{bmatrix}$, $B = \begin{bmatrix} 2 & 5 \\ 3 & 1 \end{bmatrix}$, and $C = \begin{bmatrix} 9 & 8 \\ 4 & 6 \end{bmatrix}$

Then, $(A + B) + C = \begin{bmatrix} 1+2 & 3+5 \\ 7+3 & 8+1 \end{bmatrix} + \begin{bmatrix} 9 & 8 \\ 4 & 6 \end{bmatrix} = \begin{bmatrix} 12 & 16 \\ 14 & 15 \end{bmatrix}$

$$A+(B+C) = \begin{bmatrix} 1 & 3 \\ 7 & 8 \end{bmatrix} + \begin{bmatrix} 2+9 & 5+8 \\ 3+4 & 1+6 \end{bmatrix} = \begin{bmatrix} 12 & 16 \\ 14 & 15 \end{bmatrix}$$

Therefore, $(A+B)+C = A+(B+C)$.

8.1.1.8 Product of constant (scalar) and a matrix

Let c be a constant (real number) and A a matrix of any size, then cA is determined by multiplying c with each element of matrix A.

Example 1

Let $A = \begin{bmatrix} -3 & 1 \\ 4 & 2 \end{bmatrix}$, then $5A = \begin{bmatrix} -15 & 5 \\ 20 & 10 \end{bmatrix}$

8.1.1.9 Product of two matrices

Let A be a matrix with size $m \times n$ and let B be a matrix with size $n \times r$.

The product AB is a matrix with size $m \times r$. Each row of A multiplies each column of B, the terms are added, and the result is entered in the corresponding location of AB.

Example 1

Let $A = \begin{bmatrix} 1 & 3 \\ 7 & 8 \end{bmatrix}$ and $B = \begin{bmatrix} 3 & 2 & 4 \\ -1 & 0 & 6 \end{bmatrix}$, then $AB = \begin{bmatrix} 1 & 3 \\ 7 & 8 \end{bmatrix}$

$$\begin{bmatrix} 3 & 2 & 4 \\ -1 & 0 & 6 \end{bmatrix} = \begin{bmatrix} 1\times 3+3\times(-1) & 1\times 2+3\times 0 & 1\times 4+3\times 6 \\ 7\times 3+8\times(-1) & 7\times 2+8\times 0 & 7\times 4+8\times 6 \end{bmatrix} =$$

$$\begin{bmatrix} 0 & 2 & 22 \\ 13 & 14 & 76 \end{bmatrix}$$

Size of $AB = 2 \times 3$.

Note: matrix multiplication is not commutative, that is, $AB \neq BA$.

We can use MATLAB to calculate the multiplication of matrices:

```
>> A=[1 3;7 8];
>> B=[3 2 4; −1 0 6];
>> A*B
ans =
     0     2    22
    13    14    76
```

We can use Maple to calculate the multiplication of matrices:

```
> B := Matrix([[3,2,4], [−1,0,6]]) :
> A := Matrix([[1,3], [7,8]]) :
> Multiply(A, B)
```

$$\begin{bmatrix} 0 & 2 & 22 \\ 13 & 14 & 76 \end{bmatrix}$$

8.1.1.10 Multiplicative identity matrix

Let A and I be square matrices with the same sizes. If $AI = IA = A$, then I is called the multiplicative identity matrix. I is a square matrix and has the diagonal line 1's and the rest of the matrix elements is 0's.

For example, the multiplicative identity matrix I for 2×2 and 3×3 matrices is 2×2 and 3×3, respectively.

Example 1

Let $A = \begin{bmatrix} 3 & 11 \\ -1 & 5 \end{bmatrix}$, then $I_A = \begin{bmatrix} 1 & 0 \\ 0 & 1 \end{bmatrix}$

Therefore, $AI_A = \begin{bmatrix} 3 & 11 \\ -1 & 5 \end{bmatrix}\begin{bmatrix} 1 & 0 \\ 0 & 1 \end{bmatrix} = \begin{bmatrix} 3 & 11 \\ -1 & 5 \end{bmatrix} = A$

The 2×2 identity matrix, $I_{2\times2}$, is $\begin{bmatrix} 1 & 0 \\ 0 & 1 \end{bmatrix}$

The 3×3 identity matrix, $I_{3\times3}$, $\begin{bmatrix} 1 & 0 & 0 \\ 0 & 1 & 0 \\ 0 & 0 & 1 \end{bmatrix}$

The 4×4 identity matrix, $I_{4\times4}$, $\begin{bmatrix} 1 & 0 & 0 & 0 \\ 0 & 1 & 0 & 0 \\ 0 & 0 & 1 & 0 \\ 0 & 0 & 0 & 1 \end{bmatrix}$

8.1.1.11 Multiplicative inverse matrix

Let A and A^{-1} be square matrices with the same size. If $AA^{-1} = A^{-1}A = I$, the I is called the multiplicative inverse matrix.

Example 1

Let $A = \begin{bmatrix} 2 & 1 \\ -1 & 1 \end{bmatrix}$, find A^{-1}.

Solution

Method 1. Let $A^{-1} = \begin{bmatrix} a & b \\ c & d \end{bmatrix}$, then $AA^{-1} = \begin{bmatrix} 2 & 1 \\ -1 & 1 \end{bmatrix}\begin{bmatrix} a & b \\ c & d \end{bmatrix} =$

$\begin{bmatrix} 1 & 0 \\ 0 & 1 \end{bmatrix}$

$$\begin{aligned}
2a + c = 1 &\longrightarrow 3a = 1 \longrightarrow a = 1/3 \\
2b + d = 0 &\longrightarrow 3b = -1 \longrightarrow b = -1/3 \\
-a + c = 0 &\longrightarrow a = c = 1/3. \\
-b + d = 1 &\longrightarrow d = 1 + b = 2/3
\end{aligned}$$

Therefore,

$$A^{-1} = \begin{bmatrix} 1/3 & -1/3 \\ 1/3 & 2/3 \end{bmatrix}$$

Method 2. Note, interchanging any two rows will not change the sign of the rows.

There is another method based on the determinant that can be used to find the inverse of a square matrix, which will be discussed in the next section.

8.1.1.12 Transpose of a matrix

Let A be a matrix with size $m \times n$. The transpose matrix of A is written as A^t with size $n \times m$. The transpose matrix A^t is obtained by interchanging the rows and columns of A.

Example 1

Let $A = \begin{bmatrix} 0 & 1 & 3 \\ -1 & 2 & 5 \end{bmatrix}$, then $A^t = \begin{bmatrix} 0 & -1 \\ 1 & 2 \\ 3 & 5 \end{bmatrix}$.

Using Matlab to transpose the matrix of A:

```
>> A=[0 1 3;-1 2 5];
>> A'
ans =
    0   -1
    1    2
    3    5
```

Using Maple to transpose the matrix of A:

```
> A := Matrix([[0,1,3], [−1,2,5]]) :
> Transpose(A)
```

$$\begin{bmatrix} 0 & -1 \\ 1 & 2 \\ 3 & 5 \end{bmatrix}$$

8.1.1.13 Symmetric matrix

A matrix $A = [a_{ij}]$ is called symmetric if $A^t = A$. That is, A is symmetric if it is a square matrix for which $a_{ij} = a_{ji}$. If matrix A is symmetric, then the elements of A are symmetric with respect to the main diagonal of A.

Example 1

The matrices $\begin{bmatrix} 0 & 1 & 2 \\ 1 & 3 & 4 \\ 2 & 4 & 5 \end{bmatrix}$ and $\begin{bmatrix} 1 & 0 & 0 \\ 0 & 1 & 0 \\ 0 & 0 & 1 \end{bmatrix}$ are symmetric matrices.

8.1 EXERCISES

In exercises 1−12, identify the type of matrices.

1. $\begin{bmatrix} 0 & -1 & 2 & -3 & 15 \end{bmatrix}$

2. $\begin{bmatrix} 3 \\ 1 \\ 2 \\ 5 \\ 7 \\ 8 \end{bmatrix}$

3. $\begin{bmatrix} a & b \\ c & d \end{bmatrix}$

4. $\begin{bmatrix} 0 & 0 & 0 & 0 \\ 0 & 0 & 0 & 0 \end{bmatrix}$

5. $\begin{bmatrix} 1 & 0 & 0 & 0 & 0 \\ 0 & 2 & 0 & 0 & 0 \\ 0 & 0 & 3 & 0 & 0 \\ 0 & 0 & 0 & 4 & 0 \\ 0 & 0 & 0 & 0 & 5 \end{bmatrix}$

6. $\begin{bmatrix} 1 & 0 & 0 & 0 \\ 0 & 1 & 0 & 0 \\ 0 & 0 & 1 & 0 \\ 0 & 0 & 0 & 1 \end{bmatrix}$

In exercises 7–18, find the dimensions (order) of each matrix.

7. $\begin{bmatrix} 2 & -1 & \pi \\ -3 & \frac{1}{2} & 4 \end{bmatrix}$

8. $\begin{bmatrix} -5 & 7 & 9 \\ 4 & \pi & 8 \\ 11 & -3 & 6 \end{bmatrix}$

9. $\begin{bmatrix} 12 \\ -4 \end{bmatrix}$

10. $\begin{bmatrix} 5 & -1 \end{bmatrix}$

11. $\begin{bmatrix} -\pi & \frac{1}{4} \\ 3 & 0 \\ -7 & 2 \end{bmatrix}$

12. $\begin{bmatrix} 0 & -3 \\ 3 & 0 \end{bmatrix}$

13. $\begin{bmatrix} 0 & 1 & 2 \\ 0 & 0 & 1 \\ 7 & 8 & 9 \end{bmatrix}$

14. $\begin{bmatrix} \sqrt{3} & \pi \end{bmatrix}$

15. $\begin{bmatrix} \pi \\ -2 \end{bmatrix}$

16. $\begin{bmatrix} -8 \end{bmatrix}$

17. $\begin{bmatrix} 2 & 1 & -3 & 5 & 7 \end{bmatrix}$

18. $\begin{bmatrix} 0 \\ 2 \\ -6 \\ 8 \end{bmatrix}$

In exercises 19–28, find the values of the variables in each of the following matrices.

19. $\begin{bmatrix} x & 2 & 5 \\ 1 & t & 3 \end{bmatrix} = \begin{bmatrix} 6 & 2 & 5 \\ 1 & 11 & y \end{bmatrix}$

20. $\begin{bmatrix} 9 & x & 8 \\ y & t & 3 \end{bmatrix} = \begin{bmatrix} 9 & -1 & 8 \\ 0 & 11 & 3 \end{bmatrix}$

21. $\begin{bmatrix} a & 2 \\ b & c \end{bmatrix} = \begin{bmatrix} 8 & 2 \\ 9 & 7 \end{bmatrix}$

22. $\begin{bmatrix} -9 \\ 21 \end{bmatrix} = \begin{bmatrix} -9 \\ x \end{bmatrix}$

23. $\begin{bmatrix} a & b \\ 3 & -5 \end{bmatrix} = \begin{bmatrix} 0 & 0 \\ f & g \end{bmatrix}$

24. $\begin{bmatrix} 1 & 13 & 16 \\ t & w & y \end{bmatrix} = \begin{bmatrix} 1 & u & v \\ 3 & 4 & 11 \end{bmatrix}$

25. $\begin{bmatrix} x+5 & y-3 \\ 9 & 2 \end{bmatrix} = \begin{bmatrix} 24 & 7 \\ t & 2 \end{bmatrix}$

26. $\begin{bmatrix} p-4 & q+8 \\ 0 & 11 \end{bmatrix} = \begin{bmatrix} 15 & 28 \\ 0 & k \end{bmatrix}$

27. $\begin{bmatrix} 6 & m+12 & n+3 \end{bmatrix} + \begin{bmatrix} 11 & -4 & 8 \end{bmatrix} = \begin{bmatrix} t-3 & 16 & 2n \end{bmatrix}$

28. $\begin{bmatrix} -6 & 2 \\ 8 & 4 \end{bmatrix} + \begin{bmatrix} p & 3 \\ 7 & q \end{bmatrix} = \begin{bmatrix} 11 & 5 \\ 15 & 12 \end{bmatrix}$

In exercises 29–36, let

$$A = \begin{bmatrix} 3 & 4 \\ 5 & 6 \end{bmatrix},\ B = \begin{bmatrix} -3 & -7 \\ 13 & -2 \end{bmatrix},\ \text{and } C = \begin{bmatrix} 2 & -1 \\ 1 & 4 \end{bmatrix}$$

Find each of the following matrices.

29. $A + B$
30. $B + C$
31. $A - B$
32. $B - C$
33. $2A$
34. $-3B$
35. $2A + C$
36. $-3B + A$

In exercises 37–42, the order (size) of two matrices M and N is given. Find the size of the product MN and the product NM, whenever these products exist.

37. Size of $M = 3 \times 3$, Size of $N = 3 \times 3$
38. Size of $M = 4 \times 4$, Size of $N = 4 \times 4$
39. Size of $M = 2 \times 4$, Size of $N = 4 \times 2$
40. Size of $M = 1 \times 3$, Size of $N = 3 \times 1$

41. Size of $M = 2 \times 5$, Size of $N = 4 \times 3$

42. Size of $M = 1 \times 5$, Size of $N = 2 \times 4$

In exercises 43–46, let

$$A = \begin{bmatrix} -1 & 3 \\ 1 & 4 \end{bmatrix}, B = \begin{bmatrix} 1 & 3 \\ 5 & -2 \end{bmatrix}, \text{ and } C = \begin{bmatrix} 0 & 1 & -3 \\ 4 & 2 & 5 \end{bmatrix}$$

Find each of the following products.

43. AB

44. BA

45. AC

46. CA

In exercises 47–50, find the inverse matrix A^{-1}, if it exists. Check your answer by calculating AA^{-1} and $A^{-1}A$.

47. $A = \begin{bmatrix} -2 & 1 \\ 0 & 2 \end{bmatrix}$

48. $A = \begin{bmatrix} 1 & 2 \\ 3 & 4 \end{bmatrix}$

49. $A = \begin{bmatrix} 2 & 1 & 0 \\ 1 & 1 & 1 \\ 1 & -1 & 3 \end{bmatrix}$

50. $A = \begin{bmatrix} 1 & 0 & 1 \\ 3 & 1 & 0 \\ 1 & -1 & 1 \end{bmatrix}$

51. Student populations at three high schools are presented in the below table.

Gender	High school A	High school B	High school C
Boys	260	230	215
Girls	242	219	235

Write a matrix based on the data in the table.

52. Airline distances between indicated cities in statute miles are as in the below data:
From New York to Houston 1419 mi, from New York to Boston 190 mi, from New York to Los Angeles 2451 mi, from Houston to Boston 1605 mi, from Houston to Los Angeles 1374 mi, and from Boston to Los Angeles 2596 mi. Write a matrix based on the given data.

53. A manufacturer of a certain product makes three models A, B, and C. Each model is partially made in factory X in China and then finished in factory Y in the United States. The total cost of each product consists of the manufacturing cost and the shipment cost. The costs at each factory in dollars can be described by the matrices M_1 and M_2:

$$\begin{array}{ccc} & \text{Manufacturing cost} & \text{Shipping cost} & \\ M_1 = & \left[\begin{array}{c} 43 \\ 60 \\ 80 \end{array}\right. & \left.\begin{array}{c} 51 \\ 90 \\ 30 \end{array}\right] & \begin{array}{l} \text{Model A} \\ \text{Model B} \\ \text{Model C} \end{array} \end{array}$$

$$\begin{array}{ccc} & \text{Manufacturing cost} & \text{Shipping cost} & \\ M_2 = & \left[\begin{array}{c} 44 \\ 65 \\ 150 \end{array}\right. & \left.\begin{array}{c} 65 \\ 60 \\ 35 \end{array}\right] & \begin{array}{l} \text{Model A} \\ \text{Model B} \\ \text{Model C} \end{array} \end{array}$$

Determine the total manufacturing and shipping costs for each product.

54. A certain electronic store sells brand X iPads and laptops. The following matrix A gives the sales of these items for three months; matrix B gives the sales price and dealer's cost (in dollars) of these items. Use matrix multiplication to generate a matrix that has as its entries the total sales and the total retail costs of brand X items for each of the 3 months.

$$\begin{array}{c} \quad\;\; \text{Jan. Feb. Mar.} \\ A = \begin{bmatrix} 38 & 56 & 47 \\ 35 & 41 & 25 \end{bmatrix} \begin{array}{l} \text{ipads} \\ \text{Laptops} \end{array} \end{array} \qquad \begin{array}{c} \quad\;\; \text{ipads Laptops} \\ B = \begin{bmatrix} 815 & 610 \\ 690 & 475 \end{bmatrix} \begin{array}{l} \text{Retail price} \\ \text{Dealer price} \end{array} \end{array}$$

55. A certain store has three brands of electronic devices, priced at \$125, \$134, and \$112. On a particular day, the store sells quantities of 8, 5, and 3, respectively. Write the prices and the quantities by matrices, and then determine the total dollar income from the sale of the three brands of devices on that day.

56. A certain store has three brands of electronic devices, priced at \$3210, \$870, and \$89. On a particular day, the store sells quantities of 3, 2, and 4, respectively, and on the next day 5, 5, and 3. Write the prices and the quantities by matrices, and then determine the total dollar income from the sale of the three brands of devices on each day.

8.2 DETERMINANTS

The determinant of a matrix is a tool used in many applications of engineering, sciences, and mathematics.

Definition of determinant
A determinant of a matrix is a scalar number that replaces the bracket with vertical lines.

The determinant of A is denoted by det A. A determinate differs from a matrix in that it has actual value.

A determinant of order n is a determinant with n rows and n columns.

The value of a second-order (2 × 2) determinant is calculated in form

$$\det\begin{bmatrix} a_{11} & a_{12} \\ a_{21} & a_{22} \end{bmatrix} = \begin{vmatrix} a_{11} & a_{12} \\ a_{21} & a_{22} \end{vmatrix} = a_{11}a_{22} - a_{12}a_{21}$$

By using the sign rule, each term is determined by the first row in the diagram $\begin{vmatrix} + & - & + \\ - & + & - \\ + & - & + \end{vmatrix}$.

The value of a third-order (3 × 3) determinate is calculated in form

$$\det\begin{bmatrix} a_{11} & a_{12} & a_{13} \\ a_{21} & a_{22} & a_{23} \\ a_{31} & a_{32} & a_{33} \end{bmatrix} = \begin{vmatrix} a_{11} & a_{12} & a_{13} \\ a_{21} & a_{22} & a_{23} \\ a_{31} & a_{32} & a_{33} \end{vmatrix} = a_{11}\begin{vmatrix} a_{22} & a_{23} \\ a_{32} & a_{33} \end{vmatrix} - a_{12}\begin{vmatrix} a_{21} & a_{23} \\ a_{31} & a_{33} \end{vmatrix} + a_{13}\begin{vmatrix} a_{21} & a_{22} \\ a_{31} & a_{32} \end{vmatrix}$$

Example 1
Find the value of the following determinants:

(a) $\begin{vmatrix} 2 & 3 \\ -1 & 4 \end{vmatrix}$

(b) $\begin{vmatrix} 1 & 3 & 4 \\ -2 & -1 & 2 \\ 5 & -4 & 6 \end{vmatrix}$

Solution

(a) $\begin{vmatrix} 2 & 3 \\ -1 & 4 \end{vmatrix} = 2 \times 4 - 3(-1) = 8 + 3 = 11$

(b) $$\begin{vmatrix} 1 & 3 & 4 \\ -2 & -1 & 2 \\ 5 & -4 & 6 \end{vmatrix} = 1\begin{vmatrix} -1 & 2 \\ -4 & 6 \end{vmatrix} - 3\begin{vmatrix} -2 & 2 \\ 5 & 6 \end{vmatrix} + 4\begin{vmatrix} -2 & -1 \\ 5 & -4 \end{vmatrix}$$
$$= 1(-6+8) - 3(-12-10) + 4(8+5) = 2 + 66 + 52 = 120$$

We can use determinants to find the inverse of a square matrix:

We know the fact that $AA^{-1} = I_A$, where A = square matrix

A^{-1} = inverse matrix for A
I_A = identity matrix for A

Case 1: a 2 × 2 matrix

Let $A = \begin{bmatrix} a_{11} & a_{12} \\ a_{21} & a_{22} \end{bmatrix}$ and the $\det(A) = |A| = \begin{vmatrix} a_{11} & a_{12} \\ a_{21} & a_{22} \end{vmatrix} = a_{11} \times a_{22}$
$-a_{12} \times a_{21}$

Then, $A^{-1} = \frac{1}{\det(A)}\begin{bmatrix} a_{22} & -a_{12} \\ -a_{21} & a_{11} \end{bmatrix} = \frac{1}{a_{11}a_{22} - a_{12}a_{21}}\begin{bmatrix} a_{22} & -a_{12} \\ -a_{21} & a_{11} \end{bmatrix}$

Example 2

Find the inverse matrix A^{-1} when $A = \begin{bmatrix} 2 & 3 \\ -1 & 4 \end{bmatrix}$

Solution

$$A^{-1} = \frac{1}{\det(A)}\begin{bmatrix} a_{22} & -a_{12} \\ -a_{21} & a_{11} \end{bmatrix} = \frac{1}{a_{11}a_{22} - a_{12}a_{21}}\begin{bmatrix} a_{22} & -a_{12} \\ -a_{21} & a_{11} \end{bmatrix}$$

$$\begin{vmatrix} 2 & 3 \\ -1 & 4 \end{vmatrix} = (2)(4) - (3)(-1) = 8 + 3 = 11$$

$$A^{-1} = \frac{1}{11}\begin{bmatrix} 4 & -3 \\ 1 & 2 \end{bmatrix} = \begin{bmatrix} \frac{4}{11} & \frac{-3}{11} \\ \frac{1}{11} & \frac{2}{11} \end{bmatrix}.$$

Case 2: a 3 × 3 matrix

Let $A = \begin{bmatrix} a_{11} & a_{12} & a_{13} \\ a_{21} & a_{22} & a_{23} \\ a_{31} & a_{32} & a_{33} \end{bmatrix}$, then

$$A^{-1} = \frac{1}{\det(A)} \begin{bmatrix} \begin{vmatrix} a_{22} & a_{23} \\ a_{32} & a_{33} \end{vmatrix} & \begin{vmatrix} a_{13} & a_{12} \\ a_{33} & a_{22} \end{vmatrix} & \begin{vmatrix} a_{12} & a_{13} \\ a_{22} & a_{23} \end{vmatrix} \\ \begin{vmatrix} a_{23} & a_{21} \\ a_{33} & a_{31} \end{vmatrix} & \begin{vmatrix} a_{11} & a_{13} \\ a_{31} & a_{33} \end{vmatrix} & \begin{vmatrix} a_{13} & a_{11} \\ a_{23} & a_{21} \end{vmatrix} \\ \begin{vmatrix} a_{21} & a_{22} \\ a_{31} & a_{32} \end{vmatrix} & \begin{vmatrix} a_{12} & a_{11} \\ a_{32} & a_{31} \end{vmatrix} & \begin{vmatrix} a_{11} & a_{12} \\ a_{21} & a_{22} \end{vmatrix} \end{bmatrix}$$

Example 3

Find the inverse matrix A^{-1} when $A = \begin{vmatrix} 1 & 3 & 4 \\ -2 & -1 & 2 \\ 5 & -4 & 6 \end{vmatrix}$.

Solution

$$\begin{vmatrix} 1 & 3 & 4 \\ -2 & -1 & 2 \\ 5 & -4 & 6 \end{vmatrix} = 1\begin{vmatrix} -1 & 2 \\ -4 & 6 \end{vmatrix} - 3\begin{vmatrix} -2 & 2 \\ 5 & 6 \end{vmatrix} + 4\begin{vmatrix} -2 & -1 \\ 5 & -4 \end{vmatrix}$$

$$= 1(-6+8) - 3(-12-10) + 4(8+5) = 2 + 66 + 52 = 120$$

$$A^{-1} = \frac{1}{120} \begin{bmatrix} \begin{vmatrix} -1 & 2 \\ -4 & 6 \end{vmatrix} & \begin{vmatrix} 4 & 3 \\ 6 & -1 \end{vmatrix} & \begin{vmatrix} 3 & 4 \\ -1 & 2 \end{vmatrix} \\ \begin{vmatrix} 2 & -2 \\ 6 & 5 \end{vmatrix} & \begin{vmatrix} 1 & 4 \\ 5 & 6 \end{vmatrix} & \begin{vmatrix} 4 & 1 \\ 2 & -2 \end{vmatrix} \\ \begin{vmatrix} -2 & -1 \\ 5 & -4 \end{vmatrix} & \begin{vmatrix} 3 & 1 \\ -4 & 5 \end{vmatrix} & \begin{vmatrix} 1 & 3 \\ -2 & -1 \end{vmatrix} \end{bmatrix}$$

Singular matrix

A *singular matrix* is a matrix whose determinant is zero. A nonsingular matrix is a matrix whose determinant is not zero.

Example 4

Show that the following matrices are singular.

(a) $M = \begin{bmatrix} 1 & 0 & 4 \\ 2 & 0 & -6 \\ 3 & 0 & 11 \end{bmatrix}$

(b) $N = \begin{bmatrix} 2 & -2 & 7 \\ 1 & 4 & 5 \\ 2 & 8 & 10 \end{bmatrix}$

Solution

(a) All the elements in column 2 of B are zero. Thus, $|M| = 0$.

(b) Observe that every element in row 3 of N is twice the corresponding element in row 2. We write:
(row 3) = 2 (row 2)

Row 2 and row 3 are proportional. Thus, $|N| = 0$.

We can summarize some of various matrix operations.

Let M and N be $n \times n$ matrices and k be nonzero scalar.

1) $|kM| = k^n|M|$
2) $|MN| = M||N|$
3) $|M^t| = |M|$
4) $|M^{-1}| = \frac{1}{|M|}$, assuming M^{-1} exist

Example 5

If M is a 2×2 with $|M| = 5$ and assume M^{-1} exists. Compute the following determinants.

(a) $|4M|$
(b) $|M^3|$
(c) $12|M^t|$
(d) $|5M^{-1}|$

Solution

(a) $|4M| = (4)^2|M| = 16(5) = 80$

(b) $|M^3| = |MMM| = |M||M||M| = 5(5)(5) = 125$

(c) $12|M^t| = 12|M| = 12(5) = 60$

(d) $|5M^{-1}| = (5)^2\frac{1}{|M|} = \frac{25}{5} = 5$

We can use MATLAB function det(A) to calculate the determinant of the square matrix A as:

```
>> A=[3 4;5 6];
>> det(A)
ans =
      -2.0000
```

We can use Maple to calculate the determinant of the square matrix A as:

```
> A := Matrix([[3,4], [5,6]]) :
> Determinant(A)
            -2
>
```

8.2 EXERCISES

In exercises 1–14, evaluate the determinants of the following matrices:

1. $\begin{bmatrix} 5 & 1 \\ 2 & 3 \end{bmatrix}$

2. $\begin{bmatrix} 4 & -3 \\ 2 & 6 \end{bmatrix}$

3. $\begin{bmatrix} 6 & 1 \\ -3 & 4 \end{bmatrix}$

4. $\begin{bmatrix} 7 & -6 \\ -2 & -5 \end{bmatrix}$

5. $\begin{bmatrix} 1 & 0 \\ -6 & 4 \end{bmatrix}$

6. $\begin{bmatrix} 5 & 4 \\ 2 & 1 \end{bmatrix}$

7. $\begin{bmatrix} -4 & 3 \\ 4 & -7 \end{bmatrix}$

8. $\begin{bmatrix} 0 & -3 \\ -2 & 9 \end{bmatrix}$

9. $\begin{bmatrix} 1 & 2 & 3 \\ 4 & -1 & 2 \\ 5 & -2 & 1 \end{bmatrix}$

10. $\begin{bmatrix} 2 & 1 & 3 \\ -8 & -3 & -1 \\ 7 & 1 & 1 \end{bmatrix}$

11. $\begin{bmatrix} 2 & 1 & -1 \\ 3 & 2 & 2 \\ 4 & 3 & 1 \end{bmatrix}$

12. $\begin{bmatrix} 0 & 2 & -1 \\ -2 & 6 & 1 \\ 7 & 4 & 2 \end{bmatrix}$

13. $\begin{bmatrix} 1 & 1 & 1 \\ 2 & 0 & 3 \\ 5 & -2 & 4 \end{bmatrix}$

14. $\begin{bmatrix} 0 & 0 & 4 \\ 1 & 1 & 1 \\ 3 & 3 & 3 \end{bmatrix}$

In exercises 15–18, solve the following equations for x.

15. $\begin{vmatrix} x-1 & 2 \\ 3 & x-1 \end{vmatrix} = 4$

16. $\begin{vmatrix} 4x & x \\ 5 & x+1 \end{vmatrix} = 1$

17. $\begin{vmatrix} x-2 & 0 \\ 3 & x+1 \end{vmatrix} = 5$

18. $\begin{vmatrix} 0 & 2 \\ x & x-1 \end{vmatrix} = 12$

In exercises 19–22, compute the values of x that make the following determinants zero.

19. $\begin{vmatrix} x-1 & 0 \\ x & x+1 \end{vmatrix}$

20. $\begin{vmatrix} x & 2 \\ 3 & x+2 \end{vmatrix}$

21. $\begin{vmatrix} 5 & x+2 \\ x & x-1 \end{vmatrix}$

22. $\begin{vmatrix} 0 & x-1 \\ x & x+3 \end{vmatrix}$

In exercises 23–26, determine the following matrices that are singular and give the reason.

23. $\begin{vmatrix} 0 & 0 & 0 \\ 1 & 2 & 3 \\ -4 & 5 & 9 \end{vmatrix}$

24. $\begin{vmatrix} 9 & 0 & 5 \\ 3 & 0 & 3 \\ -4 & 0 & 13 \end{vmatrix}$

25. $\begin{vmatrix} -2 & 3 & 1 \\ 5 & 8 & 3 \\ -6 & 9 & 3 \end{vmatrix}$

26. $\begin{vmatrix} 1 & 2 & 3 \\ 2 & 4 & 6 \\ -1 & 8 & 13 \end{vmatrix}$

In exercises 27–30, If M and N are 3×3 matrices and $|M| = -4$, $|N| = 3$, compute the following determinants.

27. $|MN|$

28. $|MM^t|$

29. $|M^tN|$

30. $|2MN^{-1}|$

8.3 VECTORS

Physicists and engineers have been able to use vectors in the description of nature and space. Numeric is a quantity with magnitude only (single number).

Definition of scalar

A scalar is a quantity with magnitude and no direction.

Examples of scalars are length and time. We use the term *scalar* to mean a real number $\mathbb{R}$.

Definition of vector

A vector is a quantity with both magnitude and direction.

Examples of vectors are velocity and force.

$\overrightarrow{AB}$ is a vector with initial point A and terminal point B and its length (or magnitude) is $\left\|\overrightarrow{AB}\right\|$. Thus, $\overrightarrow{AB}$ vector is a directed line segment. Also, $\overrightarrow{v}$ is a vector.

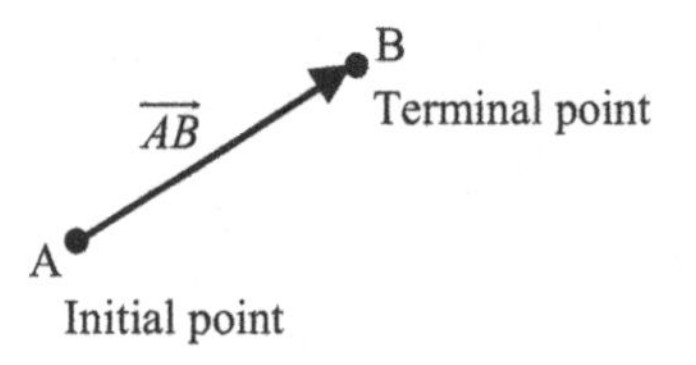

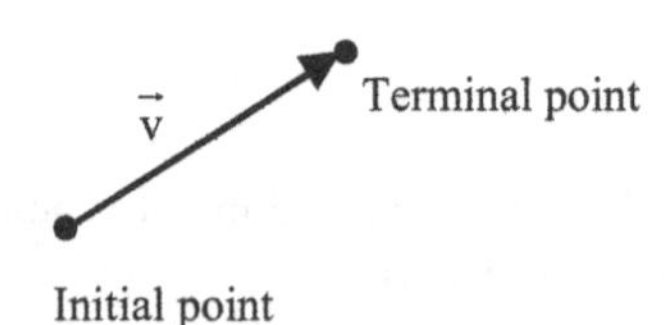

Two vectors are *equal* if both have the same magnitude and direction.

For example,

$$\overrightarrow{v} = \overrightarrow{u}$$
$$\overrightarrow{u} \neq \overrightarrow{w}$$
$$\overrightarrow{u} \neq \overrightarrow{r}$$

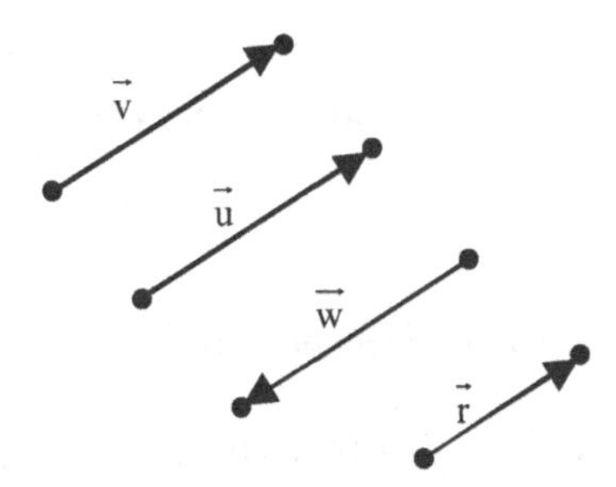

Vectors are usually presented on the rectangular coordinate (x, y) system. When the vector has the initial point at the origin (0, 0) of the rectangle coordinate (x, y) system and its angle is measured counterclockwise from the positive x-axis, we say the vector is in *standard position.*

A vector $\overrightarrow{u}$ whose initial point at the origin (0, 0) can be uniquely represented by the coordinate of its terminal point (u_1, u_2). We call this the *component form of a vector* $\overrightarrow{u}$ and it is denoted by

$$\overrightarrow{u} = \langle u_1, u_2 \rangle$$

where, u_1 and u_2 are the *components* of vector $\overrightarrow{u}$. The horizontal component is u_1 and the vertical component is u_2.

The *zero vector*, $\overrightarrow{0} = \langle 0, 0 \rangle$, is a vector whose initial point and terminal point both lie at the origin (0, 0).

The magnitude of a vector can be written in terms of magnitude of its horizontal and vertical components.

The component form for a vector $\overrightarrow{u}$ with initial point $P = (p_1, p_2)$ and terminal point $Q = (q_1, q_2)$ is

$$\overrightarrow{PQ} = \langle q_1 - p_1, q_2 - p_2 \rangle = \langle u_1, u_2 \rangle = \overrightarrow{u}.$$

The magnitude (or length) of vector $\overrightarrow{u}$ is given by

$$\|\overrightarrow{u}\| = \sqrt{(q_1 - p_1)^2 + (q_2 - p_2)^2} = \sqrt{(u_1)^2 + (u_2)^2}$$

If $\|\overrightarrow{u}\| = 1$, then $\overrightarrow{u}$ is called a *unit vector.*

Two vectors $\overrightarrow{u} = \langle u_1, u_2 \rangle$ and $\overrightarrow{v} = \langle v_1, v_2 \rangle$ are *equal* if and only if $u_1 = v_1$ and $u_2 = v_2$. The *zero vector* $\overrightarrow{0} = \langle 0, 0 \rangle$.

Example 1

Find the component form and the magnitude of the vector $\overrightarrow{u}$ that has initial point (5, −9) and the terminal point (−2, 7).

Solution

The components of vector $\overrightarrow{u} = \langle u_1, u_2 \rangle$ are given by

$$u_1 = -2 - (5) = -7$$
$$u_2 = 7 - (-9) = 16$$

Thus, $\overrightarrow{u} = \langle -7, 16 \rangle$ and the length of $\overrightarrow{u}$ is

$$\|\overrightarrow{u}\| = \sqrt{(u_1)^2 + (u_2)^2} = \sqrt{(-7)^2 + (16)^2}$$

$$\|\overrightarrow{u}\| = \sqrt{49 + 256} = \sqrt{305} = 17.46$$

8.3.1 **Vector operations**

Scalar multiplication is the product of a vector $\overrightarrow{u}$ and a scalar k, $k\overrightarrow{u}$, that is, a vector with k times as long as $\overrightarrow{u}$.

Let $\overrightarrow{u} = \langle u_1, u_2 \rangle$ be vector and let k be a scalar, then $k\overrightarrow{u} = \langle ku_1, ku_2 \rangle$.

The vector $k\overrightarrow{u}$ has the same direction as $\overrightarrow{u}$ when k is positive. The vector $k\overrightarrow{u}$ has the opposite direction of $\overrightarrow{u}$ when k is negative.

For *vector addition*,

let $\overrightarrow{u} = \langle u_1, u_2 \rangle$ and $\overrightarrow{v} = \langle v_1, v_2 \rangle$ be vector. Then the vector addition $\overrightarrow{u} + \overrightarrow{v}$ is

$$\overrightarrow{u} + \overrightarrow{v} = \langle u_1 + v_1, u_2 + v_2 \rangle.$$

The sum $\overrightarrow{u} + \overrightarrow{v}$ is formed by placing the initial point of vector $\overrightarrow{v}$ with the terminal point of vector $\overrightarrow{u}$ as shown in the figure.

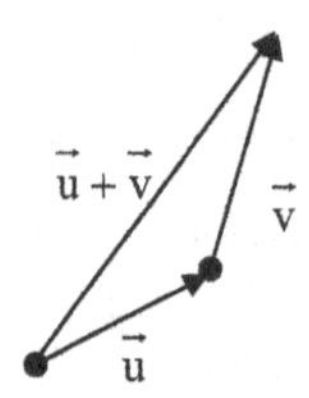

Example 1

Let $\overrightarrow{u} = \langle -3, 7 \rangle$ and $\overrightarrow{v} = \langle 4, 6 \rangle$, find the following vectors.

(a) $9\overrightarrow{u}$

(b) $\overrightarrow{u} + \overrightarrow{v}$

(c) $\overrightarrow{u} - \overrightarrow{v}$

(d) $\overrightarrow{v} + 5\overrightarrow{u}$

Solution

(a) $9\vec{u} = 9\langle -3,7\rangle = \langle -27,63\rangle$

(b) $\vec{u} + \vec{v} = \langle -3,7\rangle + \langle 4,6\rangle = \langle 1,13\rangle$

(c) $\vec{u} - \vec{v} = \langle -3,7\rangle - \langle 4,6\rangle = \langle -7,1\rangle$

(d) $\vec{v} + 5\vec{u} = \langle 4,6\rangle + 5\langle -3,7\rangle = \langle 4,6\rangle + \langle -15,35\rangle = \langle -11,41\rangle$

We summarize vector addition and scalar multiplication by the following properties.

Vector addition and scalar multiplication properties

Let $\vec{u}$, $\vec{v}$, and $\vec{w}$ be vectors and let k and c be scalars. Then,

1. $\vec{u} + \vec{v} = \vec{v} + \vec{u}$
2. $\vec{u} + (\vec{v} + \vec{w}) = (\vec{u} + \vec{v}) + \vec{w}$
3. $\vec{v} + \vec{0} = \vec{v}$
4. $\vec{v} + (-\vec{v}) = \vec{0}$
5. $\vec{0}(\vec{v}) = \vec{0}$
6. $1(\vec{v}) = \vec{v}$
7. $\|c\vec{u}\| = |c|\|\vec{u}\|$, $|c|$ is the absolute value of c
8. $k(c\vec{v}) = (kc)\vec{v}$
9. $(k + c)\vec{v} = k\vec{v} + c\vec{v}$

We can define the unit vector as:

A vector $\vec{u}$ for which $\|\vec{u}\| = 1$ is called a *unit vector*.

We defined a unit vector in the direction of a vector as:

We can define $\vec{u}$ a unit vector in the direction of $\vec{v}$ as

$$\vec{u} = \frac{\vec{v}}{\|\vec{v}\|}.$$

Example 2

Find $\vec{u}$ a unit vector in the direction of vector $\vec{v} = \langle 3, -4 \rangle$, and then show that the vector $\vec{u}$ has a length 1.

Solution

$$\vec{u} = \frac{\vec{v}}{\|\vec{v}\|} = \frac{\langle 3, -4 \rangle}{\sqrt{(3)^2 + (-4)^2}} = \frac{\langle 3, -4 \rangle}{\sqrt{25}}$$

$$\vec{u} = \left\langle \frac{3}{5}, \frac{-4}{5} \right\rangle$$

Now, we can show that $\vec{u}$ has a length 1 by

$$\|\vec{u}\| = \sqrt{\left(\frac{3}{5}\right)^2 + \left(\frac{-4}{5}\right)^2} = \sqrt{\frac{9}{25} + \frac{16}{25}} = \sqrt{\frac{25}{25}} = 1.$$

8.3.2 Standard unit vectors

Unit vectors are useful in defining the direction of any vector; we define two special unit coordinate vectors.

$\vec{i} = \langle 1, 0 \rangle$ and $\vec{j} = \langle 1, 0 \rangle$ are called *standard unit vectors.*

$\vec{i} = \langle 1, 0 \rangle$ is a unit vector in the direction of the *x*-axis.

$\vec{j} = \langle 1, 0 \rangle$ is a unit vector in the direction of the *y*-axis.

The standard unit vectors can represent any vector $\vec{u} = \langle u_1, u_2 \rangle$ as follows:

$$\vec{u} = \langle u_1, u_2 \rangle = u_1 \langle 1, 0 \rangle + u_2 \langle 1, 0 \rangle = u_1 \vec{i} + u_2 \vec{j},$$

where

$u_1 \vec{i} + u_2 \vec{j}$ is called the *linear combination of the vectors* $\vec{i}$ and $\vec{j}$
u_1 is called the *horizontal component* of $\vec{u}$
u_2 is called the *vertical component* of $\vec{u}$.

Any vector in the plane can be written as a linear combination of the standard unit vectors $\vec{i}$ and $\vec{j}$.

Example 1

Let $\vec{v}$ be the vector with initial point (3, –7) and terminal point (–2, 5). Express $\vec{v}$ as a linear combination of the standard unit vectors $\vec{i}$ and $\vec{j}$.

Solution

$$\vec{v} = \langle v_1, v_2 \rangle = \langle -2-3, 5+7 \rangle = \langle -5, 12 \rangle$$
$$\vec{v} = -5\vec{i} + 12\vec{j}$$

Example 2

Let $\vec{u} = 4\vec{i} + 11\vec{j}$ and $\vec{v} = -3\vec{i} - 5\vec{j}$. Find $5\vec{u} - 2\vec{v}$.

Solution

$$5\vec{u} - 2\vec{v} = 5(4\vec{i} + 11\vec{j}) - 2(-3\vec{i} - 5\vec{j}) = 20\vec{i} + 55\vec{j} + 6\vec{i} + 10\vec{j}\,5\vec{u} - 2\vec{v} = 26\vec{i} + 65\vec{j}$$

There are many applications to describe a vector in terms of its magnitude and direction, rather than in terms of its components.

If vector $\vec{v}$ has an angle θ (counterclockwise) form the positive x-axis to $\vec{v}$, then the terminal point of $\vec{v}$ lies on the plane and we can express the vector $\vec{v}$ as shown in the following figure.

$$\vec{v} = \|\vec{v}\|(\cos\theta\,\vec{i} + \sin\theta\,\vec{j}) = \|\vec{v}\|\langle \cos\theta, \sin\theta \rangle$$

We call θ the *direction angle of the vector* $\vec{v}$.

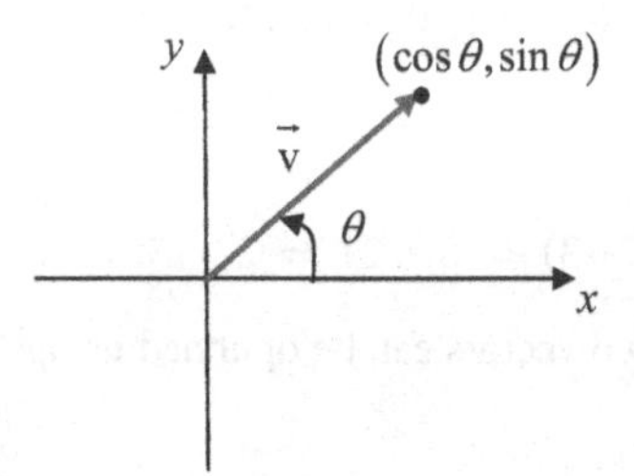

Since $\vec{v} = v_1\vec{i} + v_2\vec{j} = \|\vec{v}\|(\cos\theta\,\vec{i} + \sin\theta\,\vec{j})$, we can determine the direction angle θ for the vector $\vec{v}$ by

$$\tan\theta = \frac{\sin\theta}{\cos\theta} = \frac{\|\vec{v}\|\sin\theta}{\|\vec{v}\|\cos\theta} = \frac{v_2}{v_1}.$$

Example 3

A vector $\vec{v}$ of length 4 making an angle of 60° with the positive x-axis. Find the vector $\vec{v}$.

Solution

$$\vec{v} = \|\vec{v}\|(\cos\theta\,\vec{i} + \sin\theta\,\vec{j}) = \|\vec{v}\|\langle\cos\theta, \sin\theta\rangle$$

$$\vec{v} = 4(\cos 60^\circ\,\vec{i} + \sin 60^\circ\,\vec{j}) = 4\left(\frac{1}{2}\vec{i} + \frac{\sqrt{3}}{2}\vec{j}\right)$$

$$\vec{v} = 2\,\vec{i} + 2\sqrt{3}\,\vec{j}$$

8.3.3 The dot product

The product of two vectors has many geometric and physical applications. In this section we introduce the *dot product* (*scalar or inner product*), which is a product of two vectors to produce a real number rather than a vector.

Dot product

The *dot product* of two vectors $\vec{u} = \langle u_1, u_2\rangle$ and $\vec{v} = \langle v_1, v_2\rangle$ is

$$\vec{u}\cdot\vec{v} = u_1v_1 + u_2v_2$$

Example 1

Let $\vec{u} = \langle 6, 7\rangle$ and $\vec{v} = \langle 1, 3\rangle$, find $\vec{u}\cdot\vec{v}$.

Solution

$$\vec{u}\cdot\vec{v} = (6)(1) + (7)(3) = 6 + 21 = 27.$$

The dot product of two vectors can be optioned using Matlab commands as follows:

```
>> u=[6 7];
>> v=[1 3];
>> dot(u,v)
ans =
    27
```

The dot product of two vectors can be optioned using Maple commands as follows:

```
> u := [6,7] :
> v := [1,3] :
> DotProduct(u, v)
        27
```

Example 2

Let $\vec{u} = 9\vec{i} - 4\vec{j}$ and $\vec{v} = 2\vec{i} - 6\vec{j}$, find $\vec{u}\cdot\vec{v}$.

Solution

Since $\vec{u} = 9\vec{i} - 4\vec{j} = \langle 9, -4\rangle$ and $\vec{v} = 2\vec{i} - 6\vec{j} = \langle 2, -6\rangle$, we have

$$\vec{u}\cdot\vec{v} = (9)(2) + (-4)(-6) = 18 + 24 = 42.$$

Now, properties of dot products can follow based on the definition.

Dot product properties

Let $\vec{u}$, $\vec{v}$, and $\vec{w}$ be vectors in the plane and let k be scalars. Then,

1. $\vec{u}\cdot\vec{v} = \vec{v}\cdot\vec{u}$
2. $\vec{u}\cdot(\vec{v} + \vec{w}) = \vec{u}\cdot\vec{v} + \vec{u}\cdot\vec{w}$
3. $\vec{v}\cdot\vec{0} = 0$
4. $\vec{v}\cdot\vec{v} = \|\vec{v}\|^2$
5. $k(\vec{u}\cdot\vec{v}) = (k\vec{u})\cdot\vec{v} = \vec{u}\cdot(k\vec{v})$

Example 3

Let $\vec{u} = \langle -3, 2\rangle$, $\vec{v} = \langle 7, 4\rangle$, and $\vec{w} = \langle -5, -2\rangle$, find the following:

(a) $\|\vec{u}\|$

(b) $3\vec{u}\cdot 2\vec{v}$

(c) $-7\vec{w}$

Solution

(a) $\|\vec{u}\| = \vec{u}\cdot\vec{u} = \langle -3, 2\rangle\cdot\langle -3, 2\rangle = (-3)(-3) + (2)(2) = 123.$

(b) $3\vec{u}\cdot 2\vec{v} = 3\langle -3,2\rangle \cdot 2\langle 7,4\rangle = \langle -9,6\rangle \cdot \langle 14,8\rangle = (-9)(14)+(6)(8) = -126+48 = -78$

(c) $-7\vec{w} = -7\langle -5,-2\rangle = \langle 35,14\rangle$

The dot product can be useful in finding the angle between two vectors.

Angle between two vectors

If $\vec{u}$ and $\vec{v}$ are two nonzero vectors, then the *angle θ between the two vectors* is determined by

$$\cos\theta = \frac{\vec{u}\cdot\vec{v}}{\|\vec{u}\|\|\vec{v}\|}, \text{ where } 0 \le \theta \le \pi.$$

If the angle θ between two vectors is known, then we can find the dot product as

$$\vec{u}\cdot\vec{v} = \|\vec{u}\|\|\vec{v}\|\cos\theta$$

Example 4

Find the angle θ between the vectors $\vec{u} = \langle 3,5\rangle$ and $\vec{v} = \langle 1,-3\rangle$.

Solution

Since $\cos\theta = \frac{\vec{u}\cdot\vec{v}}{\|\vec{u}\|\|\vec{v}\|}$,

$$\cos\theta = \frac{\langle 3,5\rangle \cdot \langle 1,-3\rangle}{\sqrt{3^2+5^2}\sqrt{1^2+(-3)^2}} = \frac{3-15}{\sqrt{34}\sqrt{5}} = \frac{-12}{\sqrt{170}}$$

Therefore, the angle between the vectors $\vec{u}$ and $\vec{v}$ is

$$\theta = \arccos\left(\frac{-12}{\sqrt{170}}\right) = \cos^{-1}\theta\left(\frac{-12}{\sqrt{170}}\right) \approx 157^\circ.$$

The dot product can be used to determine if two vectors are perpendicular.

The vectors $\vec{u}$ and $\vec{v}$ are called *orthogonal* (perpendicular) if $\vec{u}\cdot\vec{v} = 0$.

The angle θ between orthogonal vectors is $\frac{\pi}{2}$.

Two vectors can be in parallel based on the scalar product and a vector.

The vectors $\vec{u}$ and $\vec{v}$ are called *parallel* if there is a nonzero real number α (or scalar) so that $\vec{u} = \alpha\vec{v}$.

Therefore, two vectors are *parallel* if they have the same or opposite directions.

Two nonzero vectors $\vec{u}$ and $\vec{v}$ have the *same direction* ($\theta = 0$) if and only if $\vec{u} = \alpha\vec{v}$, where $\alpha > 0$. Two nonzero vectors $\vec{u}$ and $\vec{v}$ have the *opposite directions* ($\theta = \pi$) if and only if $\vec{u} = \alpha\vec{v}$, where $\alpha < 0$.

Example 5

Determine the vectors $\vec{u}$ and $\vec{v}$ are orthogonal, parallel, or neither.

(a) $\vec{u} = \langle 6, -2\rangle$ and $\vec{v} = \langle 3, -1\rangle$

(b) $\vec{u} = \langle 2, -1\rangle$, $\vec{v} = \langle -4, 2\rangle$,

(c) $\vec{u} = \langle 6, -2\rangle$, $\vec{v} = \langle 1, 3\rangle$,

Solution

(a) Since $\cos\theta = \dfrac{\vec{u}\cdot\vec{v}}{\|\vec{u}\|\|\vec{v}\|}$,

$$\cos\theta = \frac{\langle 6,-2\rangle\cdot\langle 3,-1\rangle}{\sqrt{6^2+(-2)^2}\sqrt{3^2+(-1)^2}} = \frac{18+2}{\sqrt{40}\sqrt{10}} = \frac{20}{\sqrt{400}} = 1$$

The angle θ between the vectors $\vec{u}$ and $\vec{v}$ is $\theta = \cos^{-1}(1) = 0$.
Thus, the vectors $\vec{u}$ and $\vec{v}$ are parallel.

(b) Since $\cos\theta = \dfrac{\vec{u}\cdot\vec{v}}{\|\vec{u}\|\|\vec{v}\|}$,

$$\cos\theta = \frac{\langle 2,-1\rangle\cdot\langle -4,2\rangle}{\sqrt{2^2+(-1)^2}\sqrt{(-4)^2+(2)^2}} = \frac{-8-2}{\sqrt{5}\sqrt{20}} = \frac{-10}{\sqrt{100}} = -1$$

The angle θ between the vectors $\vec{u}$ and $\vec{v}$ is $\theta = \cos^{-1}(-1) = \pi$.
Thus, the vectors $\vec{u}$ and $\vec{v}$ are parallel.

(c) Since $\cos\theta = \dfrac{\vec{u}\cdot\vec{v}}{\|\vec{u}\|\|\vec{v}\|}$,

$$\cos\theta = \frac{\langle 6,-2\rangle\cdot\langle 1,3\rangle}{\sqrt{6^2+(-2)^2}\sqrt{(1)^2+(3)^2}} = \frac{6-6}{\sqrt{40}\sqrt{10}} = \frac{0}{\sqrt{400}} = \frac{0}{20} = 0$$

The angle θ between the vectors $\vec{u}$ and $\vec{v}$ is $\theta = \cos^{-1}(0) = \dfrac{\pi}{2}$.
Thus, the vectors $\vec{u}$ and $\vec{v}$ are orthogonal.

8.3.4 Work

Work (W) is performed when a constant force ($\vec{F}$) is used to move an object a certain distance ($\vec{d}$) as shown in the following figure.

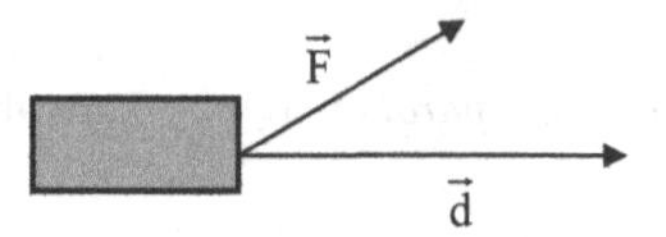

Let $\vec{F} = \vec{F_1} + \vec{F_2}$, where $\vec{F_1}$ is parallel to $\vec{d}$ and is the component of $\vec{F}$ that is oriented in the same direction of $\vec{d}$; $\vec{F_2}$ is perpendicular to $\vec{d}$ and is the component of $\vec{F}$ that is oriented in the orthogonal direction of $\vec{d}$; and θ is the angle between the force $\vec{F}$ and the direction of motion $\vec{d}$ as shown in the following figure.

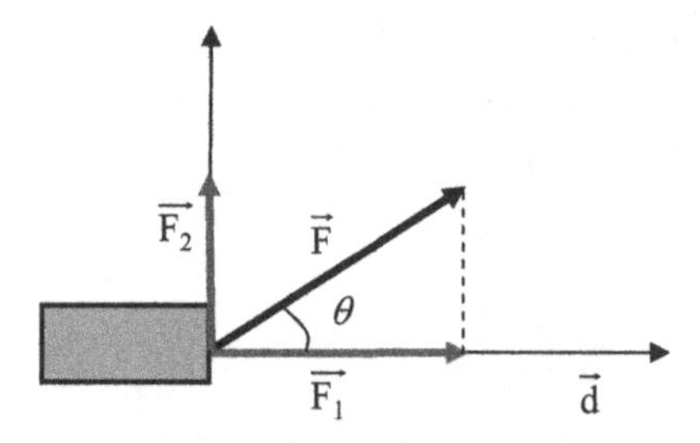

Since only the amount of the force in the direction of movement ($\vec{F_1}$) can be used in finding the work, the $\vec{F_1}$ is sometimes called the *projection of* $\vec{F}$ *onto* $\vec{d}$. Therefore, we can find the magnitude of $\vec{F_1}$ by:

$$\left\|\vec{F_1}\right\| = \left\|\vec{F}\right\|\cos\theta$$

Definition of work

If a constant force $\vec{F}$ causes motion to an object over a distance $\vec{d}$, the work done (W) is given by:

$$W = \left\|\vec{F}\right\|\left\|\vec{d}\right\|\cos\theta = \vec{F}\cdot\vec{d}$$

θ is the angle between the force $\vec{F}$ and the direction of motion $\vec{d}$.

Example 1

A vector $\vec{w}$ has magnitude 45 and is included at an angle of 60° from the horizontal. Determine the magnitudes of the horizontal and vertical components of the vector.

Solution

In the below figure, the horizontal component is labeled $\vec{u}$ and the vertical component is labeled $\vec{v}$.

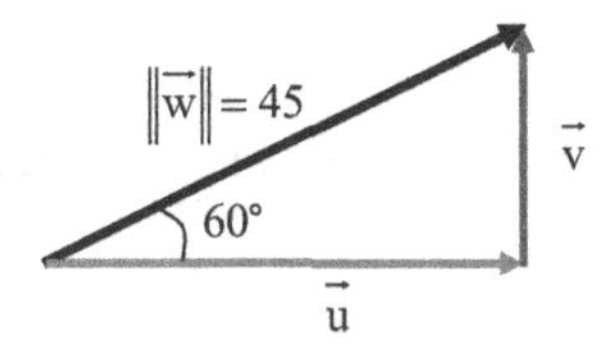

Vectors $\vec{w}$, $\vec{u}$, and $\vec{v}$ form a right triangle.

The horizontal component of $\vec{w}$ is

$$\|\vec{u}\| = \|\vec{w}\|\sin 60^\circ$$
$$\|\vec{u}\| = (45)(0.866) = 38.97,$$

and

The vertical component of $\vec{w}$ is

$$\|\vec{v}\| = \|\vec{w}\|\cos 60^\circ$$
$$\|\vec{v}\| = (45)(0.5) = 22.50$$

Example 2

A force $\vec{F} = \langle 30, -8 \rangle$ in pounds (lb) is used to push an object up a ramp. The resulting movement of the object is the distance $\vec{d} = \langle 10, 2 \rangle$ in feet (ft). Find the work done (W) by the force $\vec{F}$.

Solution

Since work is defined as

$$W = \vec{F} \cdot \vec{d}$$

Then,

$$W = \langle 30, -8 \rangle \cdot \langle 10, 2 \rangle = (30)(10) + (-8)(2) = 284 \text{ ft} - \text{lb}.$$

8.3 EXERCISES

In exercises 1–6, identify the following physical quantities as scalars or vectors.

1. acceleration
2. speed
3. frequency
4. sound
5. energy
6. gravity

In exercises 7–14, if $\vec{u} = \langle -3, 5 \rangle$ *and* $\vec{v} = \langle 2, -1 \rangle$*, calculate each quantity below.*

7. $\vec{u} + \vec{v}$
8. $\vec{u} - \vec{v}$
9. $5\vec{u} - \vec{v}$
10. $3\vec{u} + 2\vec{v}$
11. $\|\vec{u}\|$
12. $9\|\vec{v}\|$
13. $\|\vec{u}\| + \|\vec{v}\|$
14. $\|\vec{u} + \vec{v}\|$

In exercises 15–18, add the following vectors and then find the magnitude of the resultant.

15. $\vec{u} = \langle -3, 4 \rangle$, $\vec{v} = \langle 2, -5 \rangle$
16. $\vec{u} = \langle 7, -1 \rangle$, $\vec{v} = \langle -3, 6 \rangle$
17. $\vec{u} = \langle -1, -9 \rangle$, $\vec{v} = \langle 8, 2 \rangle$
18. $\vec{u} = \langle -1, 3 \rangle$, $\vec{v} = \langle -2, -8 \rangle$

In exercises 19–22, find a unit vector with the same direction as the given vector.

19. $\vec{w} = \langle 4, 3 \rangle$
20. $\vec{w} = \langle 2, 5 \rangle$
21. $\vec{w} = \langle -2, -6 \rangle$
22. $\vec{w} = \langle 1, 2 \rangle$

In exercises 23–26, express the following vectors in terms of $\vec{i}$ *and* $\vec{j}$.

23. $\vec{w} = \langle 3, -1 \rangle$

24. $\vec{w} = \langle -6, 4 \rangle$

25. $\vec{w} = \langle 1, 9 \rangle$

26. $\vec{w} = \langle -17, -25 \rangle$

In exercises 27–32, find $\vec{u} \cdot \vec{v}$.

27. $\vec{u} = \langle -3, 4 \rangle$, $\vec{v} = \langle 2, -5 \rangle$

28. $\vec{u} = \langle 7, 11 \rangle$, $\vec{v} = \langle -2, 1 \rangle$

29. $\vec{u} = \langle 4, -1 \rangle$, $\vec{v} = \langle 0, 9 \rangle$

30. $\vec{u} = \vec{i} - \vec{j}$, $\vec{v} = \vec{i}$

31. $\vec{u} = \vec{i} - 3\vec{j}$, $\vec{v} = \vec{i} + \vec{j}$

32. $\vec{u} = \vec{i} + 2\vec{j}$, $\vec{v} = -\vec{i} - \vec{j}$

In exercises 33–38, find the angle θ between the given vectors $\vec{u}$ and $\vec{v}$.

33. $\vec{u} = \langle 1, 3 \rangle$, $\vec{v} = \langle 2, 1 \rangle$

34. $\vec{u} = \langle 4, 1 \rangle$, $\vec{v} = \langle 1, -2 \rangle$

35. $\vec{u} = \langle -2, 5 \rangle$, $\vec{v} = \langle 1, 6 \rangle$

36. $\vec{u} = -2\vec{i} + \vec{j}$, $\vec{v} = \vec{i} + 3\vec{j}$

37. $\vec{u} = -5\vec{i} + 2\vec{j}$, $\vec{v} = -3\vec{j}$

38. $\vec{u} = \vec{i} - 2\vec{j}$, $\vec{v} = -\vec{i} + \vec{j}$

In exercises 39–44, determine whether the vectors $\vec{u}$ *and* $\vec{v}$ *are orthogonal, parallel, or neither.*

39. $\vec{u} = \langle -1, -1 \rangle$, $\vec{v} = \langle -2, 2 \rangle$

40. $\vec{u} = \langle 0, 1 \rangle$, $\vec{v} = \langle 1, -2 \rangle$

41. $\vec{u} = \langle 5, 0 \rangle$, $\vec{v} = \langle 3, 3 \rangle$

42. $\vec{u} = \vec{i} - 2\vec{j}$, $\vec{v} = 2\vec{i} - \vec{j}$

43. $\vec{u} = \frac{1}{2}\vec{i} - \frac{2}{3}\vec{j}$, $\vec{v} = 4\vec{i} + 3\vec{j}$

44. $\vec{u} = 2\vec{i} - 4\vec{j}$, $\vec{v} = -1\vec{i} + 2\vec{j}$

45. Determine the work done by lifting a 70-lb box from the floor to a table 4 ft high.

46. Determine the work performed by the force $\vec{F} = 22\vec{i} - 5\vec{j}$ in moving an object whose resultant motion is given by the displacement vector $\vec{d} = 40\vec{i} - 8\vec{j}$.

47. A box is pushed across a floor a distance of 60 ft by exerting a force of 20 lb downward at an angle of 30° with the horizontal. Find the work done by the force.

48. An automobile is pushed down a level street by exerting a force of 80 lb at an angle of 20° with the horizontal. Find the work done in pushing the car 120 ft.

49. A force of 50 lb on a rope attached at an angle of 60° with the horizontal to a block just overcomes friction and moves the block along the surface. Find the work done to move the block 90 ft along a level surface.

50. Find the work done by a force of 8 lb acting in the direction $\langle 1, 1 \rangle$ in moving an object 1 ft from (0,0) to (1,0).

CHAPTER 8 REVIEW EXERCISES

Perform the operations that are defined, given the following matrices in problems 1–16.

$$A = \begin{bmatrix} -3 & 2 \\ 4 & 0 \end{bmatrix}, \quad B = \begin{bmatrix} 5 & -2 \\ 1 & 6 \end{bmatrix}, \quad C = [-2 \quad 3], \quad D = \begin{bmatrix} 4 \\ -5 \end{bmatrix}, \text{ and}$$

$$E = \begin{bmatrix} 2 & 1 & 0 \\ -1 & -2 & 3 \\ 4 & 5 & 7 \end{bmatrix}$$

Find:

1. AB

2. CD

3. CB

4. AD

5. $A + B$

6. $C + D$

7. $A + C$

8. $3A - 2B$

9. $CA + C$

10. A^{-1}

11. B^t

12. I_B

13. $\det A$

14. $\det E$

15. Sizes of A, B, C, D, and E

16. The types (square, row, column) of the matrices A, B, C, D, and E.

Find the size (order) of each matrix. Show if the matrix is a row matrix, or a column matrix, or a square matrix in problems 17–24.

17. $\begin{bmatrix} 3 & 1 \\ -1 & 0 \end{bmatrix}$

18. $\begin{bmatrix} 2 & 1 & 7 \end{bmatrix}$

19. $\begin{bmatrix} 6 & 9 \end{bmatrix}$

20. $\begin{bmatrix} 0 \\ 0 \end{bmatrix}$

21. $\begin{bmatrix} 6 \\ 1 \\ 0 \\ -3 \end{bmatrix}$

22. $[5]$

23. $[2]$

24. $\begin{bmatrix} 1 & 3 \\ 5 & -2 \\ 0 & 8 \end{bmatrix}$

In problems 25 and 26 let $A = \begin{bmatrix} 0 & 2 \\ 3 & -1 \\ 5 & 1 \end{bmatrix}$, $B = \begin{bmatrix} -3 & 4 & 1 \\ 1 & 2 & 0 \end{bmatrix}$,

$C = \begin{bmatrix} -1 & 2 \\ 4 & 5 \\ 3 & -3 \end{bmatrix}$

25. Find the *dimension (size)* of each matrix

26. Compute each expression

(a) $A + C$

(b) $A - C$

(c) $-3A$

(d) BC

(e) $B + 0$

(f) A^t

27. Let matrix $A = \begin{bmatrix} 1 & -1 \\ 2 & 3 \end{bmatrix}$, find the inverse matrix of A, (A^{-1}), if it exists.

28. Find x and y so that $\begin{bmatrix} x+2y & 0 \\ -1 & 6 \end{bmatrix} = \begin{bmatrix} 2 & 0 \\ -1 & x+y \end{bmatrix}$.

29. Let $z = [z_1 \quad z_2]$ with $z_1 + z_2 = 1$, and $T = \begin{bmatrix} \frac{1}{3} & \frac{2}{3} \\ \frac{1}{5} & \frac{4}{5} \end{bmatrix}$, find z_1 and z_2 such that $zT = z$.

30. Let $M = \begin{bmatrix} c_1 & c_2 \\ 0 & 1 \end{bmatrix}$ and $N = \begin{bmatrix} -1 & 2 \\ 1 & -2 \end{bmatrix}$, find c_1 and c_2, if $M + N = \begin{bmatrix} 5 & 4 \\ 1 & -1 \end{bmatrix}$.

31. Find x and y so that

$$\begin{bmatrix} x \\ y \\ 1 \end{bmatrix} = \begin{bmatrix} 3 \\ -4 \\ 1 \end{bmatrix}$$

32. Find t and z so that

$$\begin{bmatrix} \frac{t}{2} & 7 \\ 1 & 9 \end{bmatrix} = \begin{bmatrix} 5 & 7 \\ 1 & z-3 \end{bmatrix}$$

33. Find x, y, and t so that

$$\begin{bmatrix} 1 & -3 & 0 \\ 3 & 8 & -6 \end{bmatrix} + \begin{bmatrix} 9 & x+y & x \\ 2 & t & 4 \end{bmatrix} = \begin{bmatrix} 10 & 6 & 10 \\ 5 & 2 & -2 \end{bmatrix}$$

34. Find x, y, and z so that

$$4\begin{bmatrix} 1 & -3 & 0 \\ 5 & 6 & 3 \end{bmatrix} + \begin{bmatrix} 21 & x-y & x \\ 2 & z & 4 \end{bmatrix} = \begin{bmatrix} 22 & 8 & 12 \\ 7 & 29 & 7 \end{bmatrix}$$

35. Let $A = B$ where $A = \begin{bmatrix} 1 & 0 \\ -3 & 4 \end{bmatrix}$ and $B = \begin{bmatrix} x & x+y \\ 3z & 5k+1 \end{bmatrix}$, find x, y, z, and k.

36. Let $A = \begin{bmatrix} 3 & 1 \\ 2 & 0 \end{bmatrix}$ and $A^{-1} = \begin{bmatrix} 2x+1 & 3y \\ z & k-1 \end{bmatrix}$, find the variables x, y, z, and k.

Hint: you can invert $A = \begin{bmatrix} 3 & 1 \\ 2 & 0 \end{bmatrix}$ and make it equal to $A^{-1} = \begin{bmatrix} 2x+1 & 3y \\ z & k-1 \end{bmatrix}$ or use $AA^{-1} = I$.

37. To close a sliding door, Jane pulls on a rope with a constant force of 40 lb at a constant angle of 60°. Determine the work done in moving the door 10 ft to its closed position.
38. To close a sliding door, Jane pulls on a rope with a constant force of 30 lb at a constant angle of 45°. Determine the work done in moving the door 8 ft to its closed position.

Appendix A: MAPLE

Maple is interactive mathematical and analytical software designed to perform a wide variety of mathematical calculations as well as operations on symbolic, numeric entities, and modeling. In this appendix, we give a general overview of Maple. For more information on Maple, visit the Maple Website: www.maplesoft.com.

A.1 GETTING STARTED AND WINDOWS OF MAPLE

When you double-click on the Maple icon, it opens as shown in Figure A.1. This figure shows Maple in the document mode. The worksheet mode is shown in Figure A.2, where the special [> prompt appears. This is the main area in which the user interacts with Maple. For general help, click on **Help** then **Maple Help** in the menu bar as shown in Figure A.3. Also, Maple uses the question mark (**?**), followed by the command or topic name, to get help. For example, to get help on solve, you type **?solve**. To terminate the Maple session, from the **File** menu, select **Exit**.

Start, programs, instructional software, maple 17, maple17 in red.

[>: maple window.

T: word window.

; then enter: execute the work.

?equation: learn about equation uses in maple.

Quick references.

+: addition.

−: subtraction.

(*): multiplication.

/: division.

⟶: arrow: to adjust the location.

:= means to assign name to a number.

Evalf (equation): to find the solution.

%: last expression.

%%: last two expressions.

%%%: last three expressions.

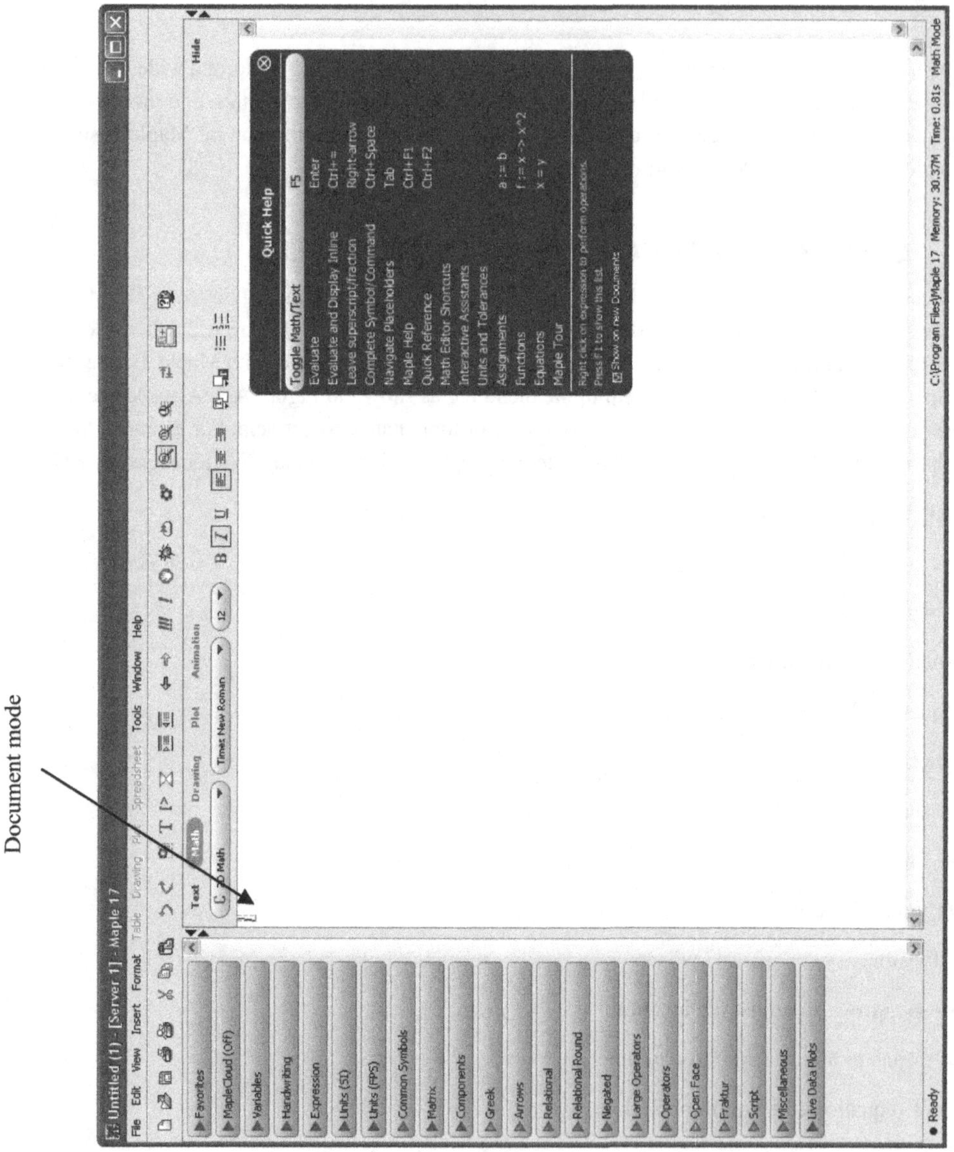

Document mode

FIGURE A.1 Default environment (document mode).

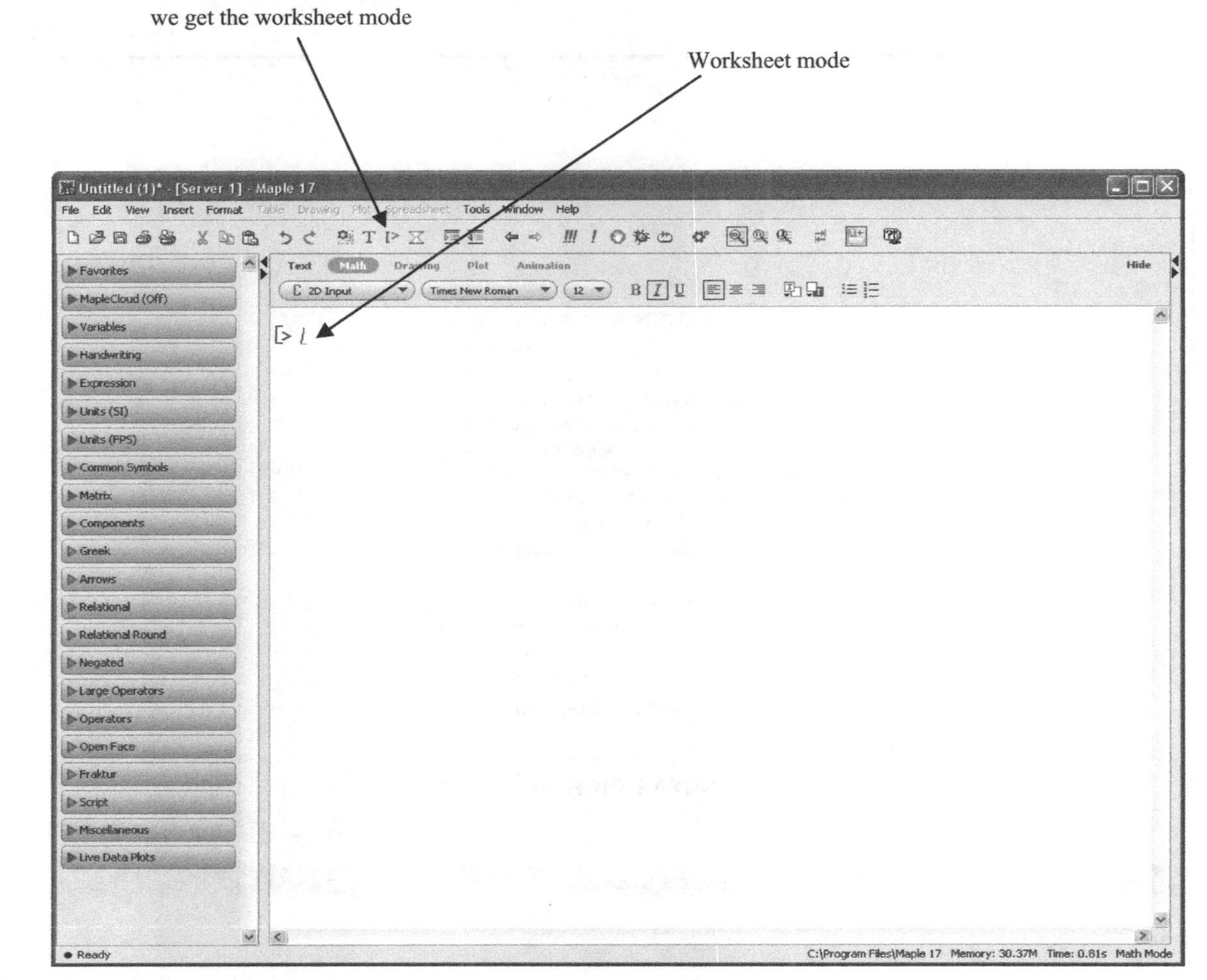

■ **FIGURE A.2** Worksheet mode.

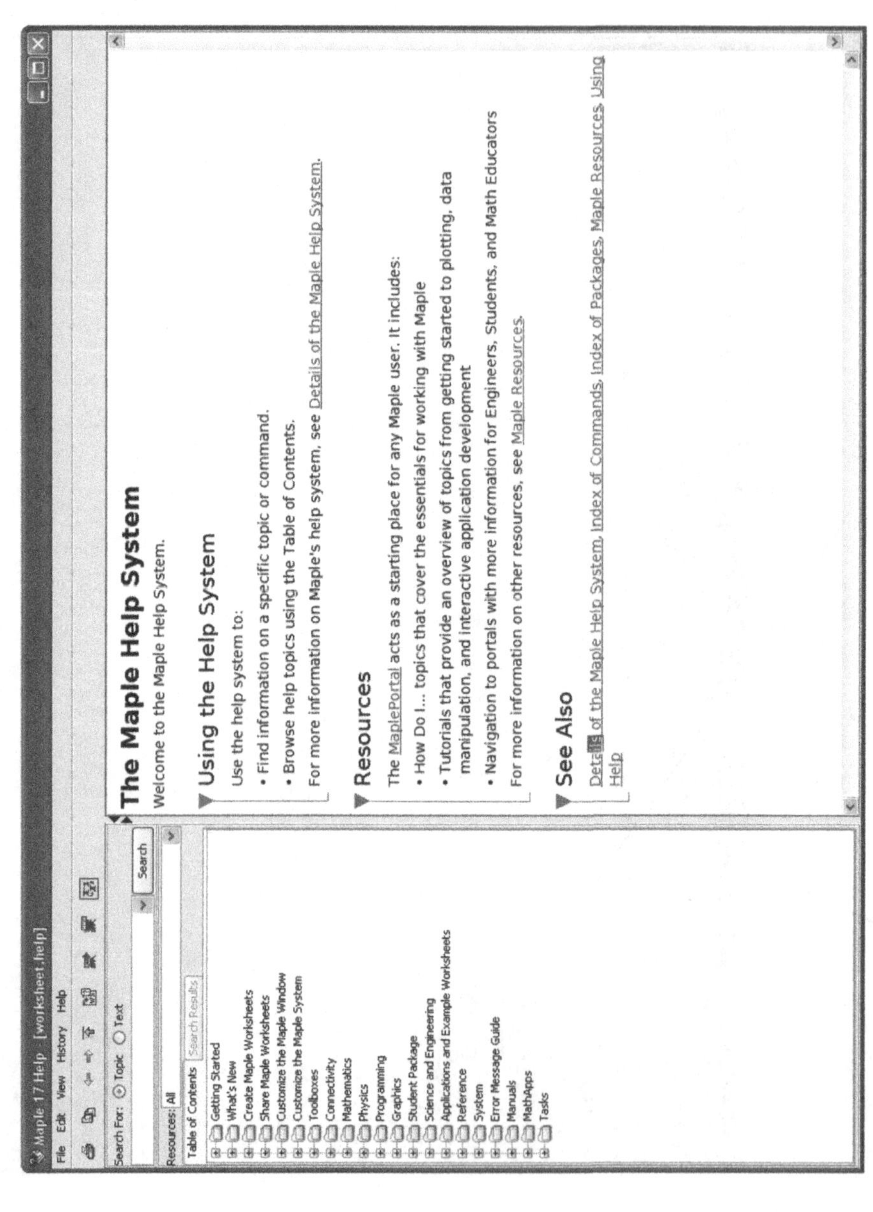

■ **FIGURE A.3** Maple help system.

A.2 ARITHMETIC

Maple can do arithmetic operations like a calculator. Table A.1 provides Maple's common arithmetic operations. To evaluate an arithmetic expression, type the expression and then press the **Enter** key.

Maple uses **pi** command to present π and **exp (1)** command to present e.

Example 1

Calculate $3^2 - \frac{(35+15)}{5} + 7 \times 4$.

Solution

$> 3^2 - \frac{(35+15)}{5} + 7 \times 4$

$$27$$

Example 2

Simple numerical calculation $9 + |-5|$.

Solution

$> 9 + |-5|$

$$14$$

Table A.1 Maple Common Arithmetic Operations

Operation	Descriptions
+	Addition
−	Subtraction
*	Multiplication
/	Division
∧	Exponentiation
!	Factorial
abs (n)	Absolute value of n
sqrt (n)	Square root of n

A.3 SYMBOLIC COMPUTATION

Maple can do a variety of symbolic calculations.

For example

$>> (x-y)^2 \cdot (x-y)^8$

$$(x-y)^{10}$$

Maple also makes simplifications to the expression when you use the command **simplify**.

For example

$> simplify\left(12 \cdot \left(\sin(\theta)^2 + \cos(\theta)^2\right)\right)$

The **expand** and **factor** commands are used to expand and factor the expression, respectively.

For example

$> expand(\sin(\beta + \Upsilon))$

$$\sin(\beta)\cos(\Upsilon) + \cos(\beta)\sin(\Upsilon)$$

$> factor(3 \cdot x^5 + 6 \cdot x^2 - 9 \cdot x \cdot y)$

$$3x(x^4 + 2x - 3y)$$

A.4 ASSIGNMENTS

To assign values to a variable, Maple uses colon equals (:=).

For example

$> x := 7$

$$x := 7$$

$> x^2 + 3 \cdot x \cdot y - z$

$$21y - z + 49$$

To clear the value of the variable x, type

$> x := 'x'$

$$x := x$$

A.5 WORKING WITH OUTPUT

One percent sign (%) refers to the output of the previous command. Two percent signs (%%) refer to the second-to-last output, and three percent signs (%%%) refer to the third-to-last output. Maple remembers the output of the last three statements you entered.

For example

$> 5 + 2$

$$7$$

$> \% \cdot 2$

$$14$$

$> \%\% + 5$

$$12$$

$> \%\%\% - 6$

$$1$$

A.6 SOLVING EQUATIONS

Maple uses **solve** command to solve equations.

For example

$> solve\left(4 \cdot y - (x-1)^2 = 8, y\right)$

$$11$$

We can solve equations with more than one variable for a specific variable.

For example

$> solve(x - \cos(y) = 11, y)$

$$\pi - \arccos(4)$$

$> solve(\{y = 2 \cdot x - 1, y = x + 2\}, \{x, y\})$

$$\{x = 3, y = 5\}$$

A.7 PLOTS WITH MAPLE

Maple uses the basic plotting command, **plot**, to plot functions, expressions, list of points, and parametric functions. For example, to plot the graph of $y = 3x^2 - x + 1$ on the interval -1 to 1, type

$> plot(3 \cdot x^3 - 2 \cdot x, x = -1...1, y = -1...1)$

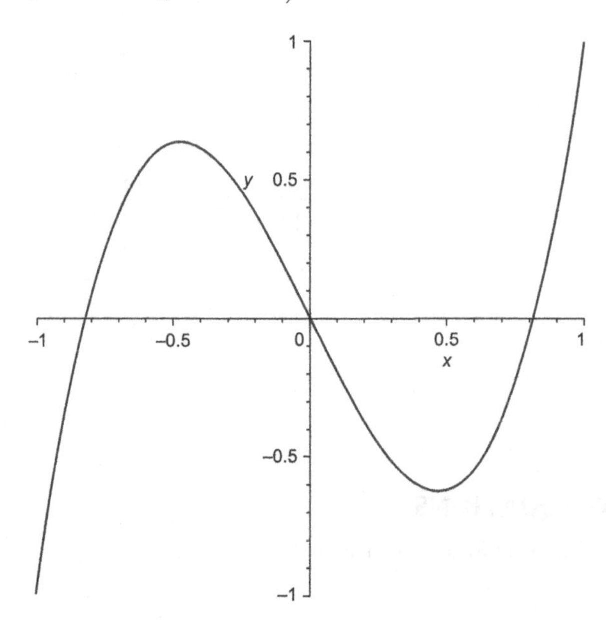

Also, we can plot several functions or expressions on same graph.

For example

$> plot(\{2 \cdot x^3 - 5, \exp(x^2), \cos(x), x + 5\}, x = -4...4, y = -3...6)$

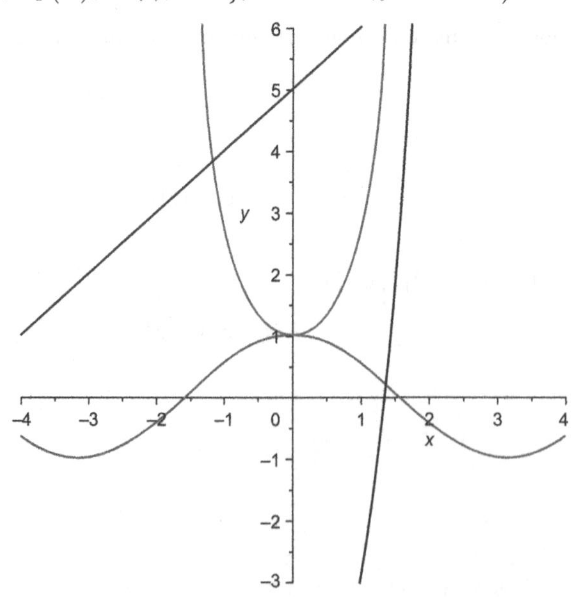

Maple allows you to annotate a plot by adding text and drawings by clicking on the plot. The **Plot** options tool bar will show up. Then click on the **Drawing** button and the drawing tool bar will show up.

For example

> $plot(\sin(x) + 0.3 \cdot x, x = -5...5);$

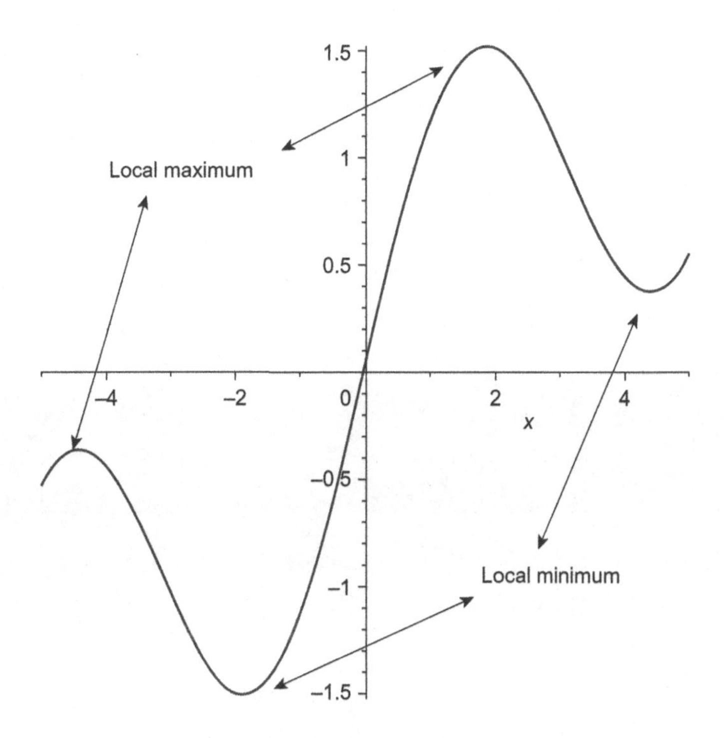

Maple allows you to annotate a plot by adding text and drawings [illegible]
The options [illegible] on the Drawing [illegible]
will show up

[illegible]

Appendix B: MATLAB

MATLAB has become a useful and dominant tool of technical professionals around the world. MATLAB is an abbreviation for Matrix Laboratory. It is a numerical computation and simulation tool that uses matrices and vectors. Also, MATLAB enables the user to solve wide analytical problems.

A copy of MATLAB software can be obtained from:

The Mathworks, Inc.
3 Apple Hill Drive
Natick, MA 01760-2098
Phone: 508-647-7000
Website: http://www.mathworks.com

This brief introduction of MATLAB (R2010b) is presented here to give a general idea about the software. MATLAB computational applications to science and engineering systems are used to solve practical problems.

B.1 GETTING STARTED AND WINDOWS OF MATLAB

When you double-click on the MATLAB icon, it opens as shown in Figure B.1. The command window, where the special >> prompt appears, is the main area in which the user interacts with MATLAB. To make the Command Window active, you need to click anywhere inside its border. To quit MATLAB, you can select **EXIT MATLAB** from the **File** menu, or by entering *quit* or *exit* at the Command Window prompt. Do not click on the X (close box) in the top right corner of the MATLAB window because it may cause problems with the operating software. Figure B.1 contains four default windows, which are Command Window, Workplace Window, Command History Window, and Current Folder Window. Table B.1 shows a list of the various windows and their purpose for MATLAB.

B.1.1 Using MATLAB in calculations

Table B.2 shows the MATLAB common arithmetic operators. The order of operations as first, parentheses ( ), the innermost are executed first for nested parentheses; second, exponentiation ^; third, multiplication * and division / (they are equal precedence); fourth, addition + and subtraction −.

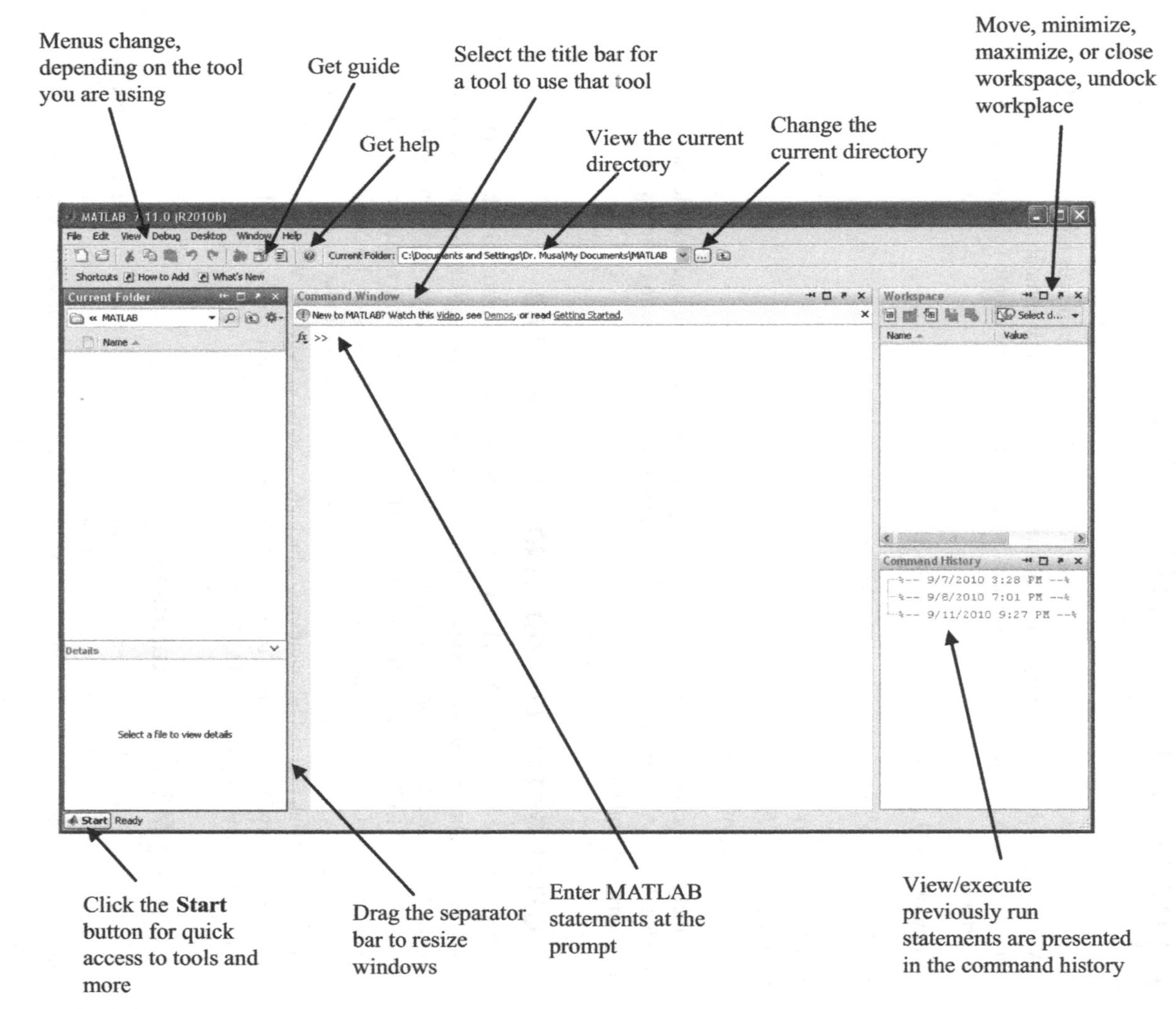

FIGURE B.1 MATALB default environment.

Table B.1 MATLAB windows

Window	Description
Command Window	Main window, enter variables, runs programs
Workplace Window	Gives information about the variable used
Command History Window	Records commands entered in the Command Window
Current Folder Window	Shows the files in current directory with details
Editor Window	Makes and debugs script and function files
Help Window	Gives help information
Figure Window	Contains output from the graphic commands
Launch Pad window	Provides access to tools, demos, and documentation

Table B.2 MATLAB common arithmetic operators

Operators symbols	Descriptions
+	Addition
−	Subtraction
*	Multiplication
/	Right division (means $\frac{a}{b}$)
\	Left division (means $\frac{b}{a}$)
∧	Exponentiation (raising to a power)
′	Converting to complex conjugate transpose
( )	Specify evaluation order

For example

```
>> a = 7; b = −2; c = 3;
>> x = 9*a + c^2 − 2

x =

   70

>> y = sqrt(x)/5

y =

   1.6733
```

Table B.3 provides common sample of MATLAB functions. You can obtain more by typing *help* in the Command Window (>> help).

Table B.3 Typical Elementary Math Functions

Function	Description
abs (x)	Absolute value or complex magnitude of x
acos (x), acosh (x)	Inverse cosine and inverse hyperbolic cosine of x (in radians)
angle (x)	Phase angle (in radians) of a complex number x
asin (x), asinh (x)	Inverse sine and inverse hyperbolic sine of x (in radians)
atan (x), atanh (x)	Inverse tangent and inverse hyperbolic tangent of x (in radians)
conj (x)	Complex conjugate of x (in radians)
cos (x), cosh (x)	Cosine and inverse hyperbolic cosine of x (in radians)
cot (x), coth (x)	Inverse cotangent and inverse hyperbolic cotangent of x (in radians)
exp (x)	Exponential of x
Fix	Round toward zero
imag (x)	Imaginary part of a complex number x
log (x)	Natural logarithm of x
$\log_2$ (x)	Natural logarithm of x to base 2
$\log_{10}$ (x)	Common logarithms (base 10) of x
real (x)	Real part of a complex number of x
sin (x), sinh (x)	Sine and inverse hyperbolic sine of x (in radians)
sqrt (x)	Square root of x
tan (x), tanh (x)	Tangent and inverse hyperbolic tangent of x (in radians)

For example

```
>> 5+3^(sin(pi/4))

ans =

   7.1746

>> y=7*sin(pi/3)

y =

   6.0622.

>> z = exp(y+2.04)

z =

   3.3017e+03.
```

In addition to operating on mathematical functions, MATLAB allows us to work easily with vectors and matrices. A vector (or one-dimensional array) is a special matrix (or two-dimensional array) with one row or one column. Arithmetic operations can apply to matrices and Table B.4 shows extra common operations that can be implemented to matrices.

Table B.4 Matrix Operations

Operations	Descriptions
A′	Transpose of matrix A
det (A)	Determinant of matrix A
inv (A)	Inverse of matrix A
eig (A)	Eigenvalues of matrix A
diag (A)	Diagonal elements of matrix A

A vector can be created by typing the elements inside brackets [ ] from a known list of numbers.

For example

```
>> A = [0 −1 2 5 6 4]

A =

  0  −1  2  5  6  4

>> B = [−1 −2 −3; 0 3 9; 1 3 6 8]

B =

  −1  −2  −3

   0   3   9

  13   6   8
```

Also, a vector can be created with constant spacing by using the command *variable-name = [a: n: b]*, where a is the first term of the vector; n is spacing; b is the last term.

For example

```
>> x = [1:0.5:5]

x =

  1.0000  1.5000  2.0000  2.5000  3.0000  3.5000  4.0000  4.5000  5.0000
```

Also, a vector can be created with constant spacing by using the command *variable-name = linspace (a, b, m)*, where a is the first element of the vector; b is the last element; m is number of elements.

For example

```
>> x=linspace(0,5*pi,6)
```

```
x =

  0   3.1416   6.2832   9.4248   12.5664   15.7080
```

Examples using Table B.4:

```
>> A = [0 1 3; 5 4 2; -6 8 9]

A =

   0  1  3

   5  4  2

  -6  8  9

>> B=A^2.

B =

  -13  28  29

    8  37  41

  -14  98  79

>> C=A'

C =

  0  5  -6

  1  4  8

  3  2  9

>> D =[-14;35];
>> inv(D)

ans =

  -0.2941  0.2353

   0.1765  0.0588

>> det(D)

ans =

  -17
```

Special constants can be used in MATLAB. Table B.5 provides special constants used in MATLAB.

Table B.5 MATLAB Named Constants

Name	Content
Pi	$\pi = 3.14159...$
i or *j*	Imaginary unit, $\sqrt{-1}$
Eps	Floating-point relative precision, 2^{-52}
Realmin	Smallest floating-point number, 2^{-1022}
Realmax	Largest floating-point number, (2-*eps*).2^{1023}
Bimax	Largest positive integer, $2^{53} - 1$
Inf or *Inf*	Infinity
nan or *NaN*	Not a number
Rand	Random element
Eye	Identity matrix
Ones	An array of 1's
Zeros	An array of 0's

For example

```
>> eye (2)

ans =

  1  0

  0  1

>> ones(2)

ans =

  1  1

  1  1

>> 1/0

ans =

  Inf

>> 0/0

ans =

  NaN
```

Arithmetic operations on arrays are done element by element. Table B.6 provides MATLAB common arithmetic operations on arrays.

Table B.6 MATLAB common arithmetic operations on arrays

Operators symbols on arrays	Descriptions
+	Addition same as matrices
−	Subtraction same as matrices
.*	Element-by-element multiplication
./	Element-by-element right division
.\	Element-by-element left division
.^	Element-by-element power
.'	Unconjugated array transpose

For example

```
>> M=[3 1 5; 2 0 4; 7 5 9]

M =

   3   1   5

   2   0   4

   7   5   9

>> M.*M

ans =

   9    1    25

   4    0    16

   49   25   81

>> A=[0 3;4 7;1 2];
>> B=[2 1;0 -1;5 6];
>> A./B

ans =

        0    3.0000
      Inf   -7.0000
   0.2000    0.3333
```

```
>> A.^2

ans =

  0   9

 16  49

  1   4
```

B.2 PLOTTING

MATLAB has nice capability for plotting two-dimensional and three-dimensional plots.

B.2.1 Two-dimensional plotting

First, we start with two-dimensional plots. The *plot* command is used to create two-dimensional plots. The simplest form of the command is *plot (x,y)*. The arguments x and y are each a vector (one-dimensional array). The vectors x and y must have the same number of elements. When the *plot* command is executed a figure will be created in the Figure Window. The *plot (x, y, 'line specifiers')* command has additional optional arguments that can be used to detail the color and style of the lines. Tables B.7–B.9 show various types of lines, points, and color types used in MATLAB.

Table B.7 MATLAB various Line Styles

Line types	MATLAB symbol
Solid (default)	-
Dashed	--
Dotted	:
Dash-dot	-.

Table B.8 MATLAB various Point Styles

Point type	MATLAB symbol
Asterisk	*
Plus sign	+
x-mark	x
Circle	o
Point	.
Square	s

Table B.9 MATLAB various Line Color Types

Color	MATLAB symbol
Black	K
Blue	B
Green	G
Red	R
Yellow	Y
Magenta	M
Cyan	C
White	W

For example

x= 0:pi/20:3*pi;%x is a vector, 0 <= x <= 2*pi, increments of pi/20.

>> y=2*sin(3*pi*x);% y is a vector.

>> plot(x,y,'--b')%creates the 2D plot with blue and dashed line

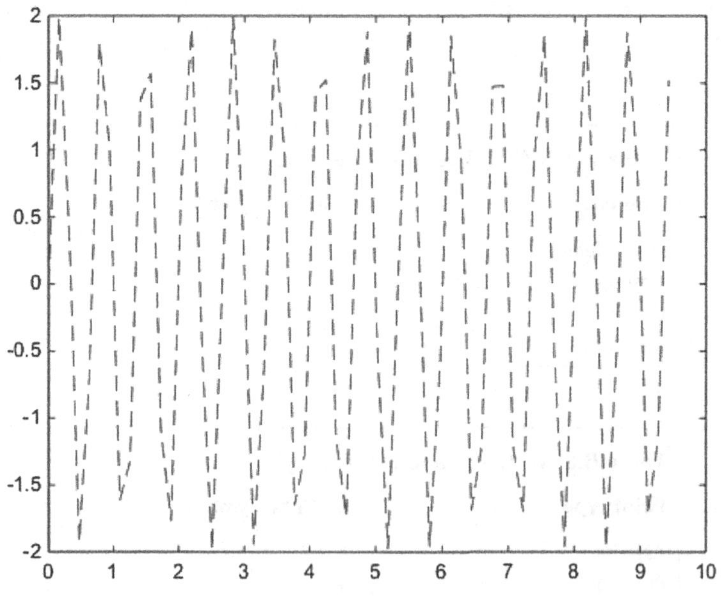

The command *fplot('function', limits, line specifiers)* is used to plot a function with form y = f(x), where the function can be typed as a string inside the command. The limits is a vector with two elements that specify the domain x [xmin, xmax], or is a vector with four elements that specifies the domain of x and the limits of the y-axis [xmin, xmax,ymin, ymax]. The line specifiers are used the same as in the plot command.

For example,

>> fplot('x^2+3*sin(2*x)-1', [−3,3],'xr')

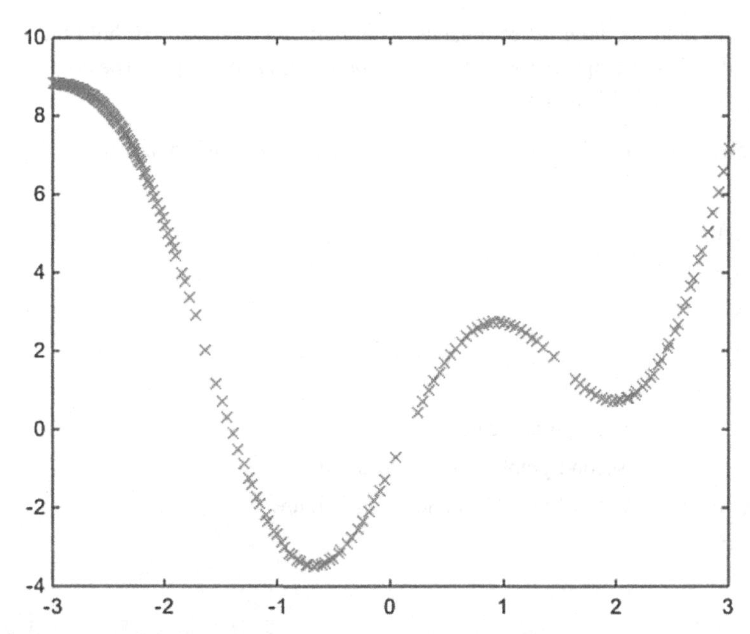

Also, we can create a plot for a function y = f(x) using the command *plot* by creating a vector of values of x for the domain that the function will be plotted, then creating y with corresponding values of f(x).

For example,

```
>> x=[0:0.:1];
>> y=cos(3*pi*x);
>> plot(x, y, 'ro:')
```

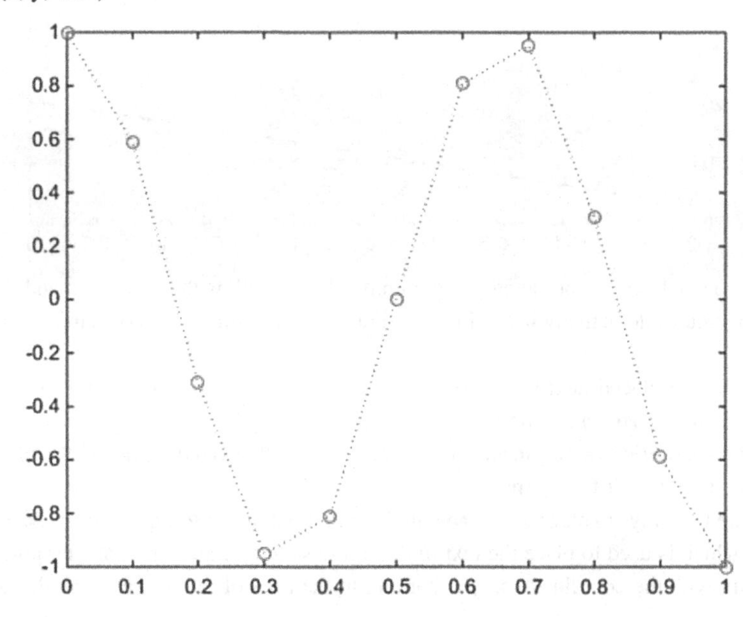

Second, using *hold on, hold off* commands. The *hold on* command will hold the first plotted graph and add to it extra figures for each time the *plot* command is typed. The *hold off* command stops the process of *hold on* command.

For example, if we use the previous example, we get the same result using the following commands:

```
>> x=[−3:0.01:6];
>> y=2*x.^3−15*x+5;
>> y=2*x.^3−15*x+5;
>> ydd=12*x;
>> plot(x,y,'-r')
>> hold on % the first graph is created
>> plot(x,yd,':b') % second graph is added to the figure
>> plot(x,ydd,'--k') % third graph is added to the figure
>> hold off
```

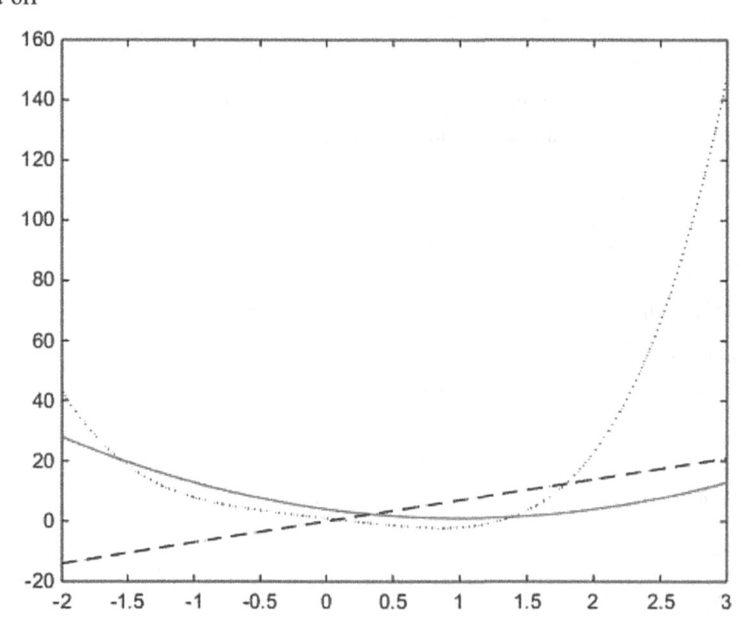

Plots in MATLAB can be formatted using commands that follow the *plot* commands, or by using the plot editor interactively in the Figure window. First, format the plot using commands as follows:

- Labels can be placed next to the axes with the *xlabel ('text as string')* for the x-axis and *ylabel ('text as string')* for the y-axis.
- The command *title ('text as string')* is a title command that can be added to the plot to place the title at the top of the figure as a text.
- There are two ways to place a text label in the plot. First, using *text (x,y,'text as string')* command, which is used to place the text in the figure such that the first character positioned at the point with the coordinates x, y according to the axes of the figure. Second, using *gtext*

(*'text as string'*) command, which is used to place the text at a position specified by the user mouse in the figure window.

- The command *legend ('string1', 'string2',..., pos)* is used to place a legend on the plot. The legend command shows a sample of line type of each graph that is plotted and places a label specified by the user beside the line sample. The *strings* in the command are the labels that are placed next to the line sample, and their order corresponds to the order that the graphs were created. The *pos* in the command is an optional number that specifies where in the figure the legend is placed. Table B.10 shows the options that can be used for *pos*.
- The command *axis* is used to change the range and the appearance of the axes of the plot, based on the minimum and maximum values of the elements of x and y. Table B.11 shows some common possible forms of *axis* command.
- The command *grid on* is used to add grid lines to the plot and the command *grid off* is used to remove grid lines from the plot.

Table B.10 Options That Can Be Used For *Pos*

Pos value	Description
−1	Place the legend outside the axes boundaries on the right side
0	Place the legend inside the axes boundaries in a location that interferes the least with graph
1	Place the legend at the upper-right corner of the plot (this is the default)
2	Place the legend at the upper-left corner of the plot
3	Place the legend at the lower-left corner of the plot
4	Place the legend at the lower-right corner of the plot

Table B.11 Some Common *axis* Commands

axis command	Description
axis ([xmin, xmax, ymin, ymax])	Sets the limits of both the x- and y-axes (xmin, xmax, ymin, ymax are numbers)
axis equal	Sets the same scale for both axes
axis tight	Sets the axis limits to the range of the data
axis square	Sets the axes region to be square

For example

```
>> x=0:pi/20:3*pi;y1=exp(−5*x);y2=sin(x*2);
>> plot(x,y1,'-b',x,y2,'--r')
>> xlabel('x')
>> ylabel('y1 , y2')
>> title('y1=exp(−5*x), y2=sin(x*2)')
>> axis([0,1,−1, 1])
```

```
>> text(5,0.5,'Comparison between y1 and y2')
>> legend('y1','y2',0)
```

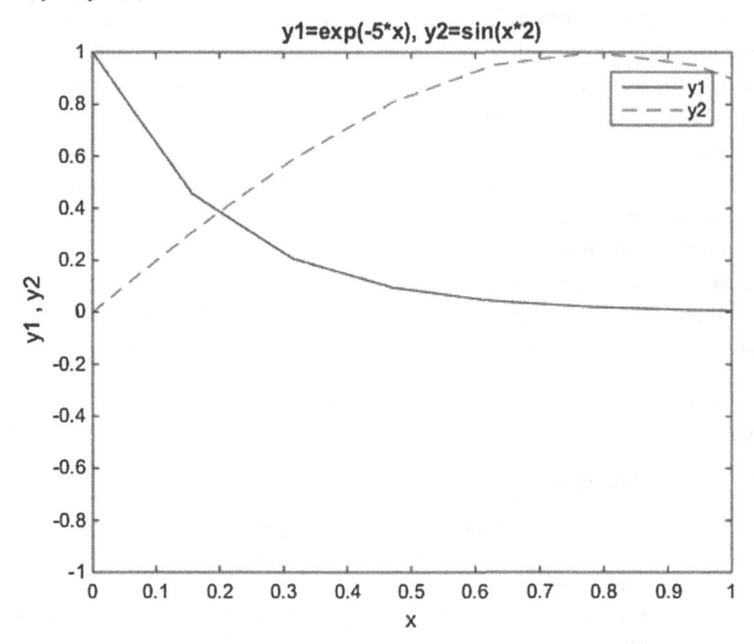

In MATLAB, users can use Greek characters in the text by typing \name of the letter within the string as in Table B.12.

Table B.12 Some Common Greek Characters

Greek characters in the string	Greek letter	Greek characters in the string	Greek letter
\alpha	α	\Phi	Φ
\beta	β	\Delta	Δ
\gamma	γ	\Gamma	Γ
\theta	θ	\Lambda	Λ
\pi	π	\Omega	Ω
\sigma	σ	\Sigma	Σ

To get a lowercase Greek letter, the name of the letter must be typed in all lowercase. To get a capital Greek letter, the name of the letter must start with a capital letter.

Second, format the plot using the plot editor interactively in the Figure window. This can be done by clicking on the plot and/or using the menus as illustrated in the following figure.

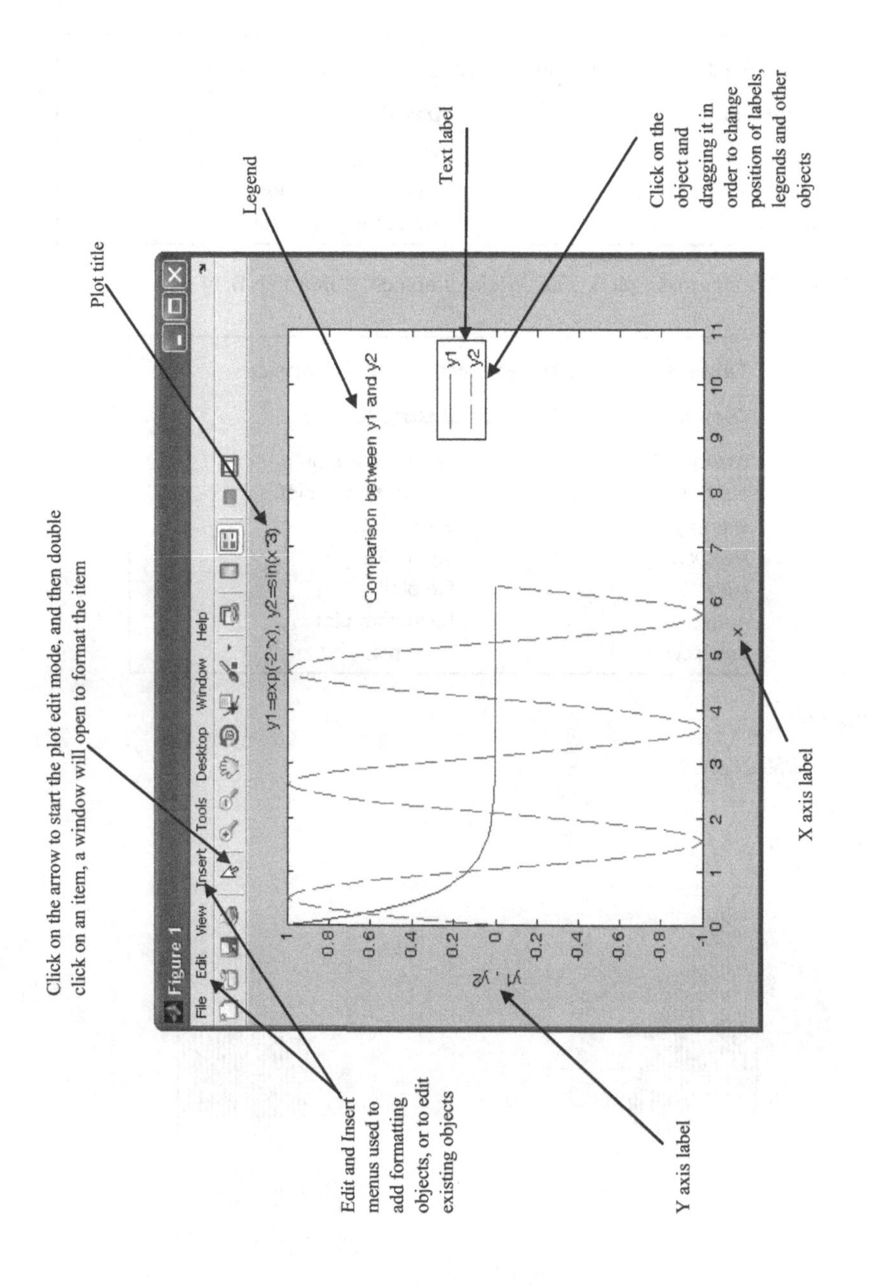
Click on the arrow to start the plot edit mode, and then double click on an item, a window will open to format the item
Plot title
Legend
Text label
Click on the object and dragging it in order to change position of labels, legends and other objects
Edit and Insert menus used to add formatting objects, or to edit existing objects
Y axis label
X axis label
Figure 1
File Edit View Insert Tools Desktop Window Help
y1=exp(-2*x), y2=sin(x.^3)
Comparison between y1 and y2
y1
y2
y1 , y2
x

MATLAB can use logarithm scaling for two-dimensional plot. Table B.13 shows MATLAB commands for logarithm scaling.

Table B.13 Two-Dimensional Graphic for Logarithm Scaling

Command	Description
Loglog	To plot log(y) versus log(x)
Semilogx	To plot y versus log(x)
Semilogy	To plot log(y) versus x

Also, MATLAB can make plots with special graphics as in Table B.14.

Table B.14 MATLAB Plots with Special Graphics

Command	Description
bar(x,y)	Vertical bar plot
barh(x,y)	Horizontal bar plot
stairs(x,y)	Stairs plot
stem(x,y)	Stem plot
pie(x)	Pie plot
hist(y)	Histogram plot
polar(x,y)	Polar plot

For example,

```
>> t=[0:pi/50:2*pi]
>> r=2+5*cos(t)
>> stem(t,r,'r.')
```

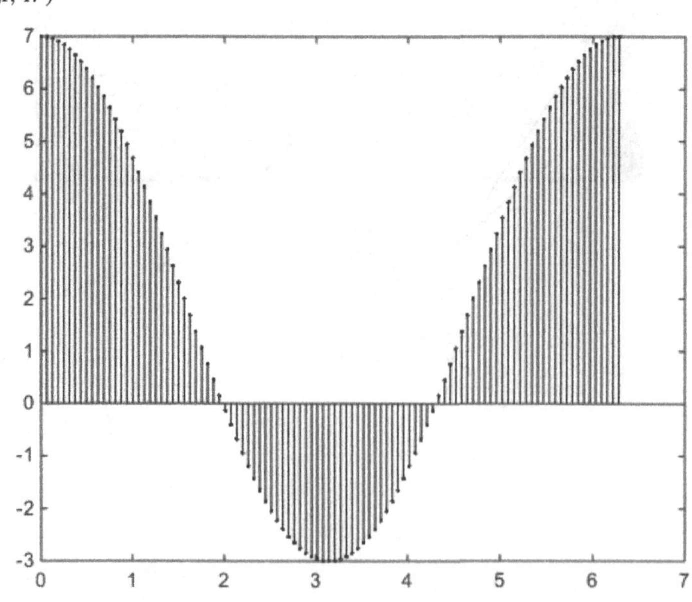

B.3 SYMBOLIC COMPUTATION

In previous sections, you learned that MATLAB can be a powerful programmable and calculator. However, basic MATLAB uses numbers as in a calculator. Most calculators and basic MATLAB lack the ability to manipulate math expressions without using numbers. In this section, you see that MATLAB can manipulate and solve symbolic expressions that make you compute with math symbols rather than numbers. This process is called *symbolic math.* You can practice some symbolic expressions in the following section.

B.3.1 Simplifying symbolic expressions

Symbolic simplification is not always straightforward; there is no universal simplification function because the meaning of a simplest representation of a symbolic expression cannot be defined clearly. MATLAB uses the *sym* or *syms* command to declare variables as symbolic variable. Then, the symbolic can be used in expressions and as arguments to many functions. For example, to rewrite a polynomial in a standard form, use the *expand* function:

```
>> syms x y; % creating a symbolic variables x and
>> x = sym('x'); y = sym('y'); % or equivalently
>> expand (cos(x + y))

ans =

  cos(x)*cos(y) – sin(x)*sin(y)
```

You can use *subs* command to substitute a numeric value for a symbolic variable or replace one symbolic variable with another.

For example

```
>> syms x;
>> f=2*x^3–5*x+2;
>> subs(f,2)

ans =

  8

>> simplify (sin(x)^2 + cos(x)^2) % Symbolic simplification

ans =

  1
```

Appendix C: Solution Manual

CHAPTER 1

Basic Concepts in Arithmetic

1.1 Exercises (Answers)

1. $1\frac{2}{3}$

2. $2\frac{1}{2}$

3. $4\frac{1}{5}$

4. $4\frac{3}{9}$

5. $\frac{3}{2}$

6. $\frac{11}{4}$

7. $\frac{17}{5}$

8. $\frac{19}{5}$

9. $\frac{1}{5}$

10. $\frac{2}{3}$

11. $\frac{1}{6}$

12. $\frac{3}{8}$

13. $\frac{10}{21}$

14. $\frac{5}{16}$

15. $\frac{2}{3}$

16. $\frac{5}{4}$

17. $\frac{35}{2}$

18. $\frac{12}{77}$

19. $\frac{2}{3}$

20. $\frac{11}{7}$

21. $\frac{3}{5}$

22. $\frac{1}{2}$

23. $\frac{6}{5}$

24. $\frac{6}{5}$

25. 7

26. $\frac{18}{5}$

27. $\frac{3}{20}$

28. $\frac{2}{21}$

29. $\frac{13}{35}$

30. $\frac{7}{10}$

31. $\frac{3}{2}$

32. $\frac{115}{2}$

33. $\frac{37}{2}$

34. $\frac{126}{5}$

35. $\frac{23}{32}$

36. $\frac{35}{8}$

37. $\frac{22}{31}$

38. $\frac{112}{45}$

39. $\frac{1}{24}$

40. $\frac{4}{5}$

41. $\frac{14}{3}$

42. $\frac{14}{5}$

43. $\frac{17}{21}$

44. $\frac{29}{12}$

45. $\frac{6}{7}$

46. $\frac{13}{6}$

47. Total amount of the bills = \$825 + \$125 + \$50 + \$272 + \$65 = \$1337.

48.
$$\begin{aligned}\text{Average rate} &= \frac{\text{Total distance}}{\text{Total time}} \\ &= \frac{45(2) + 65(4)}{6} \\ &= \frac{90 + 260}{6} = \frac{350}{6} \\ &= \frac{175}{3}\end{aligned}$$

49. Piece A = 18/56 = 9/28
Piece B = 6/56 = 3/28
Piece C = 32/56 = 4/7

50. $2 + 3(2/3) + (4/10) + (6/10) + (3/5) = 28/5 = 5\ 3/5$ h

51. $3\ 7/8 = 3\ 28/32$
the length of the Grip $= 3\ 28/32 - 2\ 4/32 = 1\ 24/32$

52. 1 space $= 4\ {}^{1}/_{2} \times 5/8 = 9/2 \times 5/8 = 45/16$
There are four spaces between the first and the fifth rivets.
Thus, $D = 4 \times 45/16 = 45/4 = 11\ {}^{1}/_{4}$ in.

53. Volume $= 6\ 3/4 \times 3\ 7/12 \times 5/6 = 27/4 + 43/12 + 5/6 = 953/120$ in.^{3}

54. $R = \frac{V}{I} = \frac{240}{0.4} = 600\ \text{V/A} = 600\ \Omega$
$P = VI = 240 \times 0.4 = 96$ W

55. 1- $I = \frac{P}{V} = \frac{1100}{110} = 10$ A
2- $R = \frac{V}{I} = \frac{110}{10} = 11\ \Omega$

56. Total $= 420 + 35 + 728 = 1183$ board feet

57. Total $= 420 + 950 + 265 = 1635$ ft

58. Work $=$ Distance (meters) $\times$ Force (newtons)
$= 100 \times 200 = 20,000$ J

59. Power = work/time

$$P = \frac{D \times F}{t}$$

$$P = \frac{600 \times 180}{30} = 3600\ \text{lb·ft/s}$$

60. Power = work/time

$$P = \frac{\text{Distance} \times \text{Force}}{\text{time}}$$

$$P = \frac{400 \times 80}{25} = 1280\ \text{W}$$

61. Torque = Force × Distance
$T = F \times D$
$T = 40 \times 15 = 600$ lb·ft

62. Moments of Force $= 20 \times 3 = 60$ N·m

63. Potential energy = mass (or weight) × rate of free fall acceleration (32.8 ft/s) × the distance the object falls.
$P = m \times g \times h$
$P = 25 \times 32.8 \times 50 = 41,000$ ft·lb

64. Potential energy = mass (or weight) × rate of free fall acceleration (32.8 ft/s) × the distance the object falls.

P = m × g × h

P = 35 × 32.8 × 40 = 45,920 ft·lb

65. $R_T = R_1 + R_2 + R_3 + R_4$

$R_T = 100\ \Omega + 300\ \Omega + 250\ \Omega + 700\ \Omega = 1350\ \Omega$

66. $$R_T = \frac{1}{\frac{1}{R_1}+\frac{1}{R_2}+\frac{1}{R_3}+\frac{1}{R_4}}$$

$$= \frac{1}{\frac{1}{1}+\frac{1}{2}+\frac{1}{8}+\frac{1}{4}} = \frac{1}{\frac{8+4+1+2}{8}} = \frac{8}{15}\ \Omega$$

67. **(a)** The total resistance is

$R_T = R_1 + R_2 + R_3 = 5 + 6 + 39 = 50\ \Omega$

(b) Using Ohm's law,

$$I = \frac{V_s}{R_T} = \frac{7}{5}\ \text{A}$$

68. $R_{eq} = 6 + 2 = 8\ \Omega$

$$V_1 = \frac{R_1}{R_{eq}}V = \frac{6}{8}(12) = 9\ \text{V}$$

$$V_2 = \frac{R_2}{R_{eq}}V = \frac{2}{8}(12) = 3\ \text{V}$$

69. $I_t = I_1 + I_2 + I_3$

$I_t = 5 + 12 + 3 = 20$ A

70. $I_t = I_1 + I_2 + I_3 + I_4$

$I_t = 4 + 10 + 7 + 9 = 30$ A

1.2 Exercises (Answers)

1. 53 hundredths
2. 8 thousandths
3. 5237 ten thousandths
4. 34 and 9 hundredths
5. 61.822
6. 44.224
7. 72.84
8. −2.34
9. 34.998
10. 0.0256
11. 48440
12. 2.641

13. $\frac{\$2.88}{6} = \$0.48$ per pound

14. $\frac{328.15}{40.5} = 8.1$ calories/min

15. 15 min = 0.25 h, distance = 65 × 0.25 = 16.25 mi

16. 8.25 × 30 = 247.5

17. $\frac{1}{4} = 0.25$

18. Perimeter = 12.3 + 3.4 + 4.1 + 5.3 + 7.2 + 8.5 = 44.8 in.

19. $I = \frac{Q}{t} = \frac{6.5}{0.3} = 21.67\ \text{A}$

20. $P = \frac{W}{t} = \frac{50,000\ \text{J}}{2 \times 60\ \text{s}} = 416.67\ \text{W}$

21. $V = 1.5 + 1.5 + 1.5 - 1.5 = 4.5$ V

22. $V = 1.5 + 1.5 - 1.5 - 1.5 = 3$ V

23. n = 6/1.2 = 5

24. n = 12/1.2 = 10

25. 156.85 hp − 102.27 = 54.58 hp

26. 126.35 hp − 101.40 = 24.95 hp

27. $3.25 × 42 = $136.5

28. 2.5 × 8.25 = $20.625

1.3 Exercises (Answers)

1. 34%
2. 15%
3. 8%
4. 2%
5. 0.43
6. 0.52
7. 0.07
8. 0.05
9. 6.08
10. 0.375
11. 35% = 35/100 = 7/20
    250/(7/20) = 714.28
12. 15% = 15/100 = 3/20
    120/(3/20) = 800.
13. We must convert to horsepower or to watts in order to make the output and input in the same units. Converting the output to W, we get

    $$\text{Output} = 1.14\ \text{hp}\left(\frac{746\ \text{W}}{\text{hp}}\right) = 850\ \text{W}$$

$$\text{Percent efficiency} = \frac{\text{output}}{\text{input}} \times 100$$

$$\text{Percent efficiency} = \frac{850}{876} \times 100 = 97\%$$

14. We must convert to horsepower or to watts in order to make the output and input in the same units. Converting the output to W, we get

$$\text{Output} = 1.25 \text{ hp}\left(\frac{746 \text{ W}}{\text{hp}}\right) = 932.5 \text{ W}$$

$$\text{Percent efficiency} = \frac{\text{output}}{\text{input}} \times 100$$

$$\text{Percent efficiency} = \frac{932.5}{975} \times 100 \cong 96\%$$

15. $75\% = 75/100 = 3/4$

$65\% = 65/100 = 13/20$

$$\frac{3}{4} \times \frac{13}{20} \times \frac{30}{1} = \frac{117}{8}.$$

16. Mechanical efficiency = (Actual work/Theoretical work) × 100

ME = (25/65) × 100 = 38.5%

17. $\frac{x}{115} = \frac{5}{100} \rightarrow 100x = 5 \times 115 \rightarrow x = 5.75$

18. $3 - 2.35 = 0.65, \frac{x}{100} = \frac{0.65}{3} \rightarrow x = \frac{0.65 \times 100}{3} = 21.7\%$ decrease

19. If 30% are eating specials, 70% are not. This means 60 people represent 70% of the number of people eating at the restaurant. So, $0.70x = 60$, or $x = \frac{60}{0.70} = 86$ people.

20. 2% = 0.02, now, 900 × 0.02 = 18

21. 0.40 × 700 = $280

22. $x = \frac{80 \times 110}{100} = 88$

Review Exercises (Chapter 1—Answers)

1. $\frac{2}{7}$

2. $\frac{7}{11}$

3. 21

4. $\frac{13}{27}$

5. $\frac{25}{3}$

6. $\frac{23}{4}$

7. $\frac{17}{2}$

8. $\frac{32}{5}$

9. ⬡ $= \frac{3}{9} = \frac{1}{3}$, △ $= \frac{4}{9}$, ▭ $= \frac{2}{9}$

10. Shade $= \frac{1}{4}$, white $= \frac{3}{4}$

11. Total weight = 20 kg + 65 kg + 100 kg + 225 kg = 410 kg.
12. Area difference = 268,820 − 52,271 = 216,579 mi^2
13. Power dissipated = 4.3 × 110 = 473 W
14. Number of revolutions = 1603 × 6 = 4809 revolutions
15. Value per share = $72,000/800 shares = $90 per share
16. Value per share = $42,000/600 shares = $70 per share

17. $I = \frac{Q}{t} = \frac{90}{20} = 4.5\ \text{C/s} = 4.5\ \text{A}$

18. $V = \frac{P}{I} = \frac{120}{20} = 6\ \text{V}$

$$R = \frac{V}{I} = \frac{6}{20} = 0.3\ \text{V/A} = 0.3\ \Omega$$

19. $R_T = R_1 + R_2 + R_3$
$= 1 + 2 + 3 = 6\ \Omega$

20. $R_T = \dfrac{1}{\frac{1}{R_1} + \frac{1}{R_2} + \frac{1}{R_3}}$

$$= \frac{1}{\frac{1}{2} + \frac{1}{4} + \frac{1}{6}} = \frac{1}{\frac{6+3+2}{12}} = \frac{12}{11}\ \Omega$$

21. $R_{eq} = 40 + 30 = 30\ \Omega$

$$V_1 = \frac{R_1}{R_{eq}} V = \frac{40}{70}(120) = \frac{48}{7}\ \text{V}$$

$$V_2 = \frac{R_2}{R_{eq}} V = \frac{30}{70}(120) = \frac{36}{7}\ \text{V}$$

22. $R_{eq} = 7 + 13 + 20 = 40\ \Omega$. Hence,

$$V_1 = \frac{R_1}{R_{eq}} V = \frac{7}{40}(120) = 21\ \text{V}$$

$$V_2 = \frac{R_2}{R_{eq}}V = \frac{13}{40}(120) = 39 \text{ V}$$
$$V_3 = \frac{R_3}{R_{eq}}V = \frac{20}{40}(120) = 60 \text{ V}$$

23. $R_{eq} = 5 + 10 + 15 = 30\ \Omega$. Hence,

$$V_1 = \frac{R_1}{R_{eq}}V = \frac{5}{30}(60) = 10 \text{ V}$$
$$V_2 = \frac{R_2}{R_{eq}}V = \frac{10}{30}(60) = 20 \text{ V}$$
$$V_3 = \frac{R_3}{R_{eq}}V = \frac{15}{30}(60) = 30 \text{ V}$$

24. $24 - 8 + 6 - V_{ab} = 0$
$V_{ab} = 24 - 8 + 6 = 22$ V

25. $10 - 100 + 50 + V_{ab} = 0$
$V_{ab} = -10 + 100 - 50 = 40$ V

26. $9\frac{1}{2} - 2\frac{3}{4} = \frac{19}{2} - \frac{11}{4} = \frac{38 - 11}{4} = \frac{27}{4} = 6\frac{3}{4}$

27. $16 - 3\frac{1}{2} = 16 - \frac{7}{2} = \frac{32 - 7}{2} = \frac{25}{2} = 12\frac{1}{2}$

28. $\frac{48}{87 + 48} = \frac{48}{135}$

29. $\frac{825,000 - 386,000}{825,000 + 386,000} = \frac{439,000}{1,211,000} = \frac{439}{1211}$

30. $3\frac{1}{7} + 2\frac{5}{14} = \frac{22}{7} + \frac{33}{14} = \frac{44 + 33}{14} = \frac{77}{14} = \frac{11}{2} = 5\frac{1}{2}$

31. $7\frac{1}{2} + 95\frac{1}{4} = \frac{15}{2} + \frac{381}{4} = \frac{30 + 381}{4} = \frac{411}{4} = 102\frac{3}{4}$

32. $21 \times 3\frac{5}{7} = \frac{21 \times 26}{7} = 78$

33. $21\frac{5}{7} \div 6 = \frac{\frac{152}{7}}{6} = \frac{152}{7} \times \frac{1}{6} = \frac{152}{42}$

34. Convert 85% to a decimal: 0.85
Multiply the value of the 40 A device rating by 0.85 = 40 A × 0.85 = 34 A

35. Convert 120% to a decimal: 1.20
Multiply the value of the 75 A load by 1.20 = 75 A × 1.20 = 90 A

36. Convert 30% to decimal form: 0.30
Add one to the decimal value: 1 + 0.30 = 1.30
Multiply 55 by the multiplier 1.30: 55 × 1.30 = 71.50

37. Convert 25% to decimal form: 0.25
    Add one to the decimal value: $1 + 0.25 = 1.25$
    Multiply 35 by the multiplier 1.25: $35 \times 1.25 = 43.75$
38. $8.25\% = 0.0825$
    $0.0825 \times \$10.00 = 0.825$
    $\$10.00 + 0.825 = \$10.825$
39. 2470 km/190 L = 13 km/L
40. 1350 mi/50 gal = 27 mi/gal

CHAPTER 2

Introduction to Algebra

2.1 Exercises (Answers)

1. $3^2 = 9$
2. $5^3 = 125$
3. $2 \times 2^4 = 2^5 = 32$
4. $3 \times 3^2 = 3^3 = 27$
5. $(mn)^4 = m^4 n^4$
6. $(vt)^3 = v^3 t^3$
7. $n = 9$
8. $n = 9$
9. $n = 6$
10. $n = 8$
11. $n = 6$
12. $n = 13$
13. $n = 0$
14. $n = 0$
15. $n = 2$
16. $n = 3/2$
17. $n = 5$
18. $n = 8$
19. $n = 6$
20. $n = 9$
21. $n = 12$
22. $n = 18$
23. 128
24. 256
25. 64
26. 729
27. 324
28. 1
29. $\frac{t}{3}$

30. $3r^2$
31. $\frac{1}{3}$
32. $\frac{5}{7}$
33. $\frac{5}{p-5}$
34. $\frac{1}{p-2}$
35. $3m + 6$
36. $2m + 16$

2.2 Exercises (Answers)

1. $V^2 + V$
2. $7V^3 + 9V^2 - 6V + 3$
3. $4I^2 + 11I + 15$
4. $-2I^2 - 14I - 8$
5. $-6t^4 - 8t^3 + 10t^2 + 12t$
6. $6t^{3/2} - 12t^3 + 9t^2 - 15t$
7. $15q^2 + 26q + 8$
8. $2q^2 + q - 3$
9. $4r^2 + 4rt + t^2$
10. $r^2 - 4rt + 4t^2$
11. $6k^3 + 11k^2 + 2k - 1$
12. $3k^3 - 5k^2 + 11k - 8$
13. $3p^2 - mp + np - 4m^2 - 2n^2 - 6nm$
14. $p^2 - 6m^2 + 6n^2 - 7np - mp - 9nm$
15. $3p^3 + p^2 - 7p - 6$
16. $9p + 2$
17. $z^2 + z + \left(\frac{6}{z+1}\right)$
18. $5z - 2 + \left(\frac{-6z + 11}{z^2 + 1}\right)$

2.3 Exercises (Answers)

1. $15{:}30 = 1/2$
2. $6/36 = 1/6$
3. $1/2{:}3/5 = 1/2 \times 5/3 = 5/6$

4. 3/7:4/6 = 3/7 × 6/4 = 9/14

5. $\frac{3}{11}$ or 3 : 11

6. $\frac{2}{9}$ or 2 : 9

7. $\frac{4}{14}$ or 4 : 14

8. $\frac{5}{15}$ or 5 : 15

9. $\frac{17}{5}$ or 17 : 5

10. $\frac{13}{2}$ or 13 : 2

11. $V = \frac{14}{3}$

12. $V = \frac{3}{4}$

13. $V = \frac{6}{5}$

14. $V = \frac{18}{11}$

15. $V = \frac{18}{3}$

16. $V = \frac{10}{13}$

17. $V = \frac{105}{2}$

18. $V = \frac{10}{9}$

19. $V = \frac{13}{5}$

20. $V = \frac{43}{4}$

21. $V = \frac{20}{13}$

22. $V = \frac{35}{3}$

23. flow rate $= \dfrac{145\text{ gal}}{5\text{ min}} = 29\text{ gal/min}$

24. flow rate $= \dfrac{180\text{ gal}}{8\text{ min}} = 22.5\text{ gal/min}$

25. the ratio is $\dfrac{6}{3850}$

26. the ratio is $\dfrac{640}{65}$

27. Inverse proportion, the slower the speed, the longer the time.
28. Direct proportion, the faster the speed, the greater the distance covered.
29. Direct proportion, the higher the temperature of gas, the greater the volume.
30. Inverse proportion, the greater the volume, the less the density.

31. $x = \dfrac{200}{9}, y = \dfrac{160}{9}$

32. $k = 7, y = 252$
33. $k = -6, y = -3$
34. $k = -15, -15$

Review Exercises (Chapter 2—Answers)

1. -2
2. -16
3. -1
4. -7
5. $\dfrac{x^3}{t}$
6. t^{11}
7. $-6z^5$
8. $27x^6y^{15}z^3$
9. $3z - 8y - 1$
10. $-6x^5 + 3x^2$
11. $4x - 3$
12. $\dfrac{-6}{x^2y^5z}$
13. $\dfrac{-2y^3}{x^8}$
14. $3x^3y^8$
15. binomial, order = 9
16. trinomial, order = 4
17. binomial, order = 4

18. monomial, order = 3
19. polynomial, order = 3
20. neither
21. $3z^2 - 4t$
22. $2z^4 - 5tz^2 + 3t^2$
23. xz^2
24. $2x + 5 - \left(\frac{6}{x+1}\right)$
25. x
26. $x + 1$
27. $18x$
28. $2t$
29. $9z$
30. $14m - 14n$
31. 5
32. 11
33. $8a$
34. $15xy$
35. x^{11}
36. $8x^3$
37. $20t^5$
38. $24z^3t^2$
39. 96
40. 31
41. The current that flows in a circuit is *directly proportional* to the applied voltage of the circuit.
42. The current that flows in a circuit is *inversely proportional* to the resistance of that circuit.

CHAPTER 3

Equations, Inequalities, and Modeling

3.1 Exercises (Answers)

1. $x = 7 + 2 = 9$
2. $x = -9 - 9 = -18$
3. $x = 6 - 11 = -5$
4. $x = 10$
5. $x = 4$
6. $x = 18/5$
7. $x = 6/7$
8. $x = -7/2$
9. $x = 11/3$

10. $x = 5$
11. $t = -1/4$
12. $m = 6/7$
13. $p = 2$
14. $v = 17/2$
15. $c = 4$
16. $t = 3/2$
17. $f = 7/3$
18. $g = 19/6$
19. $n - 5$
20. $n + 9$
21. $7n$
22. $n/2$
23. $V = IR = 4 \times 8 = 32$ V
24. Circumference $= 2\pi r$, $d = 2r$

$$50 = 2\pi r \rightarrow r = \frac{50}{2\pi} = 7.96$$

$$d = 15.92 \text{ inches}$$

25. $P = 24, L = x, W = x - 2$

$$24 = 2x + 2(x - 2)$$

$$24 = 2x + 2x - 4$$

$$24 = 4x - 4$$

$$4x = 28$$

$$x = \frac{28}{4} = 7$$

Length = 7 yards and width = 5 yards

26. Let x = width = height of box,
The length of the box = $(x + 20)$ cm
The length of the edges of the box = $2(4x) + 4(x + 20)$ cm
Hence,

$$200 = 2(4x) + 4(x + 20)$$

$$200 = 8x + 4x + 80$$

$$200 = 12x + 80$$

$$x = \frac{120}{12} = 10 \text{ cm}$$

The width of the box = 10 cm
Volume of box = length × width × height

$$= (x + 20)(x)(x) = 30 \times 10 \times 10 = 300 \text{ cm}^3$$

27. $2x + 7 = 3x - 4$
$3x - 2x = 7 + 4$
$x = 11$

28. $2x + 11 = 3x - 5$
$3x - 2x = 11 + 5$
$x = 16$

29. $9x = 71 \rightarrow x = 8$

30. $P = 4\,s$, $36 = 4\,s \rightarrow s = 36/4$

31. $14 - 8.5 = 5.5$ in.

32. $\alpha + \beta = 120°; \alpha = 2\beta$
$2\beta + \beta = 120° \rightarrow \beta = 40°$

33. $T = \frac{1}{f} = \frac{1}{60}$ s

34. $f = \frac{1}{T} = \frac{1}{200 \times 10^{-3}} = 5$ Hz

35. $y = \frac{-5}{3}x + \frac{7}{3}$

36. $y = 3/8\,x + 7/8$

37. $y = 6x + 3$

38. $y = 6x + 3$

39. $x = \frac{2 \pm \sqrt{4 - 4 \times 1 \times 1}}{2 \times 1} = 1 \pm \sqrt{0} = 1$

40. $x = \frac{-2 \pm \sqrt{4 + 4 \times 1 \times 6}}{2 \times 1} = \frac{-2 \pm \sqrt{28}}{2} = \frac{-2 \pm \sqrt{4 \times 7}}{2}$

$x = \frac{-2 \pm 2\sqrt{7}}{2} = -1 \pm \sqrt{7}$

41. $x = \frac{-4 \pm \sqrt{4^2 + 4 \times 1 \times 5}}{2 \times 1} = \frac{-4 \pm \sqrt{36}}{2} = \frac{-4 \pm 6\sqrt{1}}{2} = -2 \pm 3$

$x_1 = 1, x_2 = -5$

42. $x = \frac{2 \pm \sqrt{4 - 4 \times 3 \times 1}}{2 \times 3} = \frac{2 \pm \sqrt{-8}}{6}$

43. $x = 7$

44. $x = -6$

45. $\frac{x^2 + 2x + 2}{(x + 1)x} = 4 \rightarrow x^2 + 2x + 2 = 4x^2 + 4x \rightarrow 3x^2 + 2x - 2 = 0$

$x = \frac{-2 \pm \sqrt{4 + 4 \times 3 \times 2}}{2 \times 3} = \frac{-2 \pm \sqrt{28}}{6} = \frac{-2 \pm \sqrt{4 \times 7}}{6} = \frac{-1 \pm \sqrt{7}}{3}$

46. $\dfrac{(x+1)(x-3)+6x}{3(x-3)} = 6 \to x^2+4x-3 = 18x-54 \to x^2-14x+51 = 0$

$x = \dfrac{14 \pm \sqrt{14^2-4\times 1\times 51}}{2\times 1} = \dfrac{14\pm\sqrt{-8}}{2} = 7\pm\sqrt{-2} = 7\pm i\sqrt{2}$

47. $\dfrac{x^2+x-15}{3x} = 0 \to x^2+x-15 = 0$

$x = \dfrac{-1\pm\sqrt{1+4\times 1\times 15}}{2\times 1} = \dfrac{1\pm\sqrt{61}}{2}$

48. $\dfrac{x^2+2x-2}{2x} = 0 \to x^2+2x-2 = 0$

$x = \dfrac{-2\pm\sqrt{4+4\times 1\times 2}}{2\times 1} = \dfrac{-2\pm\sqrt{12}}{2} = -1\pm\sqrt{3}$

49. $(6\sqrt{t+6})^2 = (4\sqrt{6t-24})^2 \to 36(t+6) = 16(6t-24)$

$9(t+6) = 4(6t-24) \to 9t+54 = 24t-96$

$15t-150 = 0 \to t = 150/15 = 10$

50. $(\sqrt{t+3})^2 = (8)^2 \to t+3 = 64 \to t = 61$

51. $(\sqrt{m-3})^2 = (8)^2 \to m-3 = 64 \to m = 67$

52. $\sqrt{m} = 18+7 \to (\sqrt{m})^2 = (25)^2 \to m = 625$

53. $|x-2| = 4 \to x-2 = 4$ or $x-2 = -4$
$x = 6$ or $x = -2$

54. $|x+6| = 9 \to x+6 = 6$ or $x+6 = -9$
$x = 0$ or $x = -15$

55. $|x^2-24| = 1 \to x^2-24 = 1$ or $x^2-24 = -1$
$x^2 = 25 \to x = \pm 5$ or
$x^2 = 23 \to x = \pm\sqrt{23}$

56. $|x^2+4| = 2 \to x^2+4 = 2$ or $x^2+4 = -2$
$x^2 = -2 \to x = \pm\sqrt{-2} = \pm i\sqrt{2}$ or
$x^2 = -6 \to x = \pm\sqrt{-6} = \pm i\sqrt{6}$

3.2 Exercises (Answers)

1. 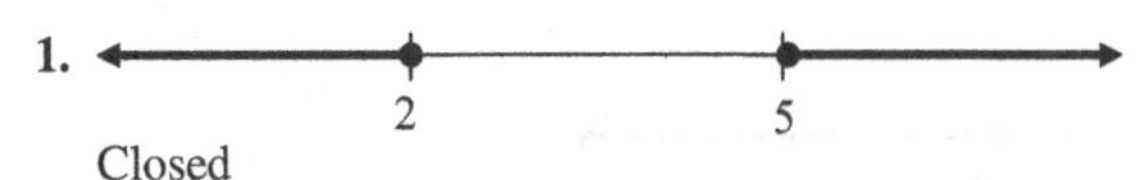

Closed

2. 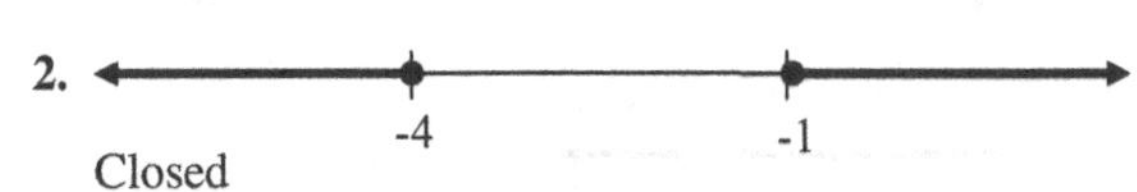

Closed

3. 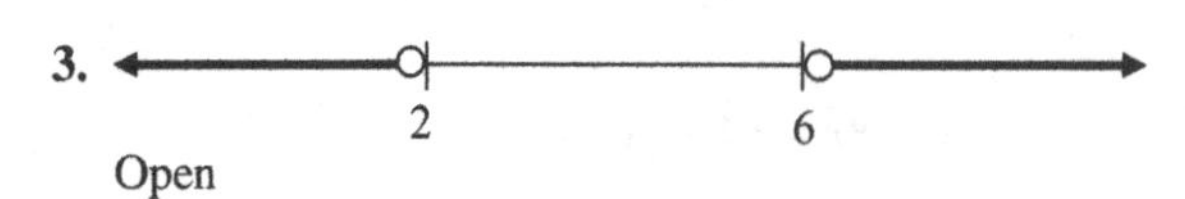

Open

4. -1 -5

Open

5. -4 0

Half-open or half-closed

6. 2 6

Half-open or half-closed

7. 3 5

Half-open or half-closed

8. 1 3

Half-open or half-closed

9. $(-\infty, -5]$

10. $(2, \infty)$

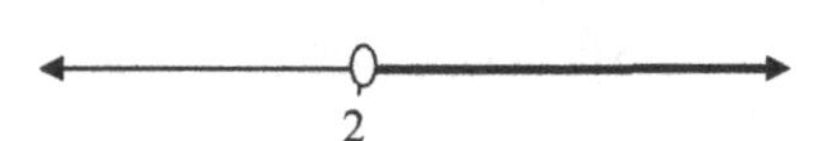

11. $(-\infty, 1)$

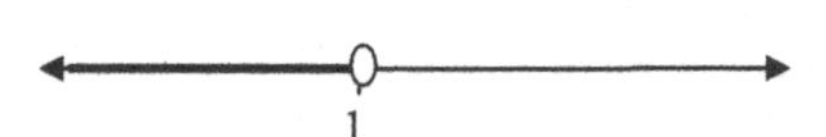

12. $[3, \infty)$

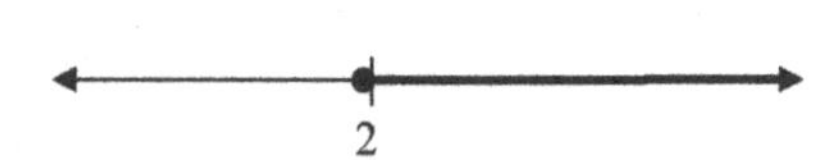

13. $y < 7 \rightarrow (-\infty, 7)$

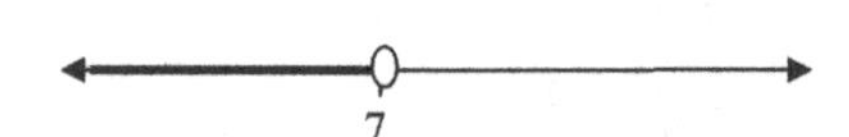

14. $y < 5 \rightarrow (-\infty, 5)$

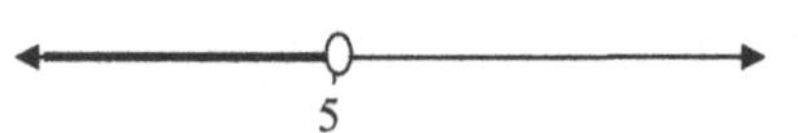

15. $t \geq -8/3 \rightarrow [-8/3, \infty)$

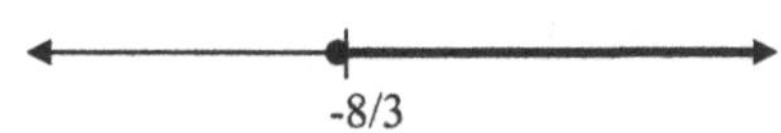

16. $m \geq -1 \rightarrow [-1, \infty)$

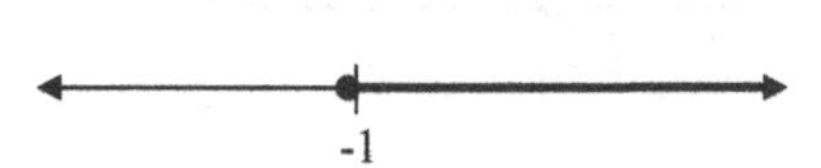

17. $n > 9 \rightarrow (9, \infty)$

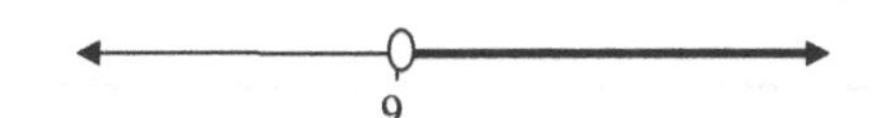

18. $v > 2 \rightarrow (2, \infty)$

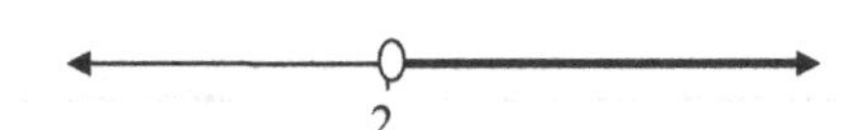

19. $p \leq -14 \rightarrow (-\infty, -14]$

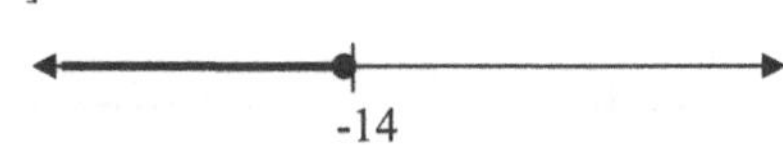

20. $v \geq 7/3 \rightarrow [7/3, \infty)$

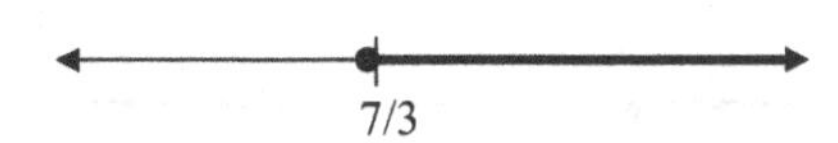

21. $x < 2 \rightarrow (-\infty, 2)$

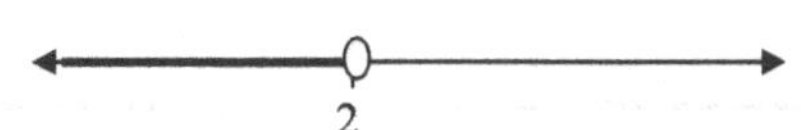

22. $x \leq 25/6 \rightarrow (-\infty, 25/6]$

23. $t > -2 \rightarrow (-2, \infty)$

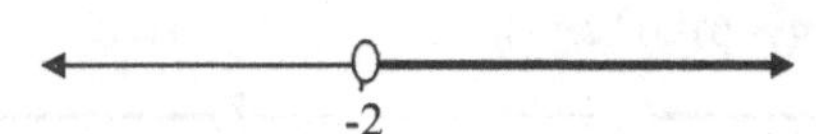

24. $x < 8 \rightarrow (-\infty, 8)$

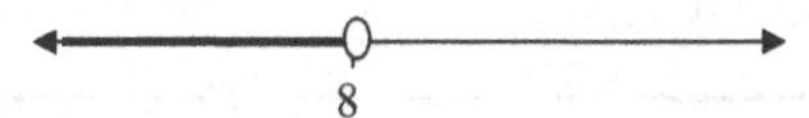

25. $m \leq 7/3 \rightarrow (-\infty, 7/3]$

26. $n \geq 9 \rightarrow [9, \infty)$

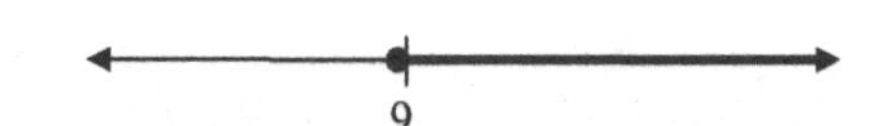

27. $-5 < t < 5$, $(-5,5)$

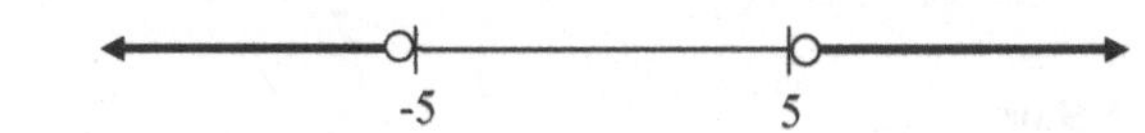

28. $-15 < t < 15$, $(-15,15)$

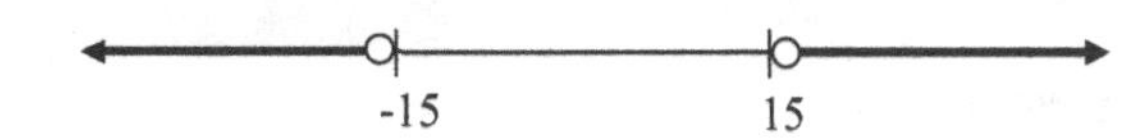

29. $u > 3$ or $u < -3$

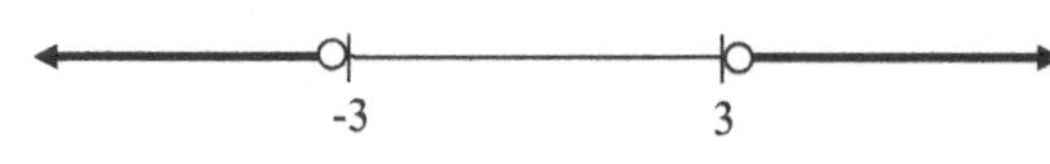

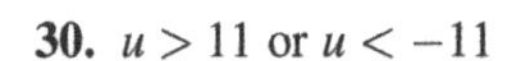
30. $u > 11$ or $u < -11$

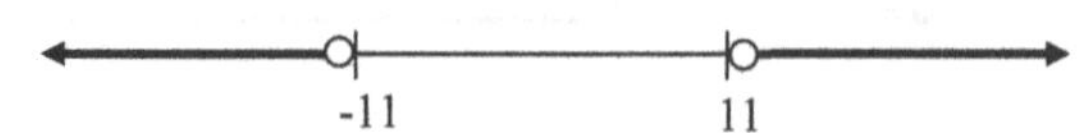

31. $-6 < x < 6$, $(-6,6)$

32. $-8 < x < 8$, $(-8,8)$

33. $-\frac{1}{2} \leq x \leq 4$, $\left[-\frac{1}{2}, 4\right]$

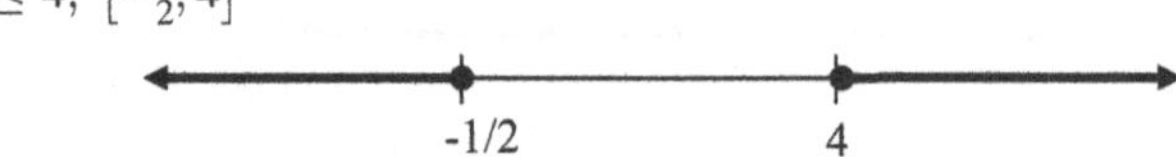

34. $-5 \leq x \leq 3$, $[-5,3]$

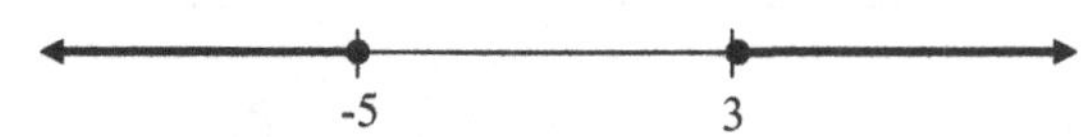

35. $x > 4$ or $x < -6$, $(-\infty,-6) \cup (4,\infty)$.

36. $x > 12$ or $x < -6$, $(-\infty,-6) \cup (12,\infty)$.

37. $x > 8$ or $x < -6$, $(-\infty,-6) \cup (8,\infty)$.

38. $x > -1$ or $x < 7 \rightarrow -1 < x < 7$, $(-1,7)$

39. $|T - 6| < 1$
$-1 < T - 6 < 1$
$5 < T < 7$

40. $|T - 7| > 2$
$T - 7 > 2 \rightarrow T > 9$ or
$T - 7 < -2 \rightarrow T < 5$

3.3 Exercises (Answers)

1. $3 + 9i$
2. $12 + 3i$
3. $9 + 2i$
4. $8 - 11i$

5. $-2-5i$
6. $-19+3i$
7. $6+i$
8. $6+4i$
9. -27
10. -24
11. $6-15i$
12. $-10+2i$
13. $2+i+6i-3=-1+7i$
14. $8-6i-20i-15=-7-26i$
15. $6+18i-2+6i=4+24i$
16. $-24-30i-28i+35=11-58i$
17. $\frac{1}{1+3i}\cdot\frac{1-3i}{1-3i}=\frac{1-3i}{1+9}=\frac{1}{10}-\frac{3}{10}i$
18. $\frac{7}{3+5i}\cdot\frac{3-5i}{3-5i}=\frac{21-35i}{9+25}=\frac{21}{34}-\frac{35}{34}i$
19. $\frac{5}{2-4i}\cdot\frac{2+4i}{2+4i}=\frac{10+20i}{4+16}=\frac{1}{2}+i$
20. $\frac{4}{7-2i}\cdot\frac{7+2i}{7+2i}=\frac{28+8i}{49+4}=\frac{28}{53}+\frac{8}{53}i$
21. $\frac{i}{2+3i}\cdot\frac{2-3i}{2-3i}=\frac{3+2i}{4+9}=\frac{3}{13}+\frac{2}{13}i$
22. $\frac{5i}{1-2i}\cdot\frac{1+2i}{1+2i}=\frac{-10+5i}{1+4}=-2+i$
23. $\frac{3+i}{2+4i}\cdot\frac{2-4i}{2-4i}=\frac{6-12i+2i+4}{4+16}=\frac{10-10i}{20}=\frac{1}{2}-\frac{1}{2}i$
24. $\frac{1-2i}{3-5i}\cdot\frac{3+5i}{3+5i}=\frac{3+5i-6i+10}{9+25}=\frac{13-i}{34}=\frac{13}{34}-\frac{1}{34}i$
25. $\frac{(3-\sqrt{-9})}{(2+\sqrt{-25})}\cdot\frac{(2-\sqrt{-25})}{(2-\sqrt{-25})}=\frac{6-15i-6i-15}{4-25}=\frac{-9-21i}{29}=\frac{-9}{29}-\frac{21}{29}i$
26. $\frac{(7+\sqrt{-4})}{(1-\sqrt{-16})}\cdot\frac{(1+\sqrt{-16})}{(1+\sqrt{-16})}=\frac{7+28i+2i-8}{1+16}=\frac{-1+30i}{17}=\frac{-1}{17}+\frac{30}{17}i$
27. $\frac{1}{2i}\cdot\frac{i}{i}=-\frac{1}{2}i$
28. $\frac{3}{4i}\cdot\frac{i}{i}=-\frac{3}{4}i$

29. $\frac{1+5i}{3i}\cdot\frac{i}{i} = \frac{5}{3} - \frac{1}{3}i$

30. $\frac{7-i}{2i}\cdot\frac{i}{i} = -\frac{1}{2} - \frac{7}{2}i$

31. $(1+3i) + 2(3-i)^2 + 11 = (1+3i) + 2(8-6i) + 11$
$= (1+3i) + (16-12i) + 11 = 28 - 9i$

32. $3(1-3i) + (2-5i)^2 - 7 = (3-9i) + (-21-10i) - 7$
$= -25 - 19i$

33. $(2-i)^2 - 3(4-i) - 8 = (3-4i) - (12-3i) = -9 - i$

34. $(3-2i)^2 - 4(5-3i) - 9 = (5-12i) - (20-12i) = -15$

35. $z^2 + 3z - 1 = (1+i)^2 + 3(1+i) - 1 = (2i) + 3 + 3i - 1 = 2 + 5i$

36. $z^2 - 2z + 3 = (2-i)^2 - 2(2-i) + 3 = (3-4i) - 4 + 2i + 3 = 2 - 2i$

37. $2z^2 - z + 4 = 2(1-i)^2 - (1-i) + 4 = -4i - 1 + i + 4 = 3 - 3i$

38. $3z^2 - 2z - 5 = 3(1+2i)^2 - (1+2i) - 5 = (-3+4i) - 1 - 2i - 5 = -9 + 2i$

39. $i^{16} = (i^2)^8 = 1,\ i^{28} = (i^2)^{14} = 1,\ i^{42} = (i^2)^{21} = -1$

40. $i^{19} = (i)(i^2)^9 = -i,\ i^{31} = (i)(i^2)^{15} = -i,\ i^{63} = (i)(i^2)^{31} = -i$

41. $\frac{1}{i^{22}} = \frac{1}{(i^2)^{11}} = -1,\ \frac{3}{i^{36}} = \frac{3}{(i^2)^{18}} = 3,\ \frac{5}{i^{54}} = \frac{5}{(i^2)^{27}} = -5$

42. $\frac{1}{i^{21}} = i^{-21} = (i)(i^{-22}) = i(i^2)^{-11} = -i,\ \frac{2}{i^{33}} = 2i^{-33} = 2(i)(i^{-34})$
$= 2(i)(i^2)^{-17} = -2i,\ \frac{4}{i^{57}} = 4i^{-57} = 4(i)(i^{-58}) = 4(i)(i^2)^{-29} = -4i$

43. $3x + 1 = 7 \rightarrow x = 2$

$2y - 5 = 8 \rightarrow y = \frac{13}{2}$

44. $x + 1 = 6 \rightarrow x = 5$

$4y + 2 = 1 \rightarrow y = \frac{-1}{4}$

45. $2x - 4 = 9 \rightarrow x = \frac{13}{2}$

$3y - 7 = 5 \rightarrow y = 4$

46. $x - 5 = 1 \rightarrow x = 6$

$2y - 6 = 3 \rightarrow y = \frac{9}{2}$

47. $2x = 4 + 3x \rightarrow x = -4$
$-3y - 5 = 9y - 5 \rightarrow y = 0$

48. $3x = 5 - 2x \rightarrow x = 1$

$y - 7 = -4y - 6 \rightarrow y = \frac{1}{5}$

49. $z = 3$

50. $z = 8$

51. $z = -6$

52. $z = -9$

53. $z = \frac{3}{4}$

54. $z = 2$

55. $z = -3$

56. $z = \frac{-11}{3}$

Review Exercises (Chapter 3—Answers)

1. $x + 5 = 13 \rightarrow x = 8$

2. $z - 6 = 3 \rightarrow z = 9$

3. $3y + 1 - 5y - 7 = 7y + 8 + y - 12 \rightarrow -2y - 6 = 8y - 4$

$10y = -2 \rightarrow y = \frac{-1}{5}$

4. $-5z = 11 \rightarrow z = \frac{-11}{5}$

5. $4\frac{y}{2} = 3 \rightarrow y = \frac{3}{2}$

6. $\frac{z}{-3} = -11 \rightarrow z = 33$

7. $7z - 3 = 2 \rightarrow z = \frac{5}{7}$

8. $5 - \frac{t}{3} = 9 \rightarrow t = -12$

9. $\frac{4t}{7} - 1 = \frac{8t}{3} + 5 \rightarrow 21\left(\frac{4t}{7}\right) = 21\left(\frac{8t}{3} + 6\right)$

$12t = 56t + 126 \rightarrow 44t = -126 \rightarrow t = \frac{-126}{44}$

10. $\frac{9s}{2} - 11 = s + 3 \rightarrow 2\left(\frac{9s}{2} - 11 = s + 3\right) \rightarrow 9s - 22 = 2s + 6$

$7s = 28 \rightarrow s = \frac{28}{7}$

11. $6(s-4) = 3 \rightarrow 6s - 24 = 3 \rightarrow s = \dfrac{27}{6}$

12. $7(t+1) - 5 = 14 \rightarrow 7t + 7 = 19 \rightarrow t = \dfrac{12}{7}$

13. $\dfrac{3z-2}{5} = z + 1 \rightarrow 3z - 2 = 5z + 5 \rightarrow z = \dfrac{-7}{2}$

14. $\dfrac{3x}{5} + \dfrac{6x}{7} = 2 \rightarrow 35\left(\dfrac{3x}{5} + \dfrac{6x}{7} = 2\right) = 21x + 30x = 70 \rightarrow x = \dfrac{70}{51}$

15. $x^2 = 16 \rightarrow x = \pm\sqrt{16} = \pm 4$

16. $x^2 = 81 \rightarrow x = \pm\sqrt{81} = \pm 9$

17. $10z^2 - 100 = 60 \rightarrow z^2 = 10 \rightarrow z = \pm\sqrt{10}$

18. $9x^2 + 5x = 0 \rightarrow x(9x + 5) = 0$

$x = 0 \text{ or } x = \dfrac{-5}{9}$

19. $4x^2 + 9x - 11 = 0 \rightarrow x = \dfrac{-9 \pm \sqrt{81 - 4 \cdot 4 \cdot (-11)}}{2(4)} = \dfrac{-9 \pm \sqrt{169}}{8}$

20. $y^2 - 4y - 5 = 0 \rightarrow y = \dfrac{4 \pm \sqrt{16 - 4 \cdot 1 \cdot (-5)}}{2(1)} = 2 \pm \dfrac{\sqrt{36}}{2}$

21. $3z^2 + 7z = 1 \rightarrow z = \dfrac{-7 \pm \sqrt{47 - 4(3)(-1)}}{2(3)} = \dfrac{-7 \pm \sqrt{59}}{6}$

22. $2z^2 + 5z = 0 \rightarrow z(2z + 5) = 0$

$z = 0 \text{ or } z = \dfrac{-5}{2}$

23. $\sqrt{x-6} = 2 \rightarrow (\sqrt{x-6})^2 = (2)^2 \rightarrow x - 6 = 4 \rightarrow x = 10$

24. $\sqrt{x^2+1} = 3 \rightarrow (\sqrt{x^2+1})^2 = (3)^2 \rightarrow x^2 + 1 = 9 \rightarrow x = \pm\sqrt{8}$

25. $\sqrt{x-2} + 2x = 5 \rightarrow (\sqrt{x-2})^2 = (5 - 2x)^2 \rightarrow x - 2 = 4x^2 - 20x + 25$

$4x^2 - 19x + 23 = 0$

26. $\sqrt{3x-1} = x - 2 \rightarrow (\sqrt{3x-1})^2 = (x-2)^2 \rightarrow 3x - 1 = x^2 - 4x + 4$

$x^2 - 7x + 5 = 0 \rightarrow x = \dfrac{7 \pm \sqrt{47 - 4(1)(5)}}{2(1)} = \dfrac{7 \pm \sqrt{27}}{2}$

27. $(\sqrt{x-1})^2 = (2\sqrt{x+5})^2 \rightarrow x - 1 = 4x + 20 \rightarrow x = \dfrac{-21}{3} = -7$

28. $\sqrt[3]{(2x+7)^2} = 9 \rightarrow \left((2x+7)^{\frac{2}{3}}\right)^{3/2} = ((3)^2)^{3/2} \rightarrow 2x + 7 = 27$

$x = 10$

29. $|x-5| = 7 \rightarrow$
$x-5 = 7 \rightarrow x = 12$ or
$x-5 = -7 \rightarrow x = -2$

30. $|3x+1|-2 = 11 \rightarrow |3x+1| = 13$

$3x+1 = 13 \rightarrow x = 4$ or

$3x+1 = -13 \rightarrow x = \frac{-14}{3}$

31. $(6,\infty)$, $x > 6$.

32. $(6,8)$, $6 < x < 8$.

33. $(-\infty,-2]$ and $(5,\infty)$, $x \leq -2$ and $x > 5$

34. $(-\infty,-3)$, $x < -3$

35. $[-7,-3)$, $x \geq -7$ and $x < -3$, $-7 \leq x < -3$

36. $(-2,6]$, $-2 < x \leq 6$

37. $(-1,3]$

38. $[0,1)$

39. $[1,15]$

40. $(-5,1)$

41. $[7,\infty)$

42. $(19,\infty)$

43. $(-\infty,43)$

44. $(-\infty,65]$

45. $\frac{x-2}{3}+\frac{4}{7} \leq 5 \rightarrow \frac{x-2}{3} \leq 5-\frac{4}{7} \rightarrow 3\left(\frac{x-2}{3}\right) \leq 3\left(\frac{31}{7}\right) \rightarrow x-2 \leq \frac{93}{7}, x \leq \frac{107}{7}$

$\left[-\infty,\frac{107}{7}\right]$

46. $\frac{2x}{3}-\frac{5}{6} \geq 1 \rightarrow \frac{2x}{3} \geq 1+\frac{5}{6} \rightarrow \frac{3}{2}\left(\frac{2x}{3}\right) \geq \frac{3}{2}\left(\frac{11}{6}\right) \rightarrow x \geq \frac{11}{4}$

$[11/4,\infty)$

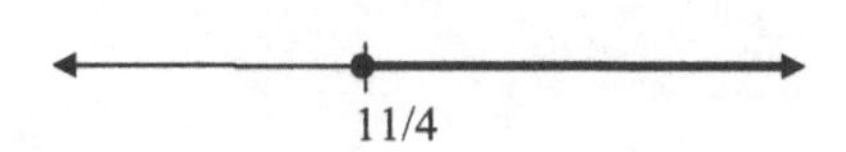

47. $(x+3)(x-2) > 0 \rightarrow x^2+x-6 > 0 \rightarrow$

for $x^2+x-6 = 0 \rightarrow x = \frac{-1}{2} \pm \frac{\sqrt{28}}{2}$

So, between $\frac{-1}{2}-\frac{\sqrt{28}}{2}$ and $\frac{-1}{2}+\frac{\sqrt{28}}{2}$, the function will either be always greater than zero or always less than zero.

We don't know which … yet! Let's pick a value in between and test it:
At $x = 0 \rightarrow x^2 + x - 6 = 0 \rightarrow 0 + 0 - 6 = -6 > 0$ is false

So between $\frac{-1}{2} - \frac{\sqrt{28}}{2}$ and $\frac{-1}{2} + \frac{\sqrt{28}}{2}$, the function is **less** than zero.
Thus,

the region $x^2 + x - 6 > 0$ in interval $\left(-\infty, \frac{-1}{2} - \frac{\sqrt{28}}{2}\right)$ or $\left(\frac{-1}{2} + \frac{\sqrt{28}}{2}, \infty\right)$

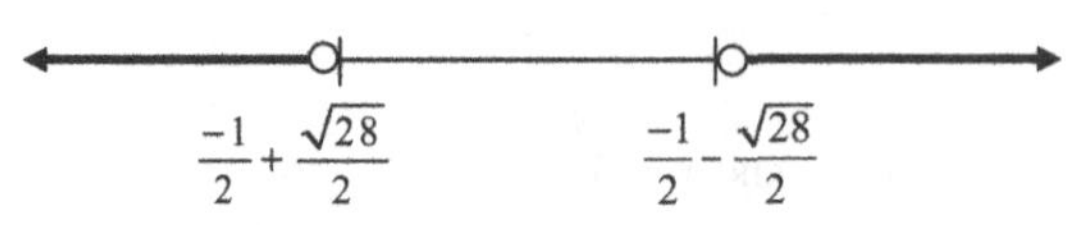

48. $2x^2 - 3x \leq 12 \rightarrow 2x^2 - 3x - 12 \leq 0$

for $2x^2 - 3x - 12 = 0 \rightarrow x = \frac{3 \pm \sqrt{9 - 4(2)(-12)}}{2(2)} = \frac{3}{4} \pm \frac{\sqrt{105}}{4}$

So, between $\frac{3}{4} - \frac{\sqrt{105}}{4}$ and $\frac{3}{4} + \frac{\sqrt{105}}{4}$ the function will either be always greater than zero or always less than zero.
We don't know which … yet! Let's pick a value in between and test it:
At $x = 0$, $2x^2 - 3x - 12 = 0 \rightarrow 0 - 0 - 12 = -12 < 0$ is true

So between $\frac{3}{4} - \frac{\sqrt{105}}{4}$ and $\frac{3}{4} + \frac{\sqrt{105}}{4}$ the function is **less** than zero.
Thus,

$$\frac{3}{4} - \frac{\sqrt{105}}{4} \leq x \leq \frac{3}{4} + \frac{\sqrt{105}}{4}$$

$\frac{3}{4} - \frac{\sqrt{105}}{4}$ $\frac{3}{4} + \frac{\sqrt{105}}{4}$

49. $\frac{x-4}{x+1} \geq 0 \rightarrow x - 4 \geq 0 \rightarrow x \geq 4$

4

50. $(x + 5)(2x - 1) \geq 0$
For

$(x + 5)(2x - 1) = 0 \rightarrow (2x - 1) = 0 \rightarrow x = \frac{1}{2}$ or

$(x + 5) = 0 \rightarrow x = -5$
So, between -5 and $1/2$, the function will either be always greater than zero or always less than zero.
We don't know which … yet! Let's pick a value in between and test it:
At $x = 0$, $(x + 5)(2x - 1) = 0 \rightarrow \rightarrow 0 - 0 - 5 = -5 > 0$ is false.

So between -5 and $1/2$, the function is less than zero.
Thus,
the region $(x + 5)(2x - 1) \geq 0$ in interval $(-\infty, -5]$ or $[1/2, \infty)$

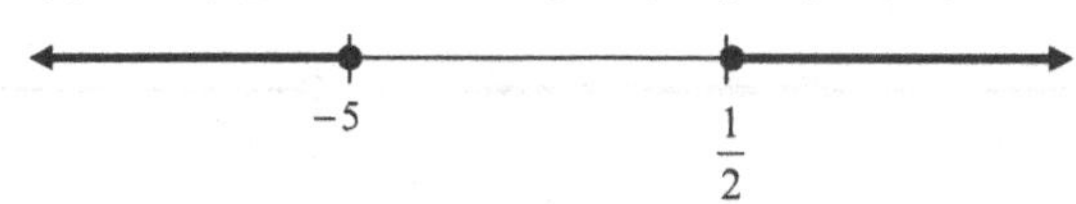

51. $x^2 - 3x \leq 14 \rightarrow x^2 - 3x - 14 \leq 0$

for $x^2 - 3x - 14 = 0 \rightarrow x = \dfrac{3 \pm \sqrt{9 - 4(1)(-14)}}{2(1)} = \dfrac{3}{2} \pm \dfrac{\sqrt{65}}{2}$

So, between $\dfrac{3}{2} - \dfrac{\sqrt{65}}{2}$ and $\dfrac{3}{2} + \dfrac{\sqrt{65}}{2}$, the function will either be always greater than zero or always less than zero.
We don't know which ... yet! Let's pick a value in between and test it:
At $x = 0$, $x^2 - 3x - 14 = 0 \rightarrow 0 - 0 - 14 = -14 < 0$ is true.

So between $\dfrac{3}{2} - \dfrac{\sqrt{65}}{2}$ and $\dfrac{3}{2} + \dfrac{\sqrt{65}}{2}$ the function is **less** than zero.
Thus,

$\dfrac{3}{2} - \dfrac{\sqrt{65}}{2} \leq x \leq \dfrac{3}{2} + \dfrac{\sqrt{65}}{2}$.

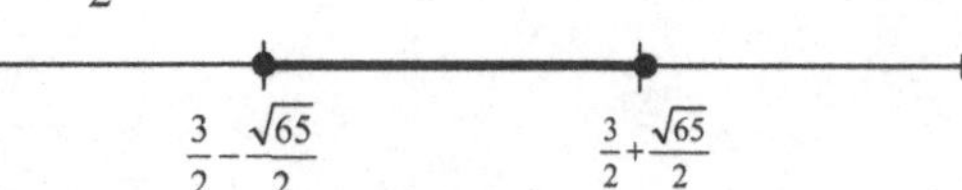

52. $x^2 - 4 \geq 0 \rightarrow x^2 \geq 4$
for $x^2 = 4 \rightarrow x = \pm 2$
So, between -2 and 2, the function will either be always greater than zero or always less than zero.
We don't know which ... yet! Let's pick a value in between and test it:
At $x = 0$, $x^2 - 4 = 0 \rightarrow 0 - 4 = -4 > 0$ is false.
So between -2 and 2, the function is **less** than zero.
Thus,
the region $x^2 - 4 \geq 0$ in interval $(-\infty, -2]$ or $[2, \infty)$.

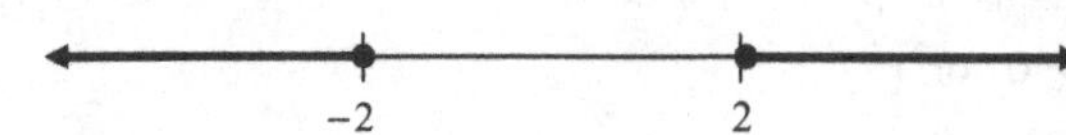

53. $x(x + 1)(x - 1) \geq 0$
For
$x(x + 1)(x - 1) = 0$
$x = 0$
$x = 1$
$x = -1$
$x \geq -1, x \geq 1, x \geq 0$
$[-1, 0] \cup [1, \infty)$

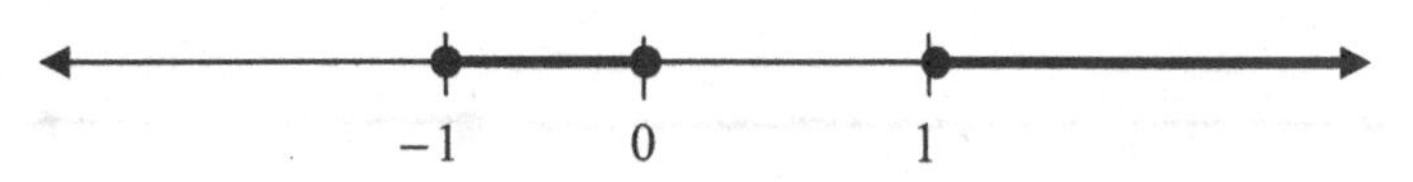

54. 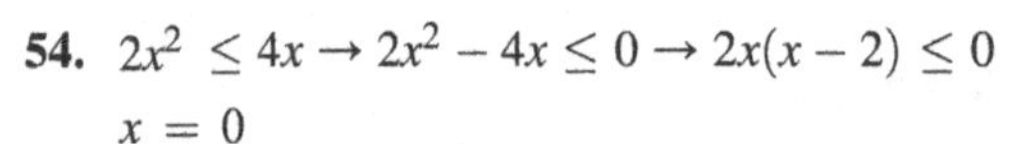
$2x^2 \le 4x \rightarrow 2x^2 - 4x \le 0 \rightarrow 2x(x-2) \le 0$

$x = 0$
$x = 2$
Thus, [0,2]

55. $5 \le x - 2 < 6 \rightarrow 7 \le x < 8$
[7,8)

56. $-12 \le 8x + 3 \le -2 \rightarrow \frac{-15}{8} \le x \le \frac{-5}{8}$
$\left[\frac{-15}{8}, \frac{-5}{8}\right]$

57. $|2x - 8| < 9 \rightarrow \ -9 < 2x - 8 < 9 \rightarrow \frac{-1}{2} < x < \frac{17}{2}$
$\left(\frac{-1}{2}, \frac{17}{2}\right)$

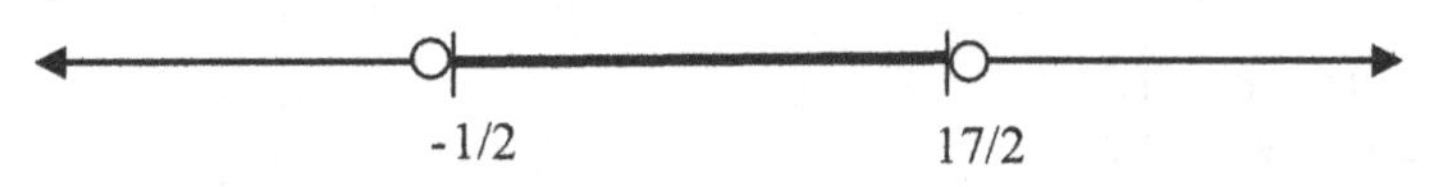

58. $|x + 3| > 6 \rightarrow x + 3 > 6 \rightarrow x > 3$ or
$x + 3 < -6 \rightarrow x < -9$
$(-\infty, -9) \cup (3, \infty)$

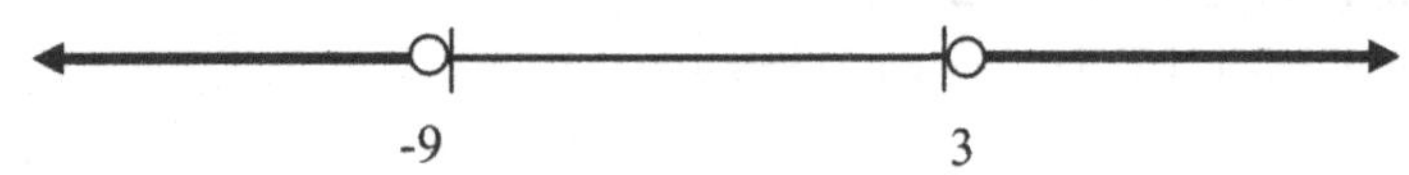

59. $|x + 4| + 3 \ge 7 \rightarrow |x + 4| \ge 4$
$x + 4 \ge 4 \rightarrow x \ge 0$ or
$x + 4 \le -4 \rightarrow x \le -8$
$(-\infty, -8] \cup [0, \infty)$

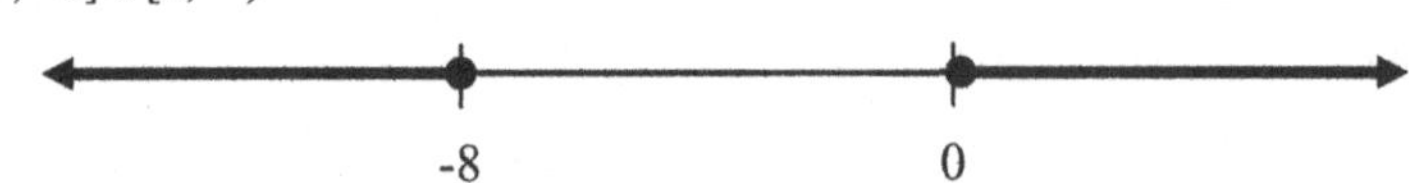

60. $|9x - 2| - 1 \le 8 \rightarrow \ -9 \le 9x - 2 \le 9 \rightarrow \frac{-7}{9} \le x \le \frac{11}{9}$
$\left[\frac{-7}{9}, \frac{11}{9}\right]$

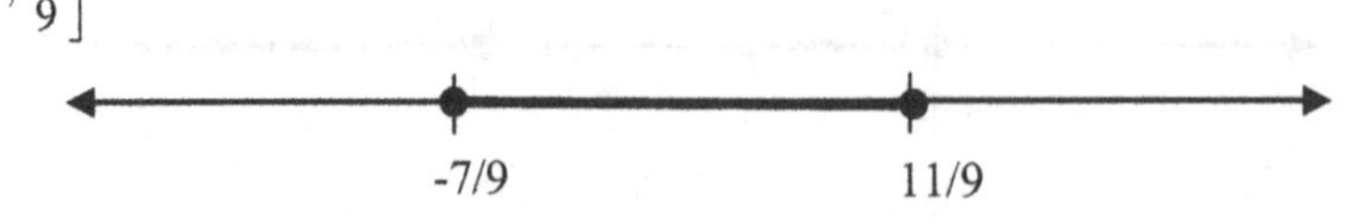

61. $x^2+9 = 0 \to x = \pm\sqrt{-9} \to x = \pm 3i$

62. $x^2+16 = 0 \to x = \pm\sqrt{-16} \to x = \pm 4i$

63. $x^2+3x+9 = 0 \to x = \dfrac{-3 \pm \sqrt{9-4(1)(9)}}{2(1)} = \dfrac{-3}{2} \pm \dfrac{\sqrt{27}}{2}i = \dfrac{-3}{2} \pm \dfrac{3\sqrt{3}}{2}$

64. $3x^2+2x+1 = 0 \to x = \dfrac{-2 \pm \sqrt{4-4(3)(1)}}{2(3)} = \dfrac{-1}{3} \pm \dfrac{\sqrt{8}}{6}i = \dfrac{-1}{3} \pm \dfrac{\sqrt{2}}{3}i$

65. $(2+4i)+(5+9i) = 7+13i$

66. $(3-i)-(1-5i) = 2+4i$

67. $4i(3-2i) = 8+12i$

68. $-3i(2-5i) = -15-6i$

69. $\dfrac{i}{2+i}\cdot\dfrac{2-i}{2-i} = \dfrac{1+2i}{5} = \dfrac{1}{5}+\dfrac{2}{5}i$

70. $\dfrac{3+2i}{1+4i}\cdot\dfrac{1-4i}{1-4i} = \dfrac{3-13i+2i+8}{1+16} = \dfrac{11-11i}{17} = \dfrac{11}{17}-\dfrac{11}{17}i$

71. $\dfrac{6-7i}{3-i}\cdot\dfrac{3+i}{3+i} = \dfrac{18+6i-21i+7}{9+1} = \dfrac{25}{10}-\dfrac{15}{10}i = \dfrac{5}{2}-\dfrac{3}{2}i$

72. $\dfrac{1-3i}{2i}\cdot\dfrac{i}{i} = \dfrac{3+i}{-2} = \dfrac{-3}{2}-\dfrac{1}{2}i$

73. $(3-2i)^2+4(5+i)-7 = (5-12i)+20+4i-7 = 15-8i$

74. $(1-i)^2+3(2+i)-9 = (-2i)+6+3i-9 = -3+i$

75. $(3-2i)(-4+7i) = -12+21i+8i+14 = 2+29i$

76. $(2+i)(3+2i) = 6+4i+3i-2 = 4+7i$

77. $i^{17} = (i)(i^2)^8 = i$

78. $i^{64} = (i^2)^{32} = 1$

79. $i^{77} = (i)(i^2)^{38} = i$

80. $i^{45} = (i)(i^2)^{22} = i$

81. $i^{-11} = (i)(i^2)^{-6} = i$

82. $i^{-13} = (i)(i^2)^{-7} = -i$

83. $7 = 3x-4 \to x = \dfrac{11}{3}$

$-11 = 2y+12 \to y = \dfrac{-23}{2}$

84. $8x-3 = 2x \to x = \dfrac{1}{2}$

$7-2y = 3y-1 \to y = \dfrac{8}{5}$

CHAPTER 4
Graphs and Functions
Section 4.1 (Answers)

1.

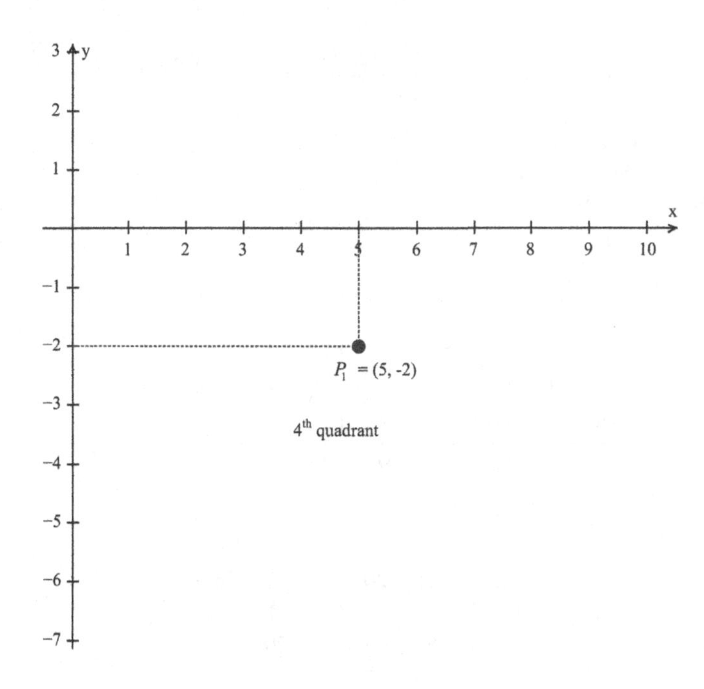

2.

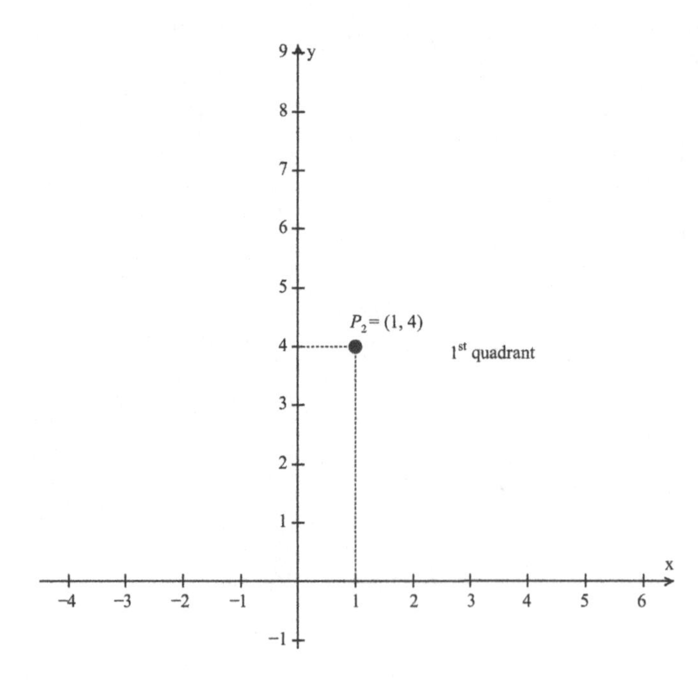

3.

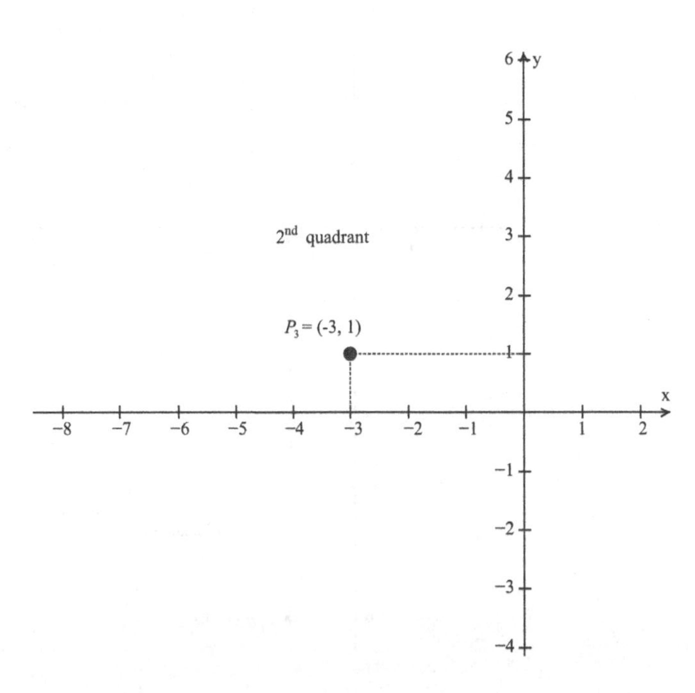
6
y
5
4
2nd quadrant
3
2
P_3 = (-3, 1)
1
x
−8
−7
−6
−5
−4
−3
−2
−1
1
2
−1
−2
−3
−4

4.

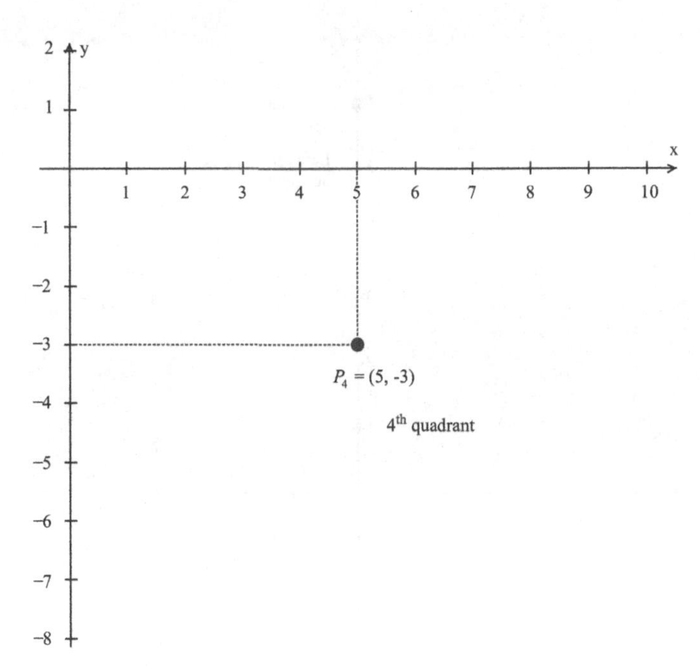
2
y
1
x
1
2
3
4
5
6
7
8
9
10
-1
-2
-3
P_4 = (5, -3)
−4
4th quadrant
−5
−6
−7
−8

5.

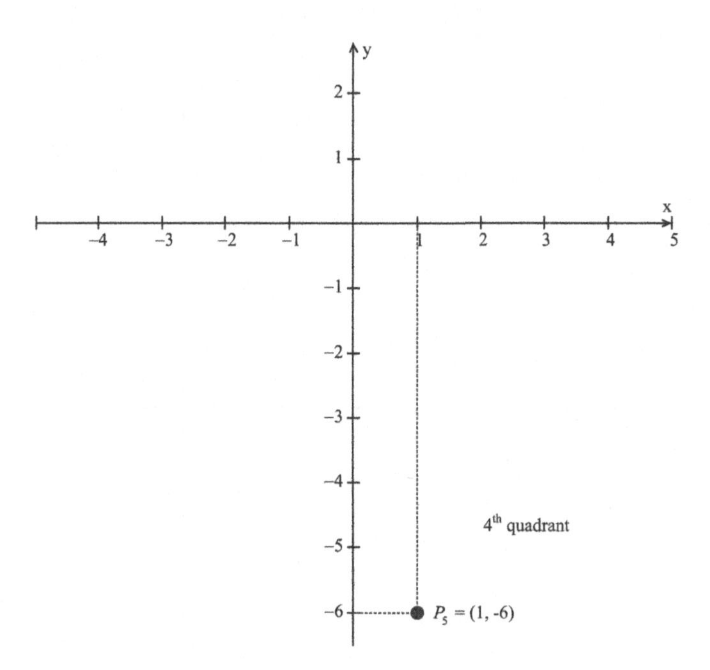
y
2
1
-4
-3
-2
-1
1
2
3
4
5
x
-1
-2
-3
-4
-5
-6
4th quadrant
P_5 = (1, -6)

6.

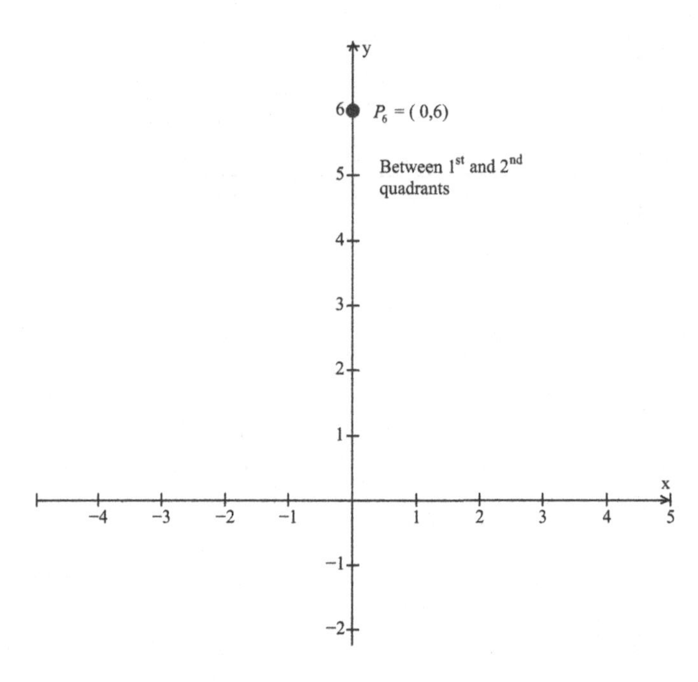
y
6
P_6 = (0,6)
Between 1st and 2nd
quadrants
5
4
3
2
1
-4
-3
-2
-1
1
2
3
4
5
x
-1
-2

7.

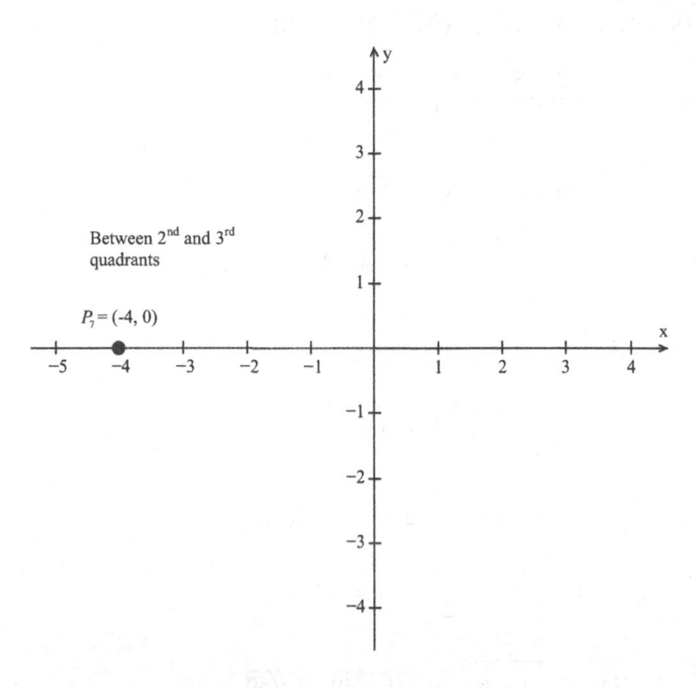
y
x
Between 2nd and 3rd quadrants
P7 = (-4, 0)
−5
−4
−3
−2
−1
1
2
3
4
4
3
2
1
−1
−2
−3
−4

8.

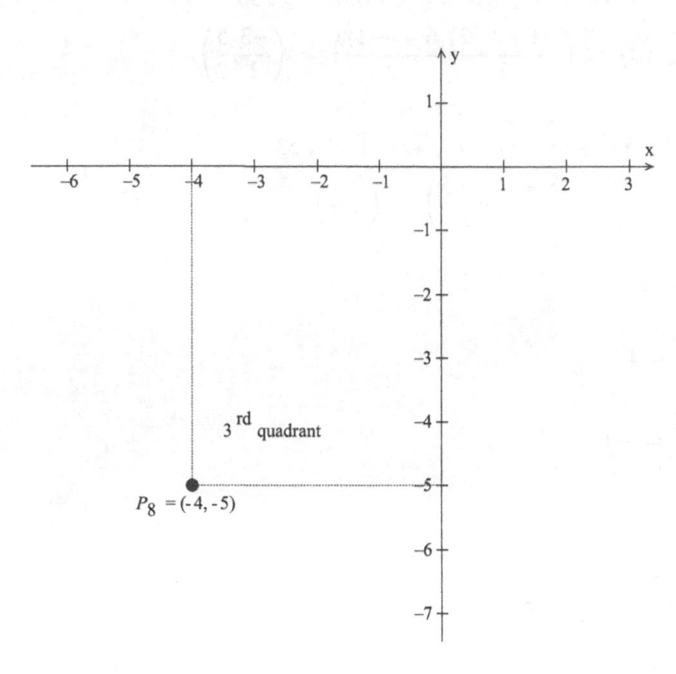
y
x
−6
−5
−4
−3
−2
−1
1
2
3
1
−1
−2
−3
−4
−5
−6
−7
3 rd quadrant
P8 = (-4, -5)

9. $d = \sqrt{(8-2)^2 + (9-4)^2} = \sqrt{36+25} = \sqrt{61}$

$$M = (x_m, y_m) = \left(\frac{2+8}{2}, \frac{4+9}{2}\right) = \left(5, \frac{13}{2}\right)$$

10. $d = \sqrt{(5-3)^2 + (0-1)^2} = \sqrt{4+1} = \sqrt{5}$

$$M = (x_m, y_m) = \left(\frac{3+5}{2}, \frac{1+0}{2}\right) = \left(4, \frac{1}{2}\right)$$

11. $d = \sqrt{(3-5)^2 + (7-6)^2} = \sqrt{4+1} = \sqrt{5}$

$$M = (x_m, y_m) = \left(\frac{5+3}{2}, \frac{6+7}{2}\right) = \left(4, \frac{13}{2}\right)$$

12. $d = \sqrt{(0-2)^2 + (-3+1)^2} = \sqrt{4+4} = \sqrt{8} = 2\sqrt{2}$

$$M = (x_m, y_m) = \left(\frac{2+0}{2}, \frac{-1+(-3)}{2}\right) = (1, -2)$$

13. $d = \sqrt{(-2+1)^2 + (-1-6)^2} = \sqrt{1+49} = \sqrt{50}$

$$M = (x_m, y_m) = \left(\frac{-1+(-2)}{2}, \frac{6+(-1)}{2}\right) = \left(\frac{-3}{2}, \frac{5}{2}\right)$$

14. $d = \sqrt{(1-8)^2 + (5-3)^2} = \sqrt{49+4} = \sqrt{53}$

$$M = (x_m, y_m) = \left(\frac{8+1}{2}, \frac{3+5}{2}\right) = \left(\frac{9}{2}, 4\right)$$

15. $m = \frac{8}{4} = 2$

16. $m = \frac{12}{3} = 4$

17. $m = \frac{-6}{6} = -1$

18. $m = \frac{-15}{-5} = 3$

19. $m = \frac{0-2}{1-5} = \frac{-2}{-4} = \frac{1}{2}$

20. $m = \frac{3-4}{1-2} = \frac{-1}{-1} = 1$

21. $m = \frac{-1-3}{1+3} = \frac{-4}{4} = -1$

22. $m = \dfrac{1-4}{6-7} = \dfrac{-3}{-1} = 3$

23.

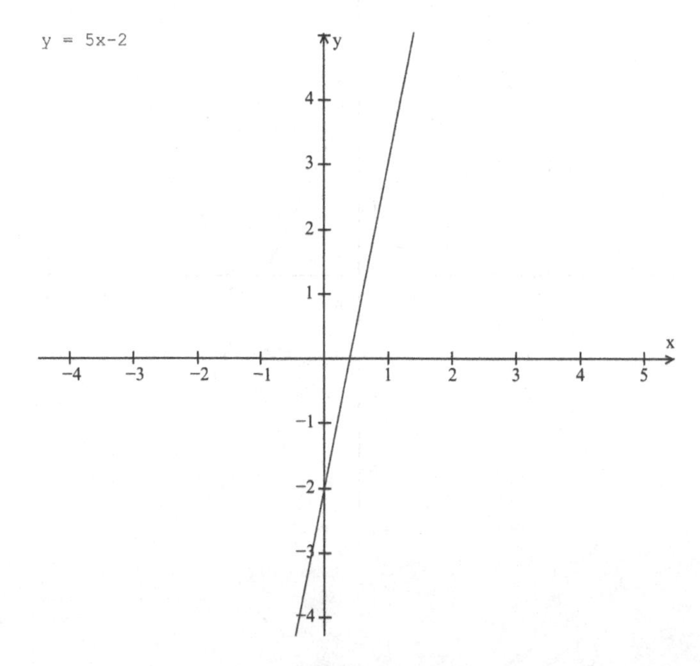

24.

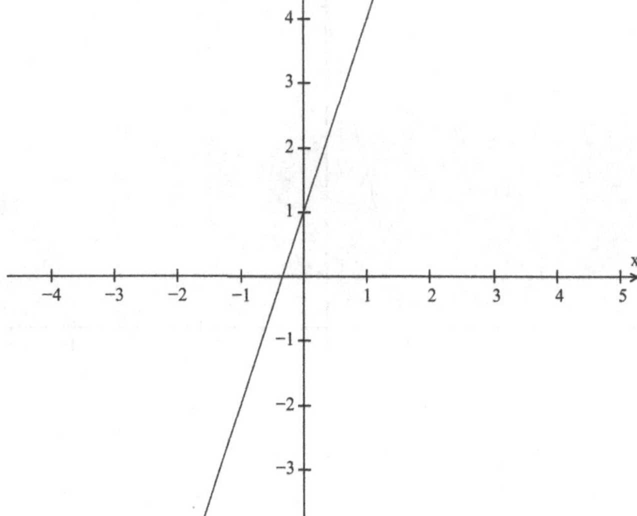

25.

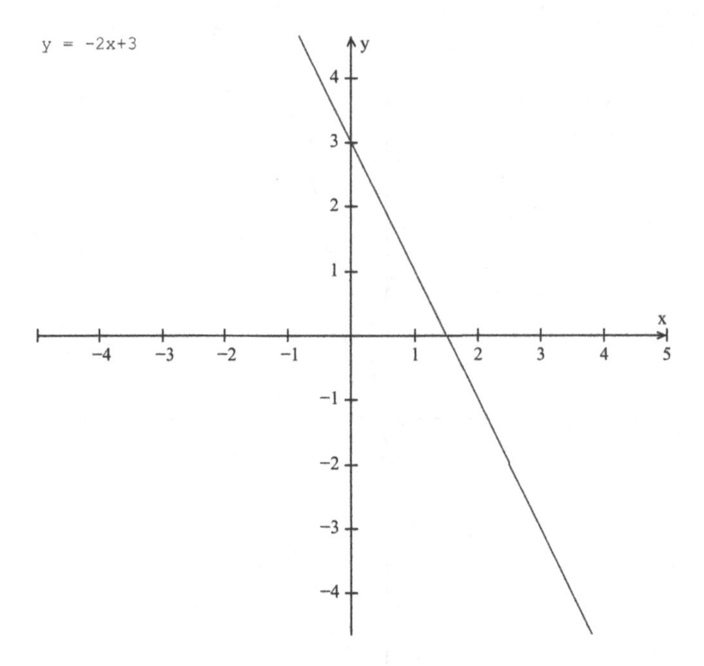

26.

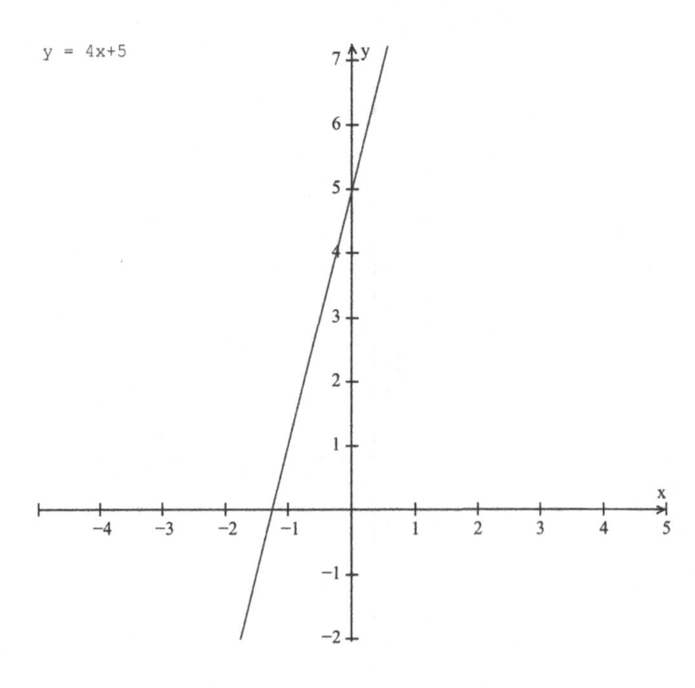

27.

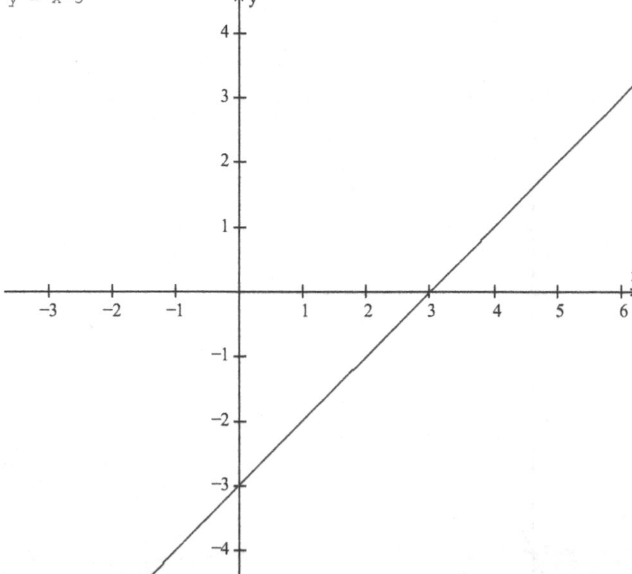

28.

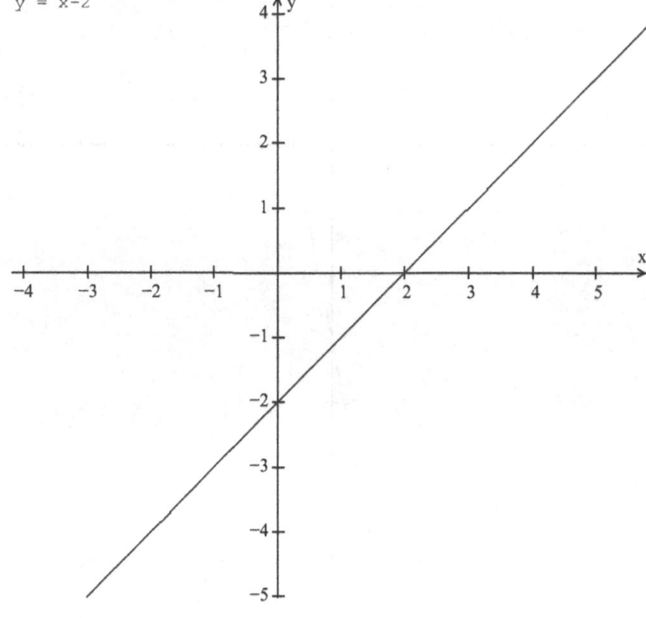

29.

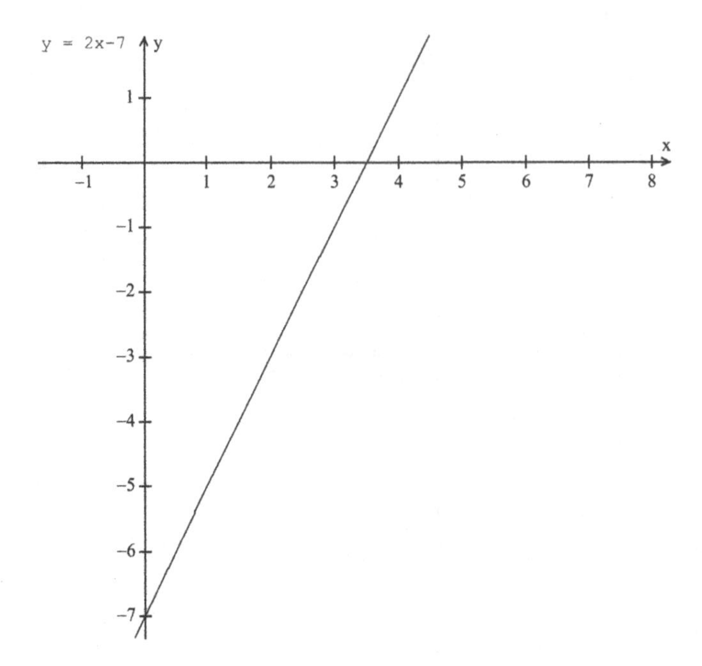

30.

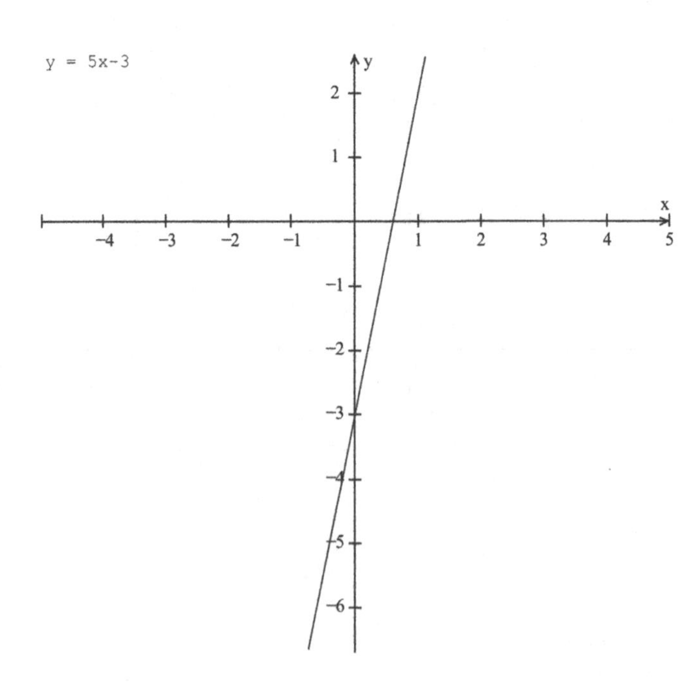

31.

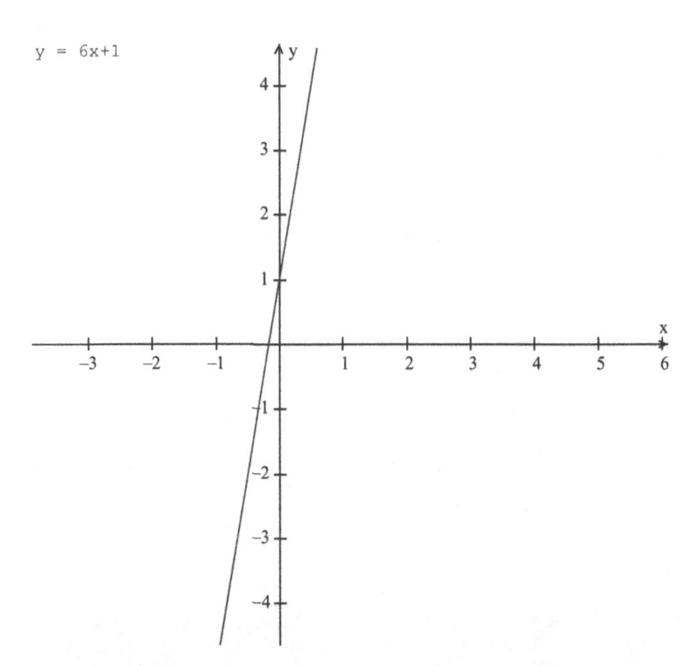

32.

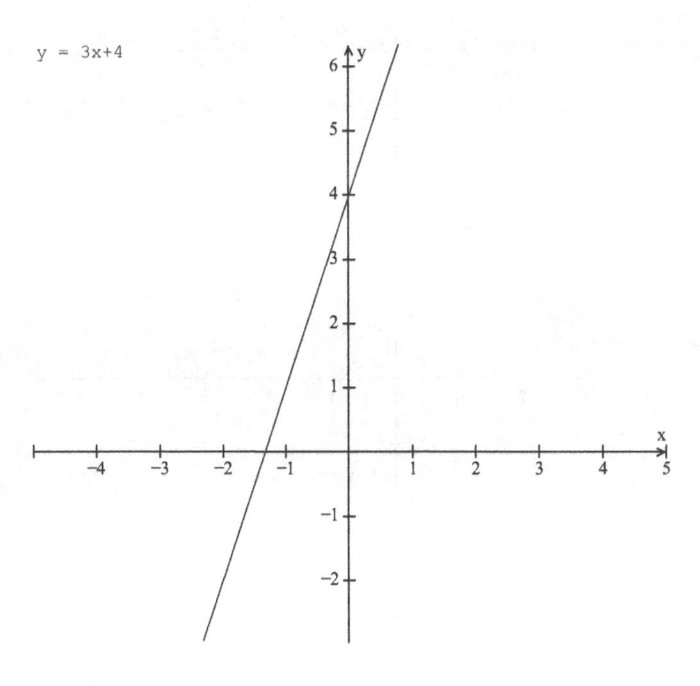

33.

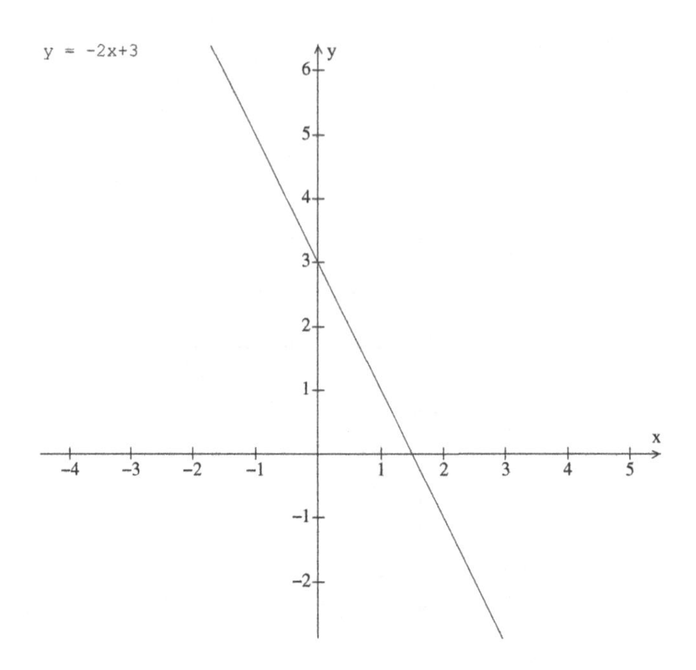

34.

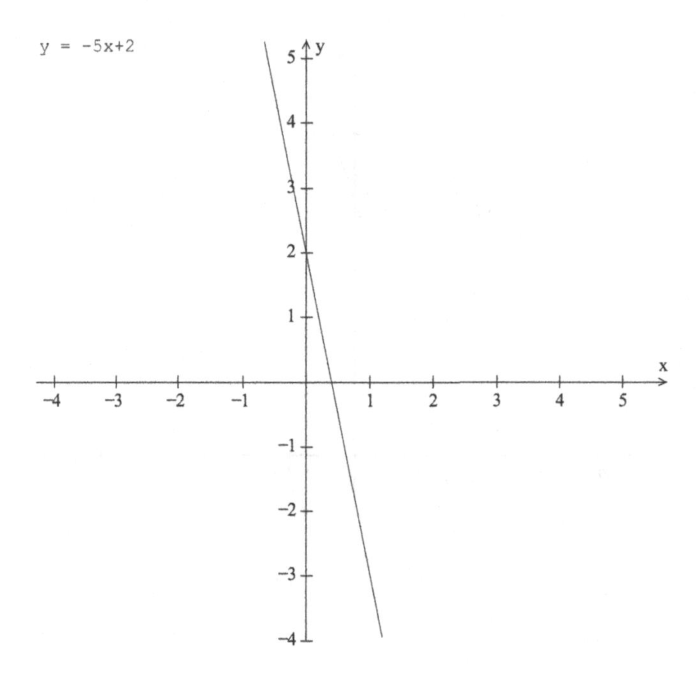

35. x-intercept $= 1$
y-intercept $= -5$
36. x-intercept $= 2/3$
y-intercept $= -2$
37. x-intercept $= 6$
y-intercept $= -12$
38. x-intercept $= 3$
y-intercept $= -9$
39. x-intercept $= -3/2$
y-intercept $= 3$
40. x-intercept $= -3/2$
y-intercept $= 3$
41. $y = 2(x + 3) + 2 \rightarrow y = 2x + 8$

42. $y = 3(x - 0) + 5 \rightarrow y = 3x + 5$

43. $y = \frac{-3}{4}(x - 2) + 1 \rightarrow y = \frac{-3}{4}x + \frac{3}{2} + 1 \rightarrow y = \frac{-3}{4}x + \frac{5}{2}$

44. $y = \frac{1}{2}(x + 3) + 2 \rightarrow y = \frac{1}{2}x + \frac{3}{2} + 2 \rightarrow y = \frac{1}{2}x + \frac{7}{2}$

45. $y = mx - 1 \rightarrow mx = 1 \rightarrow 3 = \frac{1}{m} \rightarrow m = \frac{1}{3}$
$y = \frac{1}{3}x - 1$

46. $y = mx + 6 \rightarrow mx = -6 \rightarrow 4 = \frac{-6}{m} \rightarrow m = \frac{-3}{2}$
$y = \frac{-3}{2}x + 6$

47. $y = mx + 5 \rightarrow mx = -5 \rightarrow -5 = \frac{-5}{m} \rightarrow m = 1$
$y = x + 5$

48. $y = mx - 4 \rightarrow mx = 4 \rightarrow -3 = \frac{4}{m} \rightarrow m = \frac{-4}{3}$
$y = \frac{-4}{2}x - 4$
49. This is a vertical line
$x = 3$
No slope intercept form.

50. This is a vertical line
$x = 1$
No slope intercept form.

51. This is a vertical line
$x = 2$
No slope intercept form.

52. This is a vertical line
$x = 4$
No slope intercept form.

53. $L_1 \| L_2 \leftrightarrow m_1 = m_2$
$y = 3(x + 2) + 3 \rightarrow y = 3x + 9$

54. $L_1 \| L_2 \leftrightarrow m_1 = m_2$
$y = 5(x + 1) + 1 \rightarrow y = 5x + 6$

55. $L_1 \| L_2 \leftrightarrow m_1 = m_2$
$y = -4(x + 1) + 1 \rightarrow y = -4x - 3$

56. $L_1 \| L_2 \leftrightarrow m_1 = m_2$
$y = -5(x - 2) - 1 \rightarrow y = -5x + 9$

57. $L_1 \| L_2 \leftrightarrow m_1 = m_2$
$y = -2(x + 3) - 1 \rightarrow y = -2x - 7$

58. $L_1 \| L_2 \leftrightarrow m_1 = m_2$
$y = 3(x - 0) + 0 \rightarrow y = 3x$

59. $L_1 \| L_2 \leftrightarrow m_1 = m_2$
$y = 2(x - 0) + 0 \rightarrow y = 2x$

60. $L_1 \| L_2 \leftrightarrow m_1 = m_2$
This is a vertical line
$x = 6$
No slope intercept form.

61. $L_1 \| L_2 \leftrightarrow m_1 = m_2$
This is a horizontal line. Slope = 0,
$y = 0x + 4$
$y = 0(x - 6) + 3 \rightarrow y = 3$

62. $L_1 \perp L_2 \leftrightarrow m_1 m_2 = -1$
$$\rightarrow m_2 = \frac{-1}{2}$$
$$y = \frac{-1}{2}(x + 1) + 2 \rightarrow y = \frac{-1}{2}x + \frac{3}{2}$$

63. $L_1 \perp L_2 \leftrightarrow m_1 m_2 = -1$
$$\rightarrow m_2 = \frac{-1}{3}$$
$$y = \frac{-1}{3}(x - 2) + 1 \rightarrow y = \frac{-1}{3}x + \frac{5}{3}$$

64. $L_1 \perp L_2 \leftrightarrow m_1 m_2 = -1$

$$\rightarrow m_2 = \frac{-1}{3}$$

$$y = \frac{-1}{3}(x+4) + 0 \rightarrow y = \frac{-1}{3}x - \frac{4}{3}$$

65. $L_1 \perp L_2 \leftrightarrow m_1 m_2 = -1$

$$\rightarrow m_2 = \frac{-1}{2}$$

$$y = \frac{-1}{2}(x-0) + 3 \rightarrow y = \frac{-1}{2}x + 3$$

66. $L_1 \perp L_2 \leftrightarrow m_1 m_2 = -1$

$$\rightarrow m_2 = \frac{-1}{6}$$

$$y = \frac{-1}{6}(x-0) + 1 \rightarrow y = \frac{-1}{6}x + 1$$

67. Slope of perpendicular $= 0$,
$y = m(x - x_1) + y_1$
$y = 0(x - 2) + 3$
$y = 3$

68. Slope of perpendicular is undefined.
$x = 4$
No slope–intercept form.

Section 4.2 (Answers)

1. The domain D of f is $(-\infty, \infty)$ and that is the range R of f^{-1}.
Note: $(-\infty, \infty) = \{x \mid x \in \mathbb{R}\}$.
To find the f^{-1}:
Replace $f(x)$ with y: $y = 2x - 1$
Interchange x and y: $x = 2y - 1$
Solve for y: $y = f^{-1} = \dfrac{x+1}{2}$
The domain D of f^{-1} is $(-\infty, \infty)$ and that is the range R of f.

2. The domain D of f is $(-\infty, \infty)$ and that is the range R of f^{-1}.
Note: $(-\infty, \infty) = \{x \mid x \in \mathbb{R}\}$.
To find the f^{-1}:
Replace $f(x)$ with y: $y = x + 4$
Interchange x and y: $x = y + 4$
Solve for y: $y = f^{-1} = x - 4$
The domain D of f^{-1} is $(-\infty, \infty)$ and that is the range R of f.

3. The domain D of f is all real numbers $\mathbb{R}$ except -3 and that is the range R of f^{-1}.
Note: The domain D of $f = \{x \mid x \in \mathbb{R}$, except when $x = -3\}$.
To find the f^{-1}:
Replace $f(x)$ with y: $y = \dfrac{1}{x+3}$
Interchange x and y: $x = \dfrac{1}{y+3}$
Solve for y: $y = f^{-1} = y = \dfrac{1-3x}{x}$
The domain D of f^{-1} is all real numbers $\mathbb{R}$ except 0 and that is the range R of f.

4. The domain D of f is all real numbers $\mathbb{R}$ except 7 and that is the range R of f^{-1}.
Note: The domain D of $f = \{x \mid x \in \mathbb{R}$, except when $x = 7\}$.
To find the f^{-1}:
Interchange x and y: $x = \dfrac{y}{y-7}$,
$$x(y-7) = y \rightarrow xy - 7 = y \rightarrow xy - y = 7$$
Then solve for y,
$$y(x-1) = 7 \rightarrow y = \frac{7}{x-1}$$
The domain D of f^{-1} is all real numbers $\mathbb{R}$ except 1 ($\{x|x \neq 1\}$) and that is the range R of f ($\{y|y \neq 1\}$).

5. The equation under the square root must be greater or equal to zero. By solving $x + 9 \geq 0$, we get $x \geq -9$.
Therefore the domain D of f is the interval $[-9, \infty)$ and that is the range R of f^{-1}.
Note: $[-9, \infty) = \{x \mid x \geq -9\}$.
To find the f^{-1}:
Replace $f(x)$ with y: $y = \sqrt{x+9}$
Interchange x and y: $x = \sqrt{y+9}$
Solve for y: $x^2 = y + 9$, $x \geq 0 \longrightarrow \quad y = x^2 - 9 \longrightarrow$
The domain D of f^{-1} is all real numbers $\mathbb{R}$ greater or equal to 0, and that is the range R of f.

6. The equation under the square root must be greater or equal to zero. By solving $x - 5 \geq 0$, we get $x \geq 5$.
Therefore the domain D of f is the interval $[5, \infty)$, and that is the range R of f^{-1}.
Note: $[5, \infty) = \{x|x \geq 5\}$.
To find the f^{-1}:
Replace $f(x)$ with y: $y = \sqrt{x-5}$
Interchange x and y: $x = \sqrt{y-5}$
Solve for y: $x^2 = y - 5$, $x \geq 0 \longrightarrow \quad y = x^2 + 5 \longrightarrow$
The domain D of f^{-1} is all real numbers $\mathbb{R}$ greater or equal to 0, and that is the range R of f.

7. The equation under the square root must be greater or equal to zero. We get $x \geq 0$.
Therefore the domain D of f is the interval $[0, \infty)$, and that is the range R of f^{-1}.
Note: $[0, \infty) = \{x \mid x \geq 0\}$.

To find the f^{-1}:
Replace $f(x)$ with y: $y = \sqrt{x} - 9$
Interchange x and y: $x = \sqrt{y} - 9$
Solve for y: $y = (x + 9)^2$, $x \geq -9$
The domain D of f^{-1} is all real numbers $\mathbb{R}$ greater or equal to -9, and that is the range R of f.

8. Domain: $\mathbb{R}$ (all real numbers), and that is the range R of f^{-1}.
Range: $y \geq 0$, and it is the domain D of f^{-1}

9. The domain D of f is all real numbers $\mathbb{R}$ except -2 and 1 and that is the range R of f^{-1}.
Note: The domain D of $f = \{x \mid x \in \mathbb{R}$, except when $x = -2$ and $x = 1\}$.
To find the f^{-1}:
Replace $f(x)$ with y: $y = \dfrac{3}{(x-1)(x+2)}$
Interchange x and y: $x = \dfrac{3}{(y-1)(y+2)}$
Solve for y: $y = f^{-1} = (y-1)(y+2) = \dfrac{3}{x}$
The domain D of f^{-1} is all real numbers $\mathbb{R}$ except 0 and that is the range R of f.

10. The domain D of f is all real numbers $\mathbb{R}$ except -3 and 5, and that is the range R of f^{-1}.
Note: The domain D of $f = \{x \mid x \in \mathbb{R}$, except when $x = -3$ and $x = 5\}$.
To find the f^{-1}:
Replace $f(x)$ with y: $y = \dfrac{5}{(x-5)(x+3)}$
Interchange x and y: $x = \dfrac{5}{(y-5)(y+3)}$
Solve for y: $y = f^{-1} = (y-5)(y+3) = \dfrac{5}{x}$
The domain D of f^{-1} is all real numbers $\mathbb{R}$ except 0, and that is the range R of f.

11. It is a function by the vertical line test, domain is all real numbers $\mathbb{R}$, range is $y \geq 0$.
12. Not a function, because the vertical line test crosses the graph twice.
13. A function,
Domain = {2, 3, 5}
Range = {2, 4, 8}

14. Not a function.

15. A function,
Domain = {1, 2, 4}
Range = {3, 11}

16. A function,
Domain = {1, 2, 3}
Range = {7, 8}

17. Is not one to one

18. Is one to one

19. Domain D of $f(x) = (-\infty, \infty)$
Range R of $f(x) = (-\infty, \infty)$

20. Domain D of $f(x) = (-\infty, \infty)$
Range R of $f(x) = (-\infty, \infty)$

21. To find the f^{-1}:
Replace $f(x)$ with y: $y = \sqrt{x-4}$
Interchange x and y: $x = \sqrt{y-4}$
Solve for y: $y = f^{-1} = x^2 + 4$

22. To find the f^{-1}:
Replace $f(x)$ with y: $y = \sqrt{x+3}$
Interchange x and y: $x = \sqrt{y+3}$
Solve for y: $y = f^{-1} = x^2 - 3$

23. To find the f^{-1}:
Replace $f(x)$ with y: $y = 5x$
Interchange x and y: $x = 5y$
Solve for y: $y = f^{-1} = \frac{x}{5}$

24. To find the f^{-1}:
Replace $f(x)$ with y: $y = x^5$
Interchange x and y: $x = y^5$
Solve for y: $y = f(x) = (x)^{\frac{1}{5}}$

25. To find the f^{-1}:
Replace $f(x)$ with y: $y = \dfrac{4x}{6x-1}$
Interchange x and y: $x = \dfrac{4y}{6y-1}$
Solve for y:
$y = f^{-1} \rightarrow x(6y-1) = 4y \rightarrow 6xy - x = 4y \rightarrow$
$6xy - 4y = x \rightarrow y = \dfrac{x}{(6x-4)}$

26. To find the f^{-1}:
Replace $f(x)$ with y: $y = \dfrac{x^2-9}{x+3}$
Interchange x and y: $x = \dfrac{y^2-9}{y+3}$
Solve for y:
$y = f^{-1} \rightarrow x = \dfrac{y^2-9}{y+3} \rightarrow x = \dfrac{(y-3)(y+3)}{(y+3)}$
$y = x + y$

Review Exercises (Chapter 4—Answers)

1.

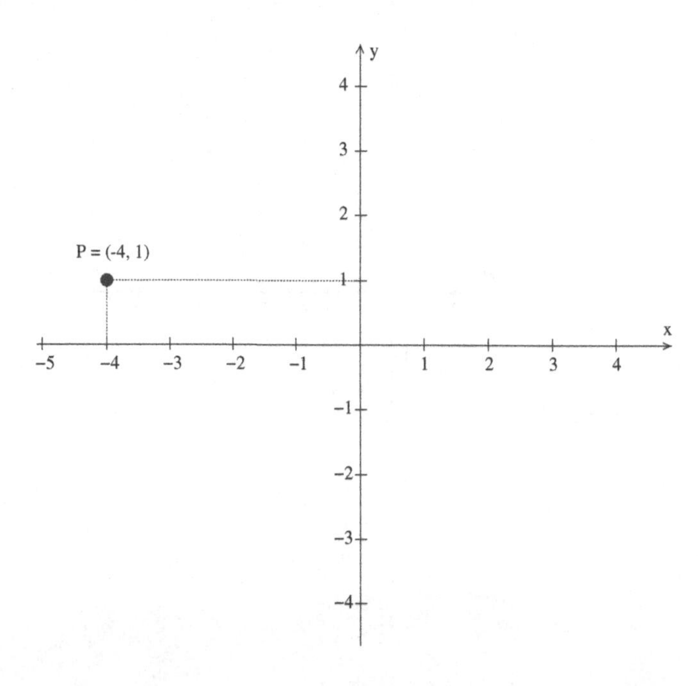

2.

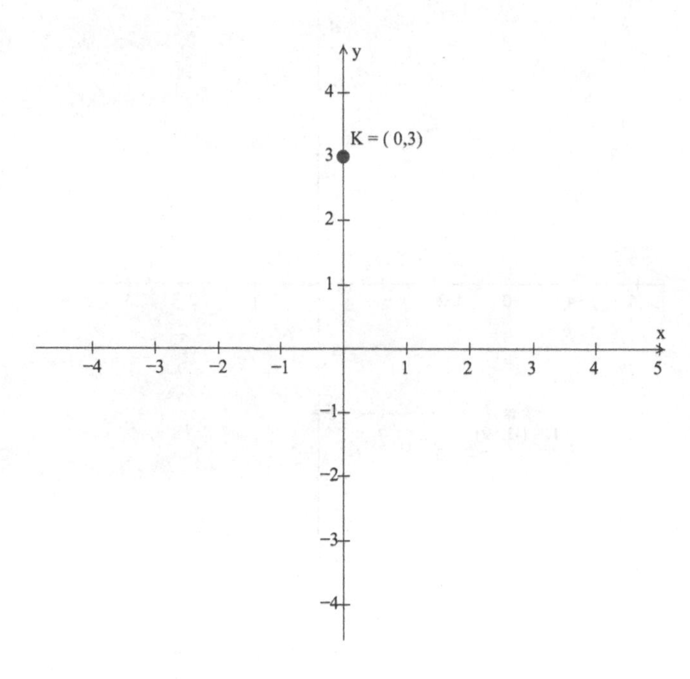

3.

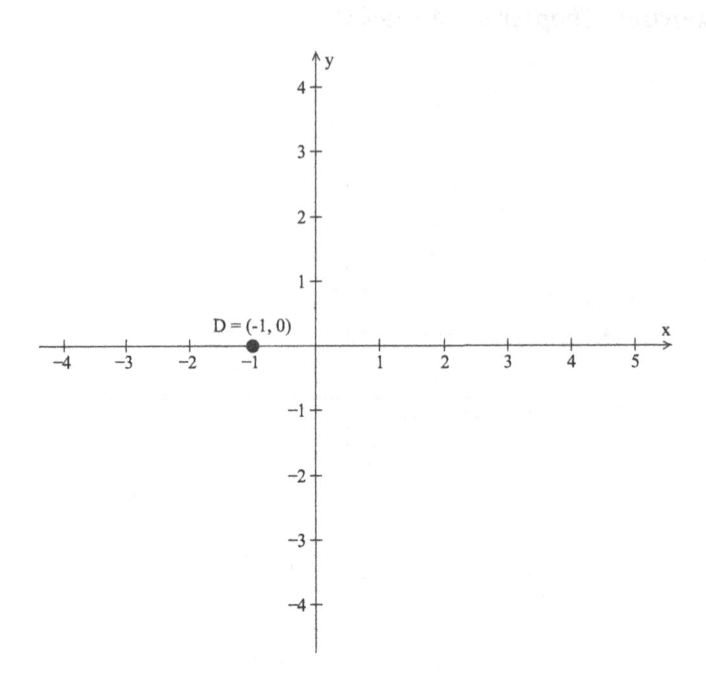

4.

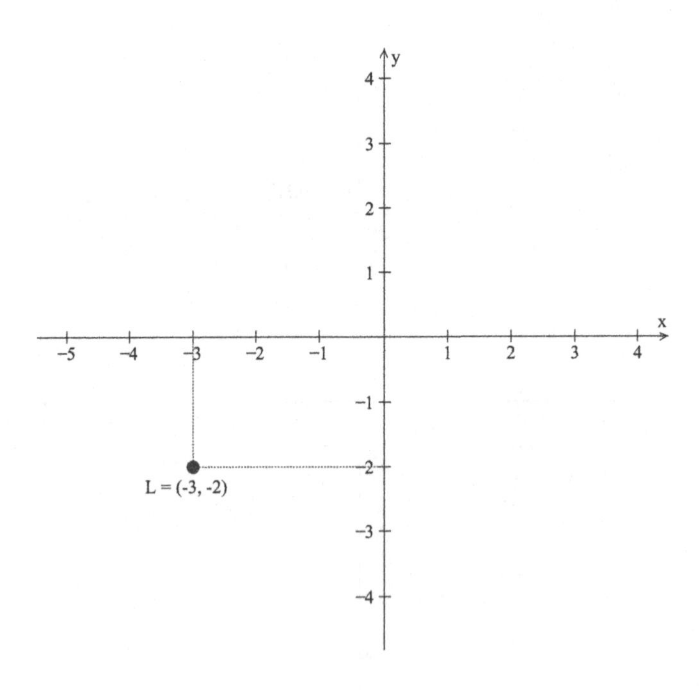

5. $d = \sqrt{(7-3)^2 + (2+5)^2} = \sqrt{16+49} = \sqrt{65}$

$$\mathrm{M} = (x_\mathrm{m}, y_\mathrm{m}) = \left(\frac{3+7}{2}, \frac{-5+2}{2}\right) = \left(5, \frac{-3}{2}\right)$$

6. $d = \sqrt{(-3-5)^2 + (-6+1)^2} = \sqrt{64+25} = \sqrt{89}$

$$\mathrm{M} = (x_\mathrm{m}, y_\mathrm{m}) = \left(\frac{5+(-3)}{2}, \frac{-1+(-6)}{2}\right) = \left(1, \frac{-7}{2}\right)$$

7. The domain D of f is $(-\infty, \infty)$ and that is the range R of f^{-1}.
Note: $(-\infty, \infty) = \{x \mid x \in \mathbb{R}\}$.
To find the f^{-1}:
Replace $f(x)$ with y: $y = 5x - 7$
Interchange x and y: $x = 5y - 7$
Solve for y: $y = f^{-1} = \dfrac{x+7}{5}$
The domain D of f^{-1} is $(-\infty, \infty)$ and that is the range R of f.

8. The domain D of f is all real numbers $\mathbb{R}$ except 9, and that is the range R of f^{-1}.
Note: The domain D of f = $\{x \mid x \in \mathbb{R}$, except when $x = 9\}$.
To find the f^{-1}:
Replace $f(x)$ with y: $y = \dfrac{5}{x-9}$
Interchange x and y: $x = \dfrac{5}{y-9}$
Solve for y: $y = f^{-1} =$

$$x = \frac{5}{y-9} \rightarrow x(y-9) = 5 \rightarrow xy - 9x = 5$$

$$y = \frac{5+9x}{x}$$

The domain D of f^{-1} is all real numbers $\mathbb{R}$ except 0, and that is the range R of f.

9. The equation under the square root must be greater or equal to zero. By solving $x + 4 \geq 0$, we get $x \geq -4$.
Therefore the domain D of f is the interval $[-4, \infty)$, and that is the range R of f^{-1}.
Note: $[-4, \infty) = \{x \mid x \geq -4\}$.
To find the f^{-1}:
Replace $f(x)$ with y: $y = \sqrt{x+4}$
Interchange x and y: $x = \sqrt{y+4}$
Solve for y: $x^2 = y + 4,\ x \geq 0 \longrightarrow y = x^2 - 4 \longrightarrow$
The domain D of f^{-1} is all real numbers $\mathbb{R}$ greater or equal to 0, and that is the range R of f.

10. The equation under the square root must be greater or equal to zero. By solving $x - 3 \geq 0$, we get $x \geq 3$.
Therefore the domain D of f is the interval $[3, \infty)$, and that is the range R of f^{-1}.

Note: $[3, \infty) = \{x|x \geq 3\}$.
To find the f^{-1}:
Replace $f(x)$ with $y : y = \sqrt{x-3}$
Interchange x and $y : x = \sqrt{y-3}$
Solve for y: $x^2 = y - 3,\ x \geq 0 \longrightarrow y = x^2 + 3 \longrightarrow$
The domain D of f^{-1} is all real numbers $\mathbb{R}$ greater or equal to 0, and that is the range R of f.

11. A function
Domain D of $f(x) = (-\infty, \infty)$
Range R of $f(x) = (-\infty, \infty)$

12. Not a function.

13. $A = \{(2,8),(1,12),(5,14)\}$
14. $B = \{(3,2),(3,4),(6,9)\}$
15. $C = \{(1,2),(3,2),(4,5)\}$
16. $D = \{(1,9),(3,5),(4,5)\}$

17. Domain D of $f(x) = [0, \infty)$
Range R of $f(x) = [0, \infty)$

18. Domain D of $f(x) = [-1,1]$
Range R of $f(x) = [-3,3]$

19. The equation under the square root must be greater or equal to zero. By solving $x - 2 \geq 0$, we get $x \geq 2$.
Therefore the domain D of f is the interval $[2, \infty)$, and that is the range R of f^{-1}.
Note: $[2, \infty) = \{x|x \geq 2\}$.
To find the f^{-1}:
Replace $f(x)$ with y: $y = \sqrt{x-2}$
Interchange x and y: $x = \sqrt{y-2}$
Solve for y: $x^2 = y - 2,\ x \geq 0 \longrightarrow y = x^2 + 2 \longrightarrow$
The domain D of f^{-1} is all real numbers $\mathbb{R}$ greater or equal to 0, and that is the range R of f.

20. The equation under the square root must be greater or equal to zero. By solving $x + 1 \geq 0$, we get $x \geq -1$.
Therefore the domain D of f is the interval $[-1, \infty)$, and that is the range R of f^{-1}.
Note: $[-1, \infty) = \{x \mid x \geq -1\}$.
To find the f^{-1}:
Replace $f(x)$ with y: $y = \sqrt{x+1}$
Interchange x and y: $x = \sqrt{y+1}$
Solve for y: $x^2 = y + 1,\ x \geq 0 \longrightarrow y = x^2 - 1 \longrightarrow$
The domain D of f^{-1} is all real numbers $\mathbb{R}$ greater or equal to 0, and that is the range R of f.

21. $y = 3x \rightarrow x = 3y \rightarrow y = f^{-1} = \frac{x}{3}$

22. $y = x^3 \rightarrow x = y^3 \rightarrow y = f^{-1} = x^{\frac{1}{3}}$

23. $y = \dfrac{2x}{3x-1} \rightarrow x = \dfrac{2y}{3y-1} \rightarrow x(3y-1) = 2y \rightarrow 3xy - x = 2y$

$3xy - 2y = x \rightarrow y = f^{-1} = \dfrac{x}{(3x-2)}$

24. $y = \dfrac{x^2-4}{x+2} \rightarrow x = \dfrac{y^2-4}{y+2} \rightarrow x = \dfrac{(y-2)(y+2)}{(y+2)}$

$x = y - 2 \rightarrow y = f^{-1} = x + 2$

CHAPTER 5

Measurement

5.1 Exercises (Answers)

1. $84\,\cancel{\text{in.}} \times \dfrac{1\text{ ft}}{12\,\cancel{\text{in.}}} = 7\text{ ft}$

2. $111\,\cancel{\text{in.}} \times \dfrac{1\text{ ft}}{12\,\cancel{\text{in.}}} = 9.25\text{ ft}$

3. $2.5\text{ ft} \times \dfrac{12\text{ in.}}{1\text{ ft}} = 30\text{ in.}$

4. $5.5\text{ ft} \times \dfrac{12\text{ in.}}{1\text{ ft}} = 66\text{ in.}$

5. $0.6\text{ yd} \times \dfrac{36\text{ in.}}{1\text{ yd}} = 21.6\text{ in.}$

6. $\left(\dfrac{13}{4}\right)\text{yd} \times \dfrac{36\text{ in.}}{1\text{ yd}} = 117\text{ in.}$

7. $\left(\dfrac{1}{9}\right)\text{ yd} \times \dfrac{36\text{ in.}}{1\text{ yd}} = 4\text{ in.}$

8. $0.40\text{ yd} \times \dfrac{36\text{ in.}}{1\text{ yd}} = 14.40\text{ in.}$

9. $4\text{ yd} \times \dfrac{3\text{ ft}}{1\text{ yd}} = 12\text{ ft}$

10. $7.2\text{ yd} \times \dfrac{3\text{ ft}}{1\text{ yd}} = 21.6\text{ ft}$

11. $33\text{ ft} \times \frac{1\text{ yd}}{3\text{ ft}} = 11\text{ yd}$

12. $54\text{ ft} \times \frac{1\text{ yd}}{3\text{ ft}} = 18\text{ yd}$

13. $27\text{ cm} \times \frac{10\text{ mm}}{1\text{ cm}} = 270\text{ mm}$

14. $12.64\text{ cm} \times \frac{10\text{ mm}}{1\text{ cm}} = 126.4\text{ mm}$

15. $223.85\text{ mm} \times \frac{1\text{ cm}}{10\text{ mm}} = 22.39\text{ cm}$

16. $83.94\text{ mm} \times \frac{1\text{ cm}}{10\text{ mm}} = 8.39\text{ cm}$

17. $0.93\text{ m} \times \frac{100\text{ cm}}{1\text{ m}} = 93\text{ cm}$

18. $0.28\text{ m} \times \frac{100\text{ cm}}{1\text{ m}} = 28\text{ cm}$

19. $145\text{ mm} \times \frac{1\text{ m}}{1000\text{ mm}} = 0.15\text{ m}$

20. $768\text{ mm} \times \frac{1\text{ m}}{1000\text{ mm}} = 0.77\text{ m}$

21. $0.83\text{ mm} \times \frac{1\text{ cm}}{10\text{ mm}} = 0.083\text{ cm}$

22. $0.0064\text{ m} \times \frac{1000\text{ mm}}{1\text{ m}} = 6.4\text{ m}$

23. 23.54 mm + 6.4 cm = 23.54 mm + 64 mm = 87.54 mm

24. 4.3 m + 87 cm = 4.3 m + 0.87 m = 5.17 m

25. 39.4 cm − 53.6 mm = 39.4 cm − 5.36 cm = 34.04 cm

26. 162.6 mm − 14.5 cm = 162.6 mm − 145 mm = 17.6 mm

27. 1.08 m − 47.3 cm = 108 cm − 47.3 cm = 60.7 cm

28. 0.68 m − 432.4 mm = 680 mm − 432.4 mm = 247.6 mm

29. 326 mm + 78.3 cm + 0.8 m = 0.326 m + 0.783 m + 0.8 m = 1.909 m

30. 0.046 m + 8.74 cm + 75.2 mm = 46 mm + 87.4 mm + 75.2 mm = 208.6 mm

31. 68.2 mm + 7.03 cm + 308.7 mm = 68.2 mm + 70.3 mm + 308.7 mm = 447.2 mm

32. 2.736 m − 624 mm = 2.736 m − 0.624 m = 2.112 m

33. $43.00\text{ mm} \times \dfrac{0.0394\text{ in.}}{1\text{ mm}} = 1.69\text{ in.}$

34. $118.40\text{ mm} \times \dfrac{0.0394\text{ in.}}{1\text{ mm}} = 4.67\text{ in.}$

35. $13.60\text{ cm} \times \dfrac{0.394\text{ in.}}{1\text{ cm}} = 5.36\text{ in.}$

36. $0.62\text{ cm} \times \dfrac{0.394\text{ in.}}{1\text{ cm}} = 0.24\text{ in.}$

37. $4.60\text{ m} \times \dfrac{39.40\text{ in.}}{1\text{ m}} = 181.24\text{ in.}$

38. $0.07\text{ m} \times \dfrac{39.40\text{ in.}}{1\text{ m}} = 2.76\text{ in.}$

39. $6.00\text{ m} \times \dfrac{3.28\text{ ft}}{1\text{ m}} = 19.68\text{ ft}$

40. $12.4\text{ m} \times \dfrac{3.28\text{ ft}}{1\text{ m}} = 40.67\text{ ft}$

41. $954.00\text{ mm} \times \dfrac{0.0394\text{ in.}}{1\text{ mm}} = 37.59\text{ in.}$

42. $76.03\text{ mm} \times \dfrac{0.0394\text{ in.}}{1\text{ mm}} = 3.00\text{ in.}$

43. $6.00\text{ m} \times \dfrac{1.09\text{ yd}}{1\text{ m}} = 6.54\text{ yd}$

44. $4.00\text{ m} \times \dfrac{1.09\text{ yd}}{1\text{ m}} = 4.36\text{ yd}$

45. $85.00\text{ cm} \times \dfrac{0.0328\text{ ft}}{1\text{ cm}} = 2.79\text{ ft}$

46. $340\text{ mm} \times \dfrac{0.00328\text{ ft}}{1\text{ mm}} = 1.12\text{ ft}$

47. $7.00\text{ in.} \times \dfrac{25.4\text{ mm}}{1\text{ in.}} = 177.8\text{ mm}$

48. $0.64\text{ in.} \times \dfrac{25.4\text{ mm}}{1\text{ in.}} = 16.26\text{ mm}$

49. $76.00\text{ in.} \times \dfrac{25.4\text{ mm}}{1\text{ in.}} = 1930.4\text{ mm}$

50. $40.65\text{ in.} \times \dfrac{2.54\text{ cm}}{1\text{ in.}} = 103.25\text{ cm}$

51. $8.00\ \text{ft} \times \dfrac{1\ \text{m}}{3.28\ \text{ft}} = 2.44\ \text{m}$

52. $0.65\ \text{ft} \times \dfrac{1\ \text{m}}{3.28\ \text{ft}} = 0.20\ \text{m}$

53. $6.50\ \text{yd} \times \dfrac{1\ \text{m}}{1.09\ \text{yd}} = 5.96\ \text{m}$

54. $2.70\ \text{yd} \times \dfrac{1\ \text{m}}{1.09\ \text{yd}} = 2.50\ \text{m}$

55. $4.58\ \text{in.} \times \dfrac{25.4\ \text{mm}}{1\ \text{in.}} = 116.33\ \text{mm}$

56. $0.54\ \text{in.} \times \dfrac{2.54\ \text{cm}}{1\ \text{in.}} = 1.37\ \text{cm}$

57. $428.00\ \text{in.} \times \dfrac{2.54\ \text{cm}}{1\ \text{in.}} = 1087.12\ \text{cm}$

58. $0.25\ \text{in.} \times \dfrac{25.4\ \text{mm}}{1\ \text{in.}} = 6.35\ \text{mm}$

59. $6.5\ \text{in.} \times \dfrac{2.54\ \text{cm}}{1\ \text{in.}} = 16.51\ \text{cm}$

60. $46.71\ \text{in.} \times \dfrac{0.0254\ \text{m}}{1\ \text{in.}} = 1.19\ \text{m}$

61. Perimeter = 25 m + 30 m + 32 m + 27 m = 114 m

62. Perimeter = 60 cm + 50 cm + (160 cm − 50 cm) + 20 cm + 160 cm = 400 cm

63. 74 mm − (3 × 10 mm) = 44 mm

64. 2240 mm − (30 × 37 mm) = 1130 mm

65. Total length = 125 mm + 40 cm + 2.5 dm = 125 mm + 400 mm + 250 mm = 775 mm

66. Total length = 60 mm + 58 mm + 80 mm = 198 mm

5.2 Exercises (Answers)

1. $185\ \text{lb} \times \dfrac{0.454\ \text{kg}}{1\ \text{lb}} = 83.99\ \text{kg}$

2. $14.5\ \text{oz} \times \dfrac{28.35\ \text{g}}{1\ \text{oz}} = 411.08\ \text{g}$

3. $600\ \text{g} \times \dfrac{1\ \text{oz}}{28.35\ \text{g}} = 21.16\ \text{oz}$

4. $70.6\ \text{kg} \times \dfrac{2.205\ \text{lb}}{1\ \text{kg}} = 155.67\ \text{lb}$

5. $645\ \text{g} \times \frac{1\ \text{kg}}{1000\ \text{g}} = 0.65\ \text{kg}$

6. $148\ \text{mg} \times \frac{10^{-3}\ \text{g}}{1\ \text{mg}} = 0.15\ \text{g}$

7. $75\ \text{g} \times \frac{1\ \text{mg}}{10^{-3}\ \text{g}} = 75{,}000\ \text{mg}$

8. $320\ \text{g} \times \frac{1\ \text{mg}}{10^{-3}\ \text{g}} = 320{,}000\ \text{mg}$

9. $3.5\ \text{kg} \times \frac{1000\ \text{g}}{1\ \text{kg}} = 3500\ \text{g}$

10. $5.4\ \text{kg} \times \frac{1000\ \text{g}}{1\ \text{kg}} = 5400\ \text{g}$

11. $645\ \mu\text{g} \times \frac{10^{-3}\ \text{mg}}{1\ \mu\text{g}} = 0.65\ \text{mg}$

12. $235\ \mu\text{g} \times \frac{10^{-3}\ \text{mg}}{1\ \mu\text{g}} = 0.24\ \text{mg}$

13. $16\ \text{mg} \times \frac{1\ \mu\text{g}}{10^{-3}\ \text{mg}} = 16{,}000\ \mu\text{g}$

14. $7.8\ \text{mg} \times \frac{1\ \mu\text{g}}{10^{-3}\ \text{mg}} = 7800\ \mu\text{g}$

15. $4.5\ \text{tons} \times \frac{907.19\ \text{kg}}{1\ \text{ton}} = 4082.40\ \text{kg}$

16. $30\ \text{tons} \times \frac{907.19\ \text{kg}}{1\ \text{ton}} = 27{,}215.7\ \text{kg}$

17. $12{,}000\ \text{kg} \times \frac{1\ \text{ton}}{907.19\ \text{kg}} = 13.23\ \text{ton}$

18. $335{,}000\ \text{kg} \times \frac{1\ \text{ton}}{907.19\ \text{kg}} = 369.27\ \text{ton}$

19. 1 lb is heavier than 1 oz

20. 1 kg is heavier than 2 g

21. 4 Gg is heavier than 2 kg

22. 1 oz is heavier than 25 g

23. 1 ton is heavier than 200 lb

24. 2 lb is heavier than 12 oz

25. 5 lb = 5 × 16 oz = 80 oz

26. 5 oz × (1 g/0.035 oz) = 142.86 oz

27. 60 kg × (2.205 lb/1 kg) = 132.3 lb

28. 25 kg × (2.205 lb/1 kg) = 55.13 lb

29. 12 oz × (1 g/0.035 oz) = 342.89 g
342.89 g/50 = 6.86 g

30. 140 lb × (0.454 kg/1 lb) = 63.56 kg
63.56 kg/8 = 7.95 kg

31. 8.35 kg × (1 lb/0.454 kg) = 18.39 lb

32. 500 kg × (1 lb/0.454 kg) = 1101.32 lb

33. $2\text{ g} \times \dfrac{1\text{ lb}}{454\text{ g}} = 4.41 \times 10^{-3}\text{ lb}$

34. $7\text{ g} \times \dfrac{1\text{ lb}}{454\text{ g}} = 0.0154\text{ lb}$

35. $5400\text{ oz} \times \dfrac{0.00751\text{ gal}}{1\text{ oz}} = 40.55\text{ gal}$

36. 5400 lb + 400 lb + 3500 lb + 2000 lb = 11,300 lb
11,300 lb × (1 ton/2000 lb) = 5.65 ton

5.3 Exercises (Answers)

1. 47 L × (1.06 qt/1 L) = 49.82 qt
2. 18.6 gal × (1 L/0.264 gal) = 70.46 L
3. 14 lb × (16 oz/1 lb) = 224 oz
4. 6 lb × (16 oz/1 lb) = 96 oz
5. 30 lb × (16 oz/1 lb) = 480 oz
6. 140 oz × (1 lb/16 oz) = 8.75 lb
7. 16 ton × (2000 lb/1 ton) = 32,000 lb
8. 700 oz × (1 lb/16 oz) = 43.75 lb
9. 15,600 lb × (1 ton/2000 lb) = 7.8 ton
10. 45 ton × (2000 lb/1 ton) = 90,000 lb

11. 384,000 oz × (0.00002835 ton/1 oz) = 10.87 ton
12. 2000 lb × (1 ton/2000 lb) = 1 ton
13. 3 ton × (1 oz/0.00002835 ton) = 105820.11 oz
14. 5 ton × (1 oz/0.00002835 ton) = 176366.84 oz
15. 1200 mL × (0.001 1 L/1 mL) = 1.32 L
16. 90 mL × (0.001 1 L/1 mL) = 0.10 L
17. 0.80 L × (1 mL/0.001 1 L) = 727.27 mL
18. 7 L × (1 mL/0.001 1 L) = 6363.64 mL
19. 2 L × (1 qt/0.946 L) = 2.11 qt
20. 1 L = 1000 mL
21. 9 qt × (1 L/1.06 qt) = 8.49 L
22. 5 qt × (1 L/1.06 qt) = 4.72 L
23. 13.5 gal = (13.5) × (3.785)L = 51.1 L
24. 125 gal = (125) × (3.785)L = 473.13 L

5.4 Exercises (Answers)

1. 6 h × (60 min/1 h) × (60 s/1 min) = 21600 s
2. 2.5 h × (60 min/1 h) × (60 s/1 min) = 9000 s
3. 14 weeks × (7 days/1 week) = 98 days
4. 17 weeks × (7 days/1 week) = 119 days
5. 20 years × (0.1 decade/1 year) = 2 decades
6. 12 years × (0.1 decade/1 year) = 1.2 decades
7. 7 h × (60 min/1 h) = 420 min
8. 5 h × (60 min/1 h) = 300 min
9. 325 weeks × (1 month/4 weeks) × (1 year/12 months) = 6.77 years
10. 120 weeks × (1 month/4 weeks) × (1 year/12 months) = 2.5 years
11. 2 years × (12 months/1 year) × (4 weeks/1 month) × (7 days/1 week) = 672 days
12. 5 years × (12 months/1 year) × (4 weeks/1 month) × (7 days/1 week) = 1680 days

13. 33 months × (1 quarter/3 months) = 11 quarters

14. 17 months × (1 quarter/3 months) = 5.67 quarters

15. 16 years × (12 months/1 year) = 192 months

16. 9 years × (12 months/1 year) = 108 months

17. 10 quarters × (3 months/1 quarter) × (1 year/12 months) = 2.5 years

18. 6 quarters × (3 months/1 quarter) × (1 year/12 months) = 1.5 years

19. 8100 decades × (1 century/10 decades) = 810 centuries

20. 6400 decades × (1 century/10 decades) = 640 centuries

21. 4200 s × (1 min/60 s) = 70 min

22. 1600 s × (1 min/60 s) = 26.67 min

23. 1 day × (24 h/1 day) × (60 min/1 h) × (60 s/1 min) = 86,400 s

24. 2 day × (24 h/1 day) × (60 min/1 h) × (60 s/1 min) = 172,800 s

25. 1 week × (7 days/1 week) × (24 h/1 day) × (60 min/1 h) × (60 s/1 min) = 604,800 s

26. 0.5 century × (100 years/1 century) × (12 months/1 year) × (4 weeks/1 month) × (7 days/1 week) × (24 h/1 day) = 403,200 h

27. 0.25 century × (100 years/1 century) = 25 years

28. (8 h/day) × (1 days/24 h) × (7 days/1 week) = 2.33 days

29. 3 h × (4/month) × (12 months/year) × (1 day/24 h) = 6 days

30. 6 h × (4/month) × (12 months/year) × (1 day/24 h) = 12 days

5.5 Exercises (Answers)

1. $F = \frac{9}{5}C + 32$

2. $C = \frac{5}{9}(F - 32)$

3. $F = \frac{9}{5}(32^\circ) + 32 = 89.6\ ^\circ F$

4. $F = \frac{9}{5}(41^\circ) + 32 = 105.8\ ^\circ F$

5. $F = \frac{9}{5}(23^\circ) + 32 = 73.4\ ^\circ F$

6. $F = \frac{9}{5}(9°) + 32 = 48.2\ °F$

7. $F = \frac{9}{5}(1°) + 32 = 33.8\ °F$

8. $F = \frac{9}{5}(12°) + 32 = 53.6\ °F$

9. $C = \frac{5}{9}(32° - 32) = 0\ °C$

10. $C = \frac{5}{9}(41° - 32) = 5\ °C$

11. $C = \frac{5}{9}(23° - 32) = -5\ °C$

12. $C = \frac{5}{9}(9° - 32) = -12.78\ °C$

13. $C = \frac{5}{9}(1° - 32) = -17.22\ °C$

14. $C = \frac{5}{9}(98° - 32) = 36.67\ °C$

15. $C = \frac{5}{9}(87° - 32) = 30.56\ °C$

16. $C = \frac{5}{9}(65° - 32) = 18.33\ °C$

17. $C = \frac{5}{9}(80° - 32) = 26.67\ °C$

18. $C = \frac{5}{9}(66° - 32) = 18.89\ °C$

19. $F = \frac{9}{5}(110°) + 32 = 230\ °F$

20. $F = \frac{9}{5}(28°) + 32 = 82.4\ °F$

5.6 Exercises (Answers)

1. mm, $10^{-3} \times m$
2. μm, 10^{-6} m
3. MW, $10^{6} \times W$
4. kW, $10^{3} \times W$

5. $900\text{ g} \times \frac{1\text{ N}}{100\text{ g}} = 9\text{ N}$

6. $4\text{ kg} \times \frac{1\text{ N}}{100\text{ g}} = 40\text{ N}$

7. $I = V/R = 18\text{ V}/6\text{ k}\Omega = 3\text{ mA}$

8. $I = V/R = 12\text{ V}/3\text{ k}\Omega = 4\text{ mA}$

9. $R = V/I = 120/3 = 40\ \Omega$

10. $G = I/V = 5\text{ mA}/120\text{ V} = 0.0417\text{ mS}$

11. $\text{Power (W)} = \frac{\text{Energy (J)}}{\text{Time (s)}} = \frac{15\text{ J}}{6\text{ s}} = 2.5\text{ W.}$

12. $E/It = \text{V} \rightarrow \text{V} = \frac{\left(\frac{20 \times 10^3}{1\text{ mi}} \times \frac{1\text{ min}}{60\text{ s}}\right)}{6\text{ A}} = 55.56\text{ V}$

13. Volume = (30 cm) (14 cm) (20 cm) = 8400 cm^3
Mass of the block = (3 g/cm^3) (8400 cm^3) = 25.2 kg

14. Volume = (25 cm) (12 cm) (10 cm) = 3000 cm^3
Mass of the block = (3 g/cm^3) (3000 cm^3) = 9 kg

15. Area of square = $(15)^2 = 225\text{ cm}^2$
$1\text{ m}^2 = 10\text{ kcm}^2$

$$\text{Pressure} = \frac{12\text{ N}}{225} = \frac{12\text{ N}}{225\text{ cm}^2} \times \frac{10\text{ kcm}^2}{1\text{ m}^2} = 533.33\text{ Pa}$$

16. Area of rectangular plate = $20 \times 200 = 4000\text{ cm}^2$
$1\text{ m}^2 = 10\text{ kcm}^2$

$$\text{Pressure} = \frac{15\text{ N}}{4000} = \frac{15\text{ N}}{4000\text{ cm}^2} \times \frac{10\text{ kcm}^2}{1\text{ m}^2} = 37.5\text{ Pa}$$

17. $I = \frac{Q}{t} = \frac{6.5\text{ C}}{0.3\text{ s}} = 21.67\text{ A}$

18. $t = \frac{4 \times 10^{-3}}{9.4} = 4.26 \times 10^{-4}\text{ A}$

19. $V = \frac{W}{Q} = \frac{42\text{ J}}{6\text{ mC}} = 7000\text{ J/C} = 7000\text{ V}$

20. $V = \frac{W}{Q} = \frac{34\text{ J}}{3\text{ mC}} = 0.0113\text{ J/C} = 0.0113\text{ V}$

Review Exercises (Chapter 5—Answers)

1. $45\ \text{m} \times \dfrac{3.28\ \text{ft}}{1\ \text{m}} = 147.6\ \text{ft}$

2. $30\ \text{m} \times \dfrac{3.28\ \text{ft}}{1\ \text{m}} = 98.4\ \text{ft}$

3. $160\ \text{ft} \times \dfrac{0.305\ \text{m}}{1\ \text{ft}} = 48.8\ \text{m}$

4. $120\ \text{ft} \times \dfrac{0.305\ \text{m}}{1\ \text{ft}} = 36.6\ \text{m}$

5. $1.6\ \text{ft} \times \dfrac{30.5\ \text{cm}}{1\ \text{ft}} = 48.8\ \text{cm}$

6. $1.9\ \text{ft} \times \dfrac{30.5\ \text{cm}}{1\ \text{ft}} = 30.5\ \text{cm}$

7. $6.4\ \text{cm} \times \dfrac{0.394\ \text{in.}}{1\ \text{cm}} = 2.522\ \text{in.}$

8. $3.8\ \text{cm} \times \dfrac{0.394\ \text{in.}}{1\ \text{cm}} = 1.497\ \text{in.}$

9. $6.20\ \text{km} \times \dfrac{0.62\ \text{mi}}{1\ \text{km}} = 3.844\ \text{mi}$

10. $8.4\ \text{km} \times \dfrac{0.62\ \text{mi}}{1\ \text{km}} = 5.208\ \text{mi}$

11. $5.2\ \text{mi} \times \dfrac{1\ \text{km}}{0.62\ \text{mi}} = 8.387\ \text{km}$

12. $6.8\ \text{mi} \times \dfrac{1\ \text{km}}{0.62\ \text{mi}} = 10.968\ \text{km}$

13. $6.8\ \text{in.} \times \dfrac{2.54\ \text{cm}}{1\ \text{in.}} = 17.272\ \text{cm}$

14. $7.5\ \text{in.} \times \dfrac{2.54\ \text{cm}}{1\ \text{in.}} = 19.05\ \text{cm}$

15. $3\ \text{m} \times \dfrac{100\ \text{cm}}{1\ \text{m}} = 300\ \text{cm}$

16. $5.5\ \text{m} \times \dfrac{100\ \text{cm}}{1\ \text{m}} = 550\ \text{cm}$

17. $4\ \text{mm} \times \dfrac{0.1\ \text{cm}}{1\ \text{m}} = 0.4\ \text{cm}$

18. $3.5\ \text{mm} \times \dfrac{0.1\ \text{cm}}{1\ \text{m}} = 0.35\ \text{cm}$

19. $7\text{ yd} \times \frac{3\text{ ft}}{1\text{ yd}} = 21\text{ ft}$

20. $5\text{ mi} \times \frac{5280\text{ ft}}{1\text{ mi}} = 26400\text{ ft}$

21. $16\text{ in.} \times \frac{1\text{ ft}}{12\text{ in.}} = 1.333\text{ ft}$

22. $8.5\text{ in.} \times \frac{1\text{ ft}}{12\text{ in.}} = 0.708\text{ ft}$

23. $4\text{ ft} \times \frac{30.5\text{ cm}}{1\text{ ft}} = 122\text{ cm}$

24. $15\text{ m} \times \frac{1\text{ ft}}{0.305\text{ m}} = 49.18$

25. $12\text{ km} \times 2.4\text{ kW} \times \frac{1\text{ W}}{10^{-3}\text{ kW}} = 2400\text{ W}$

26. $325\text{ mA} \times \frac{1\text{ A}}{1000\text{ mA}} = 0.325\text{ A}$

27. $48\text{ V} \times \frac{1\text{ kV}}{1000\text{ V}} = 0.048\text{ V}$

28. $1400\text{ k}\Omega \times \frac{1\text{ M}\Omega}{10^3\text{ k}\Omega} = 1.4\text{ M}\Omega$

29. $0.052\text{ C} \times \frac{10^3\text{ mC}}{1\text{ C}} = 52\text{ mC}$

30. $3.6\text{ mA} \times \frac{1\ \mu\text{A}}{10^{-3}\text{ mA}} = 3600\ \mu\text{A}$

31. $30,000\ \mu\text{A} \times \frac{10^{-6}\text{ A}}{1\ \mu\text{A}} = 0.03\text{ A}$

32. $620\text{ cm} \times \frac{10^{-2}\text{ m}}{1\text{ cm}} = 6.2\text{ m}$

33. mm, 10^{-3} m

34. μA, 10^{-6} A

35. ns, 10^{-9} s

36. MW, 10^{6} W

37. KV, 10^{3} V

38. ps, 10^{-12} s

39. 25 mg

40. 200 mL

41. 66 cm

42. 95 kL

43. 41 kg

44. 22 μA

45. 12 MW

46. 7 hL

47. $160\text{ mA} \times \frac{1\,A}{10^3\text{ mA}} = 0.16\text{ A}$

48. $0.25\text{ mA} \times \frac{1\ \mu\text{A}}{10^{-3}\text{ mA}} = 250\ \mu\text{A}$

49. $2.6\text{ M}\Omega \times \frac{10^3\text{ k}\Omega}{1\text{ M}\Omega} = 2600\text{ k}\Omega$

50. $1600\ \Omega \times \frac{1\text{ k}\Omega}{10^3\ \Omega} = 1.6\text{ k}\Omega$

51. $20\text{ pF} \times \frac{10^{-3}\text{ nF}}{1\text{ pF}} = 0.02\text{ nF}$

52. $0.84\text{ gHz} \times \frac{10^6\text{ kHz}}{1\text{ gHz}} = 840{,}000\text{ kHz}$

53. $3700\text{ m} \times \frac{1\text{ km}}{10^3\text{ m}} = 3.7\text{ km}$

54. $4.5 \times 10^{-4}\text{ cm} \times \frac{1\text{ mm}}{10^{-1}\text{ cm}} = 4.5 \times 10^{-3}\text{ mm}$

55. $1.8\text{ mm} \times \frac{0.0394\text{ in.}}{1\text{ mm}} = 0.07092\text{ in.}$

56. $\left(15\text{ ft/s}\right) \times \left(\frac{0.305\text{ m}}{1\text{ ft}}\right) = 4.575\text{ m/s}$

57. $2\text{ mi} \times \frac{5280\text{ ft}}{1\text{ mi}} = 10560\text{ ft}$

58. $12\text{ gal} \times \frac{3.785\text{ L}}{1\text{ gal}} = 45.42\text{ L}$

59. $170\text{ kg} \times \frac{2.205\text{ lb}}{1\text{ kg}} = 374.85\text{ lb}$

60. $3 \text{ lb} \times \frac{16 \text{ oz}}{1 \text{ lb}} = 48 \text{ oz}$

61. $0.065 \text{ oz} \times \frac{1 \text{ mg}}{10^{-3} \times 0.035 \text{ oz}} = 1842.719 \text{ mg}$

62. $65 \text{ mi/h} \times \frac{1760 \text{ yd}}{1 \text{ mi}} = 114400 \text{ yd/h}$

63. $150 \text{ g} \times \frac{0.00981 \text{ N}}{1 \text{ g}} = 1.472 \text{ N}$

64. $50 \text{ g} \times \frac{0.00981 \text{ N}}{1 \text{ g}} = 0.491 \text{ N}$

65. $30 \text{ kJ/min} = \frac{1000 \text{ J}}{60 \text{ s}} = 16.67 \text{ J/s} = 16.67 \text{ N}$

66. (70 kJ/40 s) = 1750 J/s

CHAPTER 6

Geometry

Section 6.1 (Answers)

1. *Geometry* is the study of measurements and properties of regions, figures, and solids formed by points, lines, and planes.

2. **a.** $\overline{AB} \parallel \overline{CD}$
b. $\overleftrightarrow{MN} \perp \overleftrightarrow{JK}$
c. $\overline{TJ} \perp \overline{FM}$

3. $\overline{BA}$, $\overline{AC}$, $\overline{AD}$, $\overline{BC}$

4. $\overline{BD}$, $\overline{CA}$

5. $\overline{BO}$, $\overline{BD}$, $\overline{OD}$, $\overline{OC}$, $\overline{OA}$, $\overline{CA}$

6. $\overline{OC}$, $\overline{OA}$, $\overline{AC}$, $\overline{OD}$, $\overline{OB}$, $\overline{BD}$

7. $\angle A$, $\angle B$, $\angle C$

8. $\angle A$, $\angle B$, $\angle C$, $\angle D$

9. **a.** Triangle: 3 sides
b. Quadrilateral: 4 sides
c. Pentagon: 5 sides

10. **a.** Hexagon: 6 sides
b. Octagon: 8 sides
c. Decagon: 10 sides

11. Perpendicular line segments are:
$AD \perp DC$
$AD \perp BDC$

12. Parallel line segments are:
$BC \parallel AD$
$CD \parallel BA$

13. **a.** AD: diameter
b. BC: chord
c. Point O: center
d. EO: radius

14. **a.** arc
b. secant
c. the point of tangency
d. tangent
e. chord

Section 6.2 (Answers)

1. $A = 46°$
2. $B = 54°$
3. $\theta = 45°$
4. $\phi = 30°$
5. $\alpha = 75°$
6. $\beta = 65°$

7. $34' = 34\ \cancel{\text{min}} \times \frac{1\ \text{degree}}{60\ \cancel{\text{min}}}$
$= 0.57$ degree
Thus, $25°34' = 25° + 0.57° = 25.57°$

8. $52' = 52\ \cancel{\text{min}} \times \frac{1\ \text{degree}}{60\ \cancel{\text{min}}}$
$= 0.87$ degree
Thus, $48°52' = 48° + 0.87° = 48.87°$

9. $40' = 40\ \cancel{\text{min}} \times \frac{1\ \text{degree}}{60\ \cancel{\text{min}}}$
$= 0.67$ degree
Thus, $15°40' = 15° + 0.67° = 15.67°$

10. $15' = 15\ \cancel{\text{min}} \times \frac{1\ \text{degree}}{60\ \cancel{\text{min}}}$
$= 0.25$ degree
Thus, $74°15' = 74° + 0.25° = 74.25°$

11. $12' = 12\ \cancel{\text{min}} \times \dfrac{1\ \text{degree}}{60\ \cancel{\text{min}}}$

$= 0.2$ degree

Thus, $7°12' = 7° + 0.2° = 7.2°$

12. $43' = 43\ \cancel{\text{min}} \times \dfrac{1\ \text{degree}}{60\ \cancel{\text{min}}}$

$= 0.72$ degree

Thus, $20°43' = 20° + 0.72° = 20.72°$

13. $18'30'' = \left(18' \times \dfrac{60''}{1'}\right) + 30''$

$= 1080'' + 30'' = 1110''$

$1° = 60' \times \dfrac{60''}{1'} = 3600''$

$1110'' \times \dfrac{1°}{3600''} = \dfrac{1110°}{3600} = 0.308°$

Thus, $12°18'30'' = 12° + 0.308° = 12.308°$

14. $59'24'' = \left(59' \times \dfrac{60''}{1'}\right) + 24''$

$= 3540'' + 24'' = 3564''$

$1° = 60' \times \dfrac{60''}{1'} = 3600''$

$3564'' \times \dfrac{1°}{3600''} = \dfrac{3564°}{3600} = 0.99°$

Thus, $20°59'24'' = 20° + 0.99° = 20.99°$

15. $59'17'' = \left(59' \times \dfrac{60''}{1'}\right) + 17''$

$= 3540'' + 17'' = 3557''$

$1° = 60' \times \dfrac{60''}{1'} = 3600''$

$3557'' \times \dfrac{1°}{3600''} = \dfrac{3557°}{3600} = 0.988°$

Thus, $83°59'17'' = 83° + 0.988° = 83.988°$

16. $57'7'' = \left(57' \times \dfrac{60''}{1'}\right) + 7''$

$= 3420'' + 7'' = 3427''$

$1° = 60' \times \dfrac{60''}{1'} = 3600''$

$$3427'' \times \frac{1°}{3600''} = \frac{3427°}{3600} = 0.952°$$

Thus, $115°57'7'' = 115° + 0.952° = 115.952°$

17. $$44'6'' = \left(44' \times \frac{60''}{1'}\right) + 6''$$
$$= 2640'' + 6'' = 2646''$$
$$1° = 60' \times \frac{60''}{1'} = 3600''$$
$$2646'' \times \frac{1°}{3600''} = \frac{2646°}{3600} = 0.735°$$

Thus, $146°44'6'' = 146° + 0.735° = 146.735°$

18. $$0'1'' = \left(0' \times \frac{60''}{1'}\right) + 1''$$
$$= 0'' + 1'' = 1''$$
$$1° = 60' \times \frac{60''}{1'} = 3600''$$
$$1'' \times \frac{1°}{3600''} = \frac{1°}{3600} = 0.0003°$$

Thus, $107°0'1'' = 107° + 0.000° = 107.0°$

19. $$0.6° = 0.6° \times \frac{60'}{1°} = 36'$$

$75.6° = 75°36'$

20. $$0.06° = 0.06° \times \frac{60'}{1°} = 3.6' \approx 4'$$
$12.06° = 12°4'$

21. $$0.92° = 0.92° \times \frac{60'}{1°} = 55.2' \approx 55'$$
$114.92° = 114°55'$

22. $$0.12° = 0.12° \times \frac{60'}{1°} = 7.2' \approx 7'$$
$51.12° = 51°7'$

23. $$0.13° = 0.13° \times \frac{60'}{1°} = 7.8' \approx 8'$$
$39.13° = 39°8'$

24. $$0.514° = 0.514° \times \frac{60'}{1°} = 30.84' \approx 31'$$
$0.514° = 0°31'$

25. $0.7^\circ \times \frac{60'}{1^\circ} = 42'$

$0' \times \frac{60''}{1'} = 0''$

$89^\circ + 42' + 0'' = 89^\circ 42' 0''$

Thus, $89.7^\circ = 89^\circ 42' 0''$.

26. $0.18^\circ \times \frac{60'}{1^\circ} = 10.8'$

$0.8' \times \frac{60''}{1'} = 48''$

$102^\circ + 10' + 48'' = 102^\circ 10' 48''$

Thus, $102.18^\circ = 102^\circ 10' 48''$.

27. $0.34^\circ \times \frac{60'}{1^\circ} = 20.4'$

$0.4' \times \frac{60''}{1'} = 24''$

$23^\circ + 20' + 24'' = 23^\circ 20' 24''$

Thus, $23.34^\circ = 23^\circ 20' 24''$.

28. $0.03^\circ \times \frac{60'}{1^\circ} = 1.8'$

$0.8' \times \frac{60''}{1'} = 48''$

$241^\circ + 1' + 48'' = 241^\circ 1' 48''$

Thus, $241.03^\circ = 241^\circ 1' 48''$.

29. $0.275^\circ \times \frac{60'}{1^\circ} = 16.5'$

$0.5' \times \frac{60''}{1'} = 30''$

$68^\circ + 16' + 30'' = 68^\circ 16' 30''$

Thus, $68.275^\circ = 68^\circ 16' 30''$.

30. $0.5384^\circ \times \frac{60'}{1^\circ} = 32.304'$

$0.304' \times \frac{60''}{1'} = 18.24'' \approx 18''$

$18^\circ + 32' + 18'' = 18^\circ 32' 18''$

Thus, $18.5384^\circ = 18^\circ 32' 18''$.

31.

$$\begin{array}{r} 11^\circ\ 5' \\ +\quad 36^\circ\ 17' \\ \hline 47^\circ\ 22'\ 0'' \end{array}$$

Thus, the sum of the two angles is $47^\circ 22' 0''$.

32.

$$\begin{array}{r} 60^\circ\ 33'\ 41'' \\ +\ \ 78^\circ\ 29'\ 52'' \\ \hline 138^\circ\ 62'\ 93'' \end{array}$$

So we can say by using Table 6.1 that
$138^\circ 62' 93'' = 139^\circ 3' 33''$
Thus, the sum of the two angles is $139^\circ 3' 33''$

Section 6.3 (Answers)

1. Perimeter $= 4 + 5 + 8 + 2 + 10 = 29$
2. Perimeter $= 21 + 15 + 22 + 28 + 35 = 121$
3. Perimeter $= 11 + 7 + 4 + 15 = 37$ ft
4. Perimeter $= 115 + 65 + 86 + 125 + 92 = 483$ mm
5. Perimeter $= 2(7 + 19) = 52$ in.
6. Perimeter for rhombus $= 4(5.4) = 21.6$ in.
7. Perimeter of square $= 4(3.5) = 14$ ft
8. Perimeter for parallelogram $= 2(8.5 + 4.6) = 26.2$ in.
9. Perimeter of triangle $= 6 + 8 + 10 = 24$ in.
10. Perimeter of triangle $= 45 + 50 + 35 = 130$ in.
11. Perimeter for equilateral triangle $= P_{\substack{equilateral \\ triangle}} = 3a = 3(11) = 33$ cm
12. Perimeter for isosceles triangle $= P_{\substack{isosceles \\ triangle}} = a + 2b = 9 + 2(15) = 39$ in.
13. Circumference of a fly wheel $= C = \pi \times d = \pi \times 20 = 62.83$ in.
14. Circumference of a fly wheel $= C = \pi \times d = \pi \times 30 = 94.25$ mm
15. The perimeter of the pasture $= P_{rectangle} = 2a + 2b = 2(a + b) = 2(155 + 64) = 438$ yd.
16. The circumference of the circle $= C = \pi \times d = \pi \times 24 = 75.40$ in.

Section 6.4 (Answers)

1. $A_{square} = s^2 = (7)^2 = 49 \text{ in.}^2$
2. $A_{square} = s^2 = (16)^2 = 256 \text{ in.}^2$
3. $A_{rectangle} = l \times w$
   $A = 25 \times 12 = 300 \text{ ft}^2$

4. $A_{rectangle} = l \times w$
$A = 11 \times 6 = 66\ \text{cm}^2$

5. 1 yd = 3 ft
Area $= 3 \times 3 = 9\ \text{ft}^2$

6. $w = 16\ \text{cm} = 0.16\ \text{m},\ L = 2.75\ \text{m}$
$A = (0.16) \times (2.75) = 0.44\ \text{m}^2$

7. $w = 35\ \text{cm} = 0.35\ \text{m},\ L = 5.25\ \text{m}$
$A = (0.35) \times (5.25) = 1.84\ \text{m}^2$

8. $A_{rectangle} = l \times w$
$A = 25 \times 46 = 1150\ \text{cm}^2$

9. $A_{parallelogram} = b \times h$
$A = 14 \times 8 = 112\ \text{cm}^2$

10. $A_{parallelogram} = b \times h$
$A = 9 \times 4 = 36\ \text{cm}^2$

11. $A_{trapezoid} = \frac{1}{2}h(b_1 + b_2)$

$$A = \frac{1}{2}(18)(22 + 37)$$

$$A = 531\ \text{m}^2$$

12. $A_{trapezoid} = \frac{1}{2}h(b_1 + b_2)$

$$A = \frac{1}{2}(7)(13 + 19)$$

$$A = 112\ \text{m}^2$$

13. $A_{triangle} = \frac{1}{2}bh$

$$A = \frac{1}{2}(13)(4) = 26\ \text{in.}^2$$

14. $A_{triangle} = \frac{1}{2}bh$

$$A = \frac{1}{2}(10)(18) = 90\ \text{in.}^2$$

15. $A = \pi\, r^2 = \pi\, d^2/4$

$$A = \pi\left(\frac{24^2}{4}\right) = 452.39\ \text{in.}^2$$

16. $A = \pi\, r^2 = \pi\, d^2/4$

$$A = \pi\left(\frac{12^2}{4}\right) = 113.01\ \text{in.}^2$$

17. $A_{circle} = \pi\left(\frac{d^2}{4}\right)$

Area of large circle $= A_{circle} = \pi\left(\frac{9^2}{4}\right) = 63.62 \text{ in.}^2$

Area of small circle $= A_{circle} = \pi\left(\frac{5^2}{4}\right) = 19.64 \text{ in.}^2$

Area of ring $= 63.62 \text{ in.}^2 - 19.64 \text{ in.}^2 = 43.98 \text{ in.}^2$

18. $A_{circle} = \pi r^2$
Area of large circle $= \pi(15)^2 = 706.86 \text{ in.}^2$
Area of small circle $= \pi(5)^2 = 78.54 \text{ in.}^2$
Area of ring $= 706.86 - 78.54 = 628.32 \text{ in.}^2$

19. $A_{sector} = \pi r^2\left(\frac{\theta}{360°}\right)$

$A = \pi(6)^2\left(\frac{30}{360}\right) = 9.42 \text{ in.}^2$

20. $A_{sector} = \pi r^2\left(\frac{\theta}{360°}\right)$

$A = \pi(7)^2\left(\frac{75}{360}\right) = 32.10 \text{ in.}^2$

Section 6.5 (Answers)

1. $V = l \times w \times h$
$V = 8 \text{ cm} \times 6 \text{ cm} \times 5 \text{ cm} = 240 \text{ cm}^3$

2. $V = l \times w \times h$
$V = 14 \text{ cm} \times 8 \text{ cm} \times 7 \text{ cm} = 784 \text{ cm}^3$

3. $14 \text{ ft } 8 \text{ in.} = 14\frac{8}{12} = 14.667 \text{ ft}$

$12 \text{ ft } 4 \text{ in.} = 12\frac{4}{12} = 12.333 \text{ ft}$

$10 \text{ ft } 2 \text{ in.} = 10\frac{2}{12} = 10.167 \text{ ft}$
$V = l \times w \times h$
$V = 14.667 \times 12.333 \times 10.167 = 1839.09 \text{ ft}^3$

4. $18 \text{ ft } 4 \text{ in.} = 18\frac{4}{12} = 6 \text{ ft}$

$16 \text{ ft } 2 \text{ in.} = 16\frac{2}{12} = 2.67 \text{ ft}$

$12 \text{ ft } 6 \text{ in.} = 12\frac{6}{12} = 6 \text{ ft}$

$V = l \times w \times h$

$V = 6 \times 2.67 \times 6 = 96.12 \text{ ft}^3$

5. The volume of a crate = 3.60 ft × 5.60 ft × 7.80 ft = 157.248 ft^3

6. The volume of a crate = 5.20 ft × 7.30 ft × 8.50 ft = 322.66 ft^3

7. $V_{\text{cube}} = s^3$

$V = (3)^3 = 27 \text{ cm}^3$

8. $V_{\text{cube}} = s^3$

$V = (5)^3 = 125 \text{ cm}^3$

9. $V_{\text{cylinder}} = \pi r^2 h$

$V = \pi(10)^2(12) = 3769.91 \text{ in.}^3$

10. $V_{\text{cylinder}} = \pi r^2 h$

$V = \pi(9)^2(7) = 1781.28 \text{ in.}^3$

11. $V = \frac{1}{3}\pi r^2 h$

$V = \frac{1}{3}\pi(6)^2(24) = 904.78 \text{ ft}^2$

12. $V = \frac{1}{3}\pi r^2 h$

$V = \frac{1}{3}\pi(8)^2(28) = 1876.58 \text{ ft}^2$

13. $V_{pyramid} = \frac{1}{3}Bh$

$V_{pyramid} = \frac{1}{3}(4)(6) = 8 \text{ in.}^3$

14. $V_{pyramid} = \frac{1}{3}Bh$

$V_{pyramid} = \frac{1}{3}(5)(9) = 15 \text{ m}^3$

15. $V_{sphere} = \frac{4}{3}\pi(4.8)^3 = 52.64 \text{ m}^3$

16. $V_{sphere} = \frac{4}{3}\pi(12)^3 = 7238.23 \text{ m}^3$

Review Exercises (Chapter 6—Answers)

1. Ray, $\overrightarrow{HT}$
2. Line, $\overleftrightarrow{HT}$ or $\overleftrightarrow{TH}$
3.

$$\begin{array}{r} 12^\circ\ 6' \\ 38^\circ\ 18' \\ \hline \theta = 50^\circ\ 24' \end{array}$$

4.

$$\begin{array}{r} 61^\circ\ 30'\ 42'' \\ 77^\circ\ 28'\ 50'' \\ \hline \theta = 138^\circ\ 59'\ 32'' \end{array}$$

5. Subdividing the figure into two rectangles as below

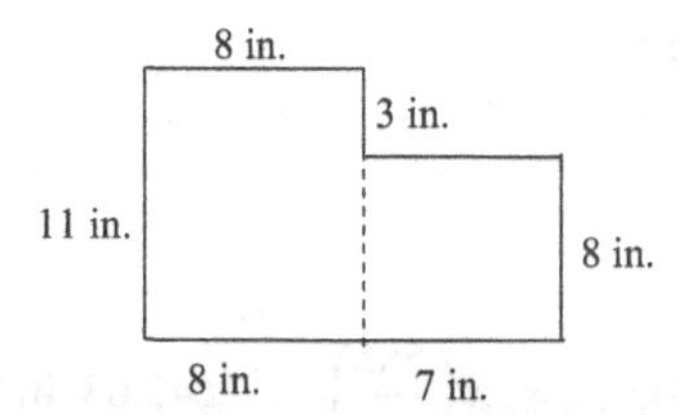

Perimeter = 11 + (8 + 7) + 8 + 3 + 3 = 40 in.

6.

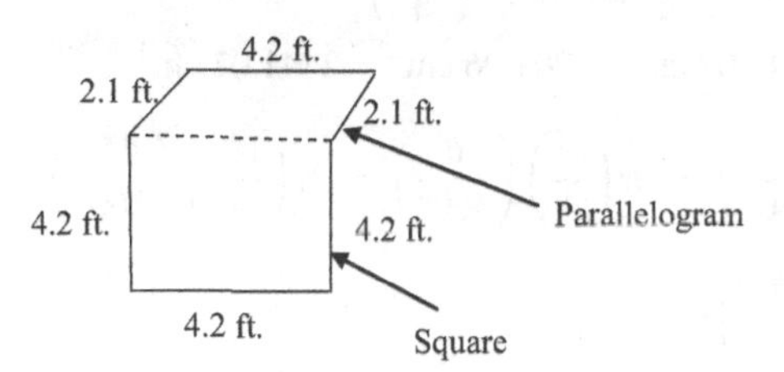

Perimeter = (4) (4.2) + (2) (2.1) = 21

7. d = C/3.14 in.
   C = 87.46 in.
   d = 87.46/3.14 = 27.85 in.

8. $P_{parallelogram} = 2a + 2b = 2(a + b)$
   $P = 2(83.2\text{ ft} + 176.5\text{ ft}) = 519.4\text{ ft}$

9. $A_{rectangle} = l \times w$
   $A = 20 \times 46 = 920\text{ ft}^2$

10. $A_{\text{square}} = s^2$
$s = \sqrt{481} = 21.93$ in.

11. $A_{rectangle} = l \times w$
$108 = l \times 12 \rightarrow l = \dfrac{108}{12} = 9$ in.
$P_{rectangle} = 2a + 2b = 2(a + b)$
$P = 2(9 + 12) = 42$ in.

12. $w = \dfrac{A}{l} = \dfrac{1932}{84} = 23$ in.

13. $A = 24 \times 18 = 432$ in.^{2}

14. $A = lw = 7 \times 5 = 35$ yd^2
the fabric cost = \$15 × 35 = \$525

15. $A = lw = 12 \times 17 = 204$ ft^2
the tile cost = \$4.50 × 204 = \$918

16. $A_{circle} = \pi\left(\dfrac{d^2}{4}\right)$

Area of large circle $= A_{circle} = \pi\left(\dfrac{58^2}{4}\right) = 2642.08$ in.^{2}

Area of small circle $= A_{circle} = \pi\left(\dfrac{16^2}{4}\right) = 201.06$ in.^{2}

Area of ring = 2642.08 cm^2 – 201.06 cm^2 = 2441.02 cm^2

17. $A_{sector} = \pi r^2\left(\dfrac{\theta}{360°}\right) = \pi\left(\dfrac{d^2}{4}\right)\left(\dfrac{\theta}{360°}\right) = \pi\left(\dfrac{18^2}{4}\right)\left(\dfrac{45°}{360°}\right) = 31.81$ in.^{2}

18. $A_{sector} = \pi r^2\left(\dfrac{\theta}{360°}\right)$

$A = \pi(30)^2\left(\dfrac{120}{360}\right) = 942.48$ yd^2

19. $V_{\text{rectangular}} = l \times w \times h$
$V = 7 \times 5 \times 2 = 70$ m^3

20. $V_{\text{rectangular}} = l \times w \times h$
$V = 12 \times 9 \times 3 = 324$ m^3

21. $V_{\text{rectangular}} = l \times w \times h$
$V = 11 \times 9 \times 10 = 990$ ft^3

22. $V_{\text{rectangular}} = l \times w \times h$
$V = 70 \times 7.7 \times 2.65 = 1428.35$ ft^3

23. $V_{\text{cube}} = s^3$
$V = 2^3 = 8\ \text{cm}^3$

24. $V_{\text{cube}} = s^3$
$V = 7^3 = 343\ \text{cm}^3$

25. $V_{\text{cylinder}} = \pi r^2 h$
$V = \pi \times (22)^2 \times 76 = 115560.34\ \text{ft}^3$

26. $V_{\text{cylinder}} = \pi r^2 h = \pi\left(\dfrac{d^2}{4}\right)h = \pi\left(\dfrac{40^2}{4}\right)98 = 123150.43\ \text{cm}^3$

27. $V_{\text{cone}} = \dfrac{1}{3}\pi r^2 h = \dfrac{1}{3}\pi\left(\dfrac{d^2}{4}\right)h = \dfrac{1}{3}\pi\left(\dfrac{26^2}{4}\right)12 = 2123.72\ \text{ft}^3$

28. $V_{\text{cone}} = \dfrac{1}{3}\pi r^2 h = \dfrac{1}{3}\pi\left(\dfrac{d^2}{4}\right)h = \dfrac{1}{3}\pi\left(\dfrac{(6.04)^2}{4}\right)1.60 = 15.28\ \text{in.}^3$

29. $V_{pyramid} = \dfrac{1}{3}Bh = \dfrac{1}{3}(26)8 = 69.33\ \text{ft}^3$

30. $180 = \dfrac{1}{3}Bh = \dfrac{1}{3}(B)8 \rightarrow B = \dfrac{(180)(3)}{8} = 67.5\ \text{cm}^2$

31. $V_{sphere} = \dfrac{4}{3}\pi r^3 = \dfrac{4}{3}\pi(4.2)^3 = 310.34\ \text{m}^3$

32. $V_{sphere} = \dfrac{4}{3}\pi r^3 = \dfrac{4}{3}\pi\left(\dfrac{d^3}{8}\right) = \dfrac{4}{3}\pi\left(\dfrac{1.6^3}{8}\right) = 2.14\ \text{cm}^3$

CHAPTER 7

Trigonometry

Section 7.1 (Answers)

1. $75° \times \dfrac{60'}{1°} = 4500'$

2. $540' \times \dfrac{1°}{60'} = 9°$

3. $15000'' \times \dfrac{1'}{60''} = 250'$

4. $90° \times \dfrac{\pi \text{ radian}}{180°} = \dfrac{\pi}{2}\ \text{rad}$

5. $\dfrac{2\cancel{\pi}}{5} \times \dfrac{180°}{\cancel{\pi}} = 72°$

6. $\frac{\not\pi}{4} \times \frac{180^\circ}{\not\pi} = 45^\circ$

7. 1 circle = 360°
   2 circles = 720°

8. $2\text{ circles} \times \frac{360^\circ}{1\text{ circle}} \times \frac{60'}{1^\circ} = 43200'$

9. $2\text{ circles} \times \frac{360^\circ}{1\text{ circle}} \times \frac{60'}{1^\circ} \times \frac{60''}{1'} = 2,592,000''$

10. $1^\circ \times \frac{\pi\text{rad}}{180^\circ} = \frac{\pi}{180}\text{rad}$

11. $30^\circ \times \frac{\pi\text{rad}}{180^\circ} = \frac{\pi}{6}\text{rad}$

12. $60' \times \frac{1^\circ}{60'} \times \frac{\pi}{180^\circ}\text{rad} = \frac{\pi}{180}\text{rad}$

13. $3\text{ h} \times \frac{60'}{1\text{ h}} \times \frac{1^\circ}{60'} = 3^\circ$

14. $360^\circ \times \frac{\pi}{180^\circ}\text{rad} = 2\pi\text{ rad}$

Section 7.2 (Answers)

1. Acute angle
2. Right angle
3. Acute angle
4. Obtuse angle
5. Obtuse angle
6. Zero angle
7. Straight angle
8. Acute angle
9. Acute angle
10. Obtuse angle
11. **(a)** $\theta_1 + \theta_2 = 90^\circ$
    $\theta = 90^\circ - 65^\circ = 25^\circ$

    (b) $\theta_1 + \theta_2 = 90^\circ$
    $\theta = 90^\circ - 90^\circ = 0^\circ$

(c) $\theta_1 + \theta_2 = 90°$
$\theta = 90° - 45° = 45°$

(d) $\theta_1 + \theta_2 = 90°$
$\theta = 90° - 85° = 5°$

12. (a) $\theta_1 + \theta_2 = 180°$
$\theta = 180° - 125° = 55°$

(b) $\theta_1 + \theta_2 = 180°$
$\theta = 180° - 145° = 35°$

(c) $\theta_1 + \theta_2 = 180°$
$\theta = 180° - 115° = 65°$

(d) $\theta_1 + \theta_2 = 180°$
$\theta = 180° - 165° = 15°$

13. $c^2 = a^2 + b^2$, therefore, $c = \sqrt{a^2 + b^2}$
$c = \sqrt{2^2 + 3^2} = \sqrt{4 + 9} = \sqrt{13}$

14. $c^2 = a^2 + b^2$,
$b = \sqrt{c^2 - a^2} = \sqrt{13 - 9} = \sqrt{4} = 2$

15. $c = \sqrt{4^2 + 6^2} = \sqrt{16 + 36} = \sqrt{52}$

16. $c = \sqrt{12^2 + 20^2} = \sqrt{144 + 400} = \sqrt{544}$

17. $c = \sqrt{5^2 + 8^2} = \sqrt{25 + 64} = \sqrt{89}$

18. $c = \sqrt{5^2 + 10^2} = \sqrt{25 + 100} = \sqrt{125}$

19. $a = \sqrt{24^2 - 8^2} = \sqrt{576 - 64} = \sqrt{512}$

20. $a = \sqrt{14^2 - 6^2} = \sqrt{196 - 36} = \sqrt{160}$

21. $b = \sqrt{15^2 - 3^2} = \sqrt{225 - 9} = \sqrt{216}$

22. $b = \sqrt{10^2 - 4^2} = \sqrt{100 - 16} = \sqrt{84}$

23. False, because it is used only for any right triangle

24. The length of $AB = \sqrt{40^2 - 12^2} = \sqrt{1600 - 144} = \sqrt{1456} = 38.16$ ft

25. $d = \sqrt{18^2 + 12^2} = \sqrt{324 + 144} = \sqrt{456} = 21.63$ in.

26. $d = \sqrt{9^2 + 4^2} = \sqrt{81 + 16} = \sqrt{97} = 9.85$ ft

27. $AB = \sqrt{7^2 - 5^2} = \sqrt{49 - 25} = \sqrt{24} = 4.90$ cm
$BC = \sqrt{12^2 - 5^2} = \sqrt{144 - 25} = \sqrt{119} = 10.91$ cm
The length $(AC) = 4.90 + 10.91 = 15.81$ cm

28. $AB = \sqrt{10^2 - 6^2} = \sqrt{100 - 36} = \sqrt{64} = 8$ cm
$BC = \sqrt{16^2 - 6^2} = \sqrt{256 - 36} = \sqrt{220} = 14.83$ cm
The length $(AC) = 8 + 14.83 = 22.83$ cm

29. $AB = \sqrt{30^2 - 25^2} = \sqrt{900 - 625} = \sqrt{275} = 16.58$ cm
$BC = AB = \sqrt{30^2 - 25^2} = \sqrt{900 - 625} = \sqrt{275} = 16.58$ cm
The length $(AC) = 2(16.58) = 33.16$ cm

30. $AB = \sqrt{60^2 - 50^2} = \sqrt{3600 - 2500} = \sqrt{1100} = 33.17$ ft

31. Applied voltage $= \sqrt{86^2 + 92^2} = \sqrt{7396 + 8464} = \sqrt{15860} = 125.94$ volts

32. Voltage across the resistance $= \sqrt{620^2 - 345^2} = \sqrt{384400 - 119025} = \sqrt{265375}$
$= 515.15$ volts

Section 7.3 (Answers)

1. $\sin\theta = \dfrac{\text{length of } \textit{opposite} \text{ side of } \theta}{\text{length of hypotenuse}} = \dfrac{9}{16.64} = 0.541 \text{ m},$

$\cos\theta = \dfrac{\text{length of adjacent side of } \theta}{\text{length of hypotenuse}} = \dfrac{14}{16.64} = 0.84 \text{ m},$

$\tan\theta = \dfrac{\text{length of } \textit{opposite} \text{ side of } \theta}{\text{length of adjacent side of } \theta} = \dfrac{9}{14} = 0.64 \text{ m},$

$\csc\theta = \dfrac{\text{length of hypotenuse}}{\text{length of } \textit{opposite} \text{ side of } \theta} = \dfrac{16.64}{9} = 1.85 \text{ m},$

$\sec\theta = \dfrac{\text{length of hypotenuse}}{\text{length of adjacent side of } \theta} = \dfrac{16.64}{14} = 1.19 \text{ m},$

$\cot\theta = \dfrac{\text{length of adjacent side of } \theta}{\text{length of } \textit{opposite} \text{ side of } \theta} = \dfrac{14}{9} = 4.56 \text{ m}$

2. $\sin\theta = \dfrac{\text{length of } \textit{opposite} \text{ side of } \theta}{\text{length of hypotenuse}} = \dfrac{16}{18.87} = 0.85 \text{ m},$

$\cos\theta = \dfrac{\text{length of adjacent side of } \theta}{\text{length of hypotenuse}} = \dfrac{10}{18.87} = 0.53 \text{ m},$

$\tan\theta = \dfrac{\text{length of } \textit{opposite} \text{ side of } \theta}{\text{length of adjacent side of } \theta} = \dfrac{16}{10} = 1.6 \text{ m},$

$\csc\theta = \dfrac{\text{length of hypotenuse}}{\text{length of } \textit{opposite} \text{ side of } \theta} = \dfrac{18.87}{16} = 1.18 \text{ m},$

$\sec\theta = \dfrac{\text{length of hypotenuse}}{\text{length of adjacent side of } \theta} = \dfrac{18.87}{10} = 1.89 \text{ m},$

$\cot\theta = \dfrac{\text{length of adjacent side of } \theta}{\text{length of } \textit{opposite} \text{ side of } \theta} = \dfrac{10}{16} = 0.63 \text{ m}$

3. $a = \sqrt{294^2 - 214^2} = \sqrt{86436 - 45796} = \sqrt{40640} = 201.60$

$\sin\theta = \dfrac{201.6}{294} = 0.69$

$\csc\phi = \dfrac{1}{\sin\phi} = \dfrac{294}{214} = 1.37$

4. $b = \sqrt{0.082^2 - 0.068^2} = \sqrt{6.724 \times 10^{-3} - 4.624 \times 10^{-3}} = \sqrt{2.1 \times 10^{-3}} = 0.046$

$\sin\phi = \dfrac{0.046}{0.082} = 0.561$

$\sec\theta = \dfrac{1}{\cos\theta} = \dfrac{0.082}{0.068} = 1.21$

5. $\tan\theta = \dfrac{3}{2.4} = 1.25$

$c = \sqrt{2.4^2 + 3^2} = \sqrt{14.76} = 3.84$

$\sin\phi = \dfrac{2.4}{3.84} = 0.625$

6. $a = \sqrt{8.64^2 - 4.2^2} = \sqrt{74.65 - 17.64} = \sqrt{57.01} = 7.55$

$\csc\theta = \dfrac{1}{\sin\theta} = \dfrac{8.64}{7.55} = 1.14$

$\cot\phi = \dfrac{7.55}{4.2} = 1.80$

7. $\tan\theta = \dfrac{6}{8} = 0.75$

$c = \sqrt{6^2 + 8^2} = \sqrt{36 + 64} = 10$

$\sin\phi = \dfrac{8}{10} = 0.8$

8. $\cot\theta = \dfrac{1}{1} = 1$

$c = \sqrt{2}$

$\cos\phi = \dfrac{1}{\sqrt{2}}$

9. $a = \sqrt{19^2 - 10^2} = \sqrt{361 - 100} = \sqrt{261} = 16.16$

$\csc\theta = \dfrac{1}{\sin\theta} = \dfrac{19}{16.16} = 1.18$

$\cos\phi = \dfrac{16.16}{19} = 0.85$

10. $b = \sqrt{28^2 - 9^2} = \sqrt{784 - 81} = \sqrt{703} = 26.51$

$\cos\theta = \dfrac{26.51}{28} = 0.95$

$\cot\phi = \dfrac{26.51}{9} = 2.95$

11. $\sin 60° = \cos(90 - 60) = \cos 30°$

12. $\tan 25° = \cot(90 - 25) = \cot 65°$

13. $\sec 30° = \csc (90 - 30) = \csc 60°$

14. $\csc 40° = \sec (90 - 40) = \sec 50°$

15. $\cot 10° = \tan (90 - 10) = \tan 80°$

16. $\cos 80° = \sin (90 - 80) = \sin 10°$

17. $\sin 125° = \cos (180 - 125) = \cos 55°$

18. $\cos 105° = \sin (180 - 105) = \sin 75°$

19. $\tan 170° = \cot (180 - 170) = \cot 10°$

20. $\csc 130° = \sec (180 - 130) = \sec 50°$

21. $\cot 145° = \tan (180 - 145) = \tan 35°$

22. $\sec 160° = \csc (180 - 160) = \csc 20°$

23. $\sin \theta = \dfrac{1}{1.15} = 0.87$

24. $\cos \phi = \dfrac{1}{1.67} = 0.60$

25. $\tan \beta = \dfrac{1}{0.76} = 1.32$

26. $\cot \alpha = \dfrac{1}{0.35} = 2.86$

Section 7.4 (Answers)

1. **(a)** $\csc 180° = \dfrac{1}{\sin 180°} = \dfrac{1}{0} = \infty$

(b) $\sec 180° = \dfrac{1}{\cos 180°} = \dfrac{1}{-1} = -1$

(c) $\cot 180° = \dfrac{\cos 180°}{\sin 180°} = \dfrac{-1}{0} = \infty$

(d) $\sin 270° = \dfrac{1}{-1} = -1$

2. **(a)** $\csc 115° = \csc(180 - 115) = \csc 65° = \dfrac{1}{\sin 65°} = \dfrac{1}{0.91} = 1.10$

(b) $\sec 130° = -\sec(180 - 130) = -\sec 50° = -\dfrac{1}{\cos 50°} = -\dfrac{1}{0.64} = -1.56$

(c) $\cot 90° = \dfrac{\cos 90°}{\sin 90°} = \dfrac{0}{1} = 0$

(d) $\sin 147° = \sin(180 - 147) = \sin 33° = 0.54$

3. $r = \sqrt{x^2+y^2} = \sqrt{3^2+5^2} = 5.83$. So now we know, $x=3$, $y=5$, $r=5.83$, we can find the values of the trigonometric functions of the angle α

$$\sin\alpha = \frac{5}{5.83} = 0.86, \quad \cos\alpha = \frac{3}{5.83} = 0.51$$

$$\tan\alpha = \frac{5}{3} = 1.67, \quad \cot\alpha = \frac{3}{5} = 0.6$$

$$\sec\alpha = \frac{5.83}{3} = 1.94, \quad \csc\alpha = \frac{5.83}{5} = 1.17$$

4. $r = \sqrt{x^2+y^2} = \sqrt{3^2+(-1)^2} = \sqrt{10} = 3.16$. So now we know, $x=3$, $y=-1$, $r=3.16$, we can find the values of the trigonometric functions of the angle α

$$\sin\alpha = \frac{-1}{3.16} = 0.32, \quad \cos\alpha = \frac{3}{3.16} = 0.95$$

$$\tan\alpha = \frac{-1}{3} = -0.33, \quad \cot\alpha = \frac{3}{-1} = -3$$

$$\sec\alpha = \frac{3.16}{3} = 1.05, \quad \csc\alpha = -\frac{3.16}{1} = -3.16$$

5. $$\frac{\sin 44^\circ}{24} = \frac{\sin\theta_1}{14} \rightarrow \sin\theta_1 = 0.405 \rightarrow \theta_1 = \sin^{-1}(0.405) = 23.89^\circ$$

$$\theta_2 = 180^\circ - (44^\circ + 23.89^\circ) = 180^\circ - 67.89^\circ = 112.11^\circ$$

$$\frac{\sin 44^\circ}{24} = \frac{\sin 112.11^\circ}{b} \rightarrow b = \frac{(\sin 112.11^\circ)24}{\sin 44^\circ} = \frac{22.24}{0.69} = 32.23$$

6. $\theta = 180^\circ - (38^\circ + 75^\circ) = 180^\circ - 113^\circ = 67^\circ$

$$\frac{\sin 67^\circ}{24} = \frac{\sin 38^\circ}{b} \rightarrow b = \frac{(\sin 38^\circ)24}{\sin 67^\circ} = \frac{14.78}{0.92} = 16.07$$

$$\frac{\sin 67^\circ}{24} = \frac{\sin 75^\circ}{a} \rightarrow a = \frac{(\sin 75^\circ)24}{\sin 67^\circ} = \frac{23.18}{0.92} = 25.20$$

7. $\theta = 180^\circ - (45^\circ + 55^\circ) = 180^\circ - 100^\circ = 80^\circ$

$$\frac{\sin 80^\circ}{30} = \frac{\sin 55^\circ}{a_1} \rightarrow a_1 = \frac{(\sin 55^\circ)30}{\sin 80^\circ} = \frac{24.57}{0.98} = 25.07$$

$$\frac{\sin 80^\circ}{30} = \frac{\sin 45^\circ}{a_2} \rightarrow a_2 = \frac{(\sin 45^\circ)30}{\sin 80^\circ} = \frac{21.21}{0.98} = 21.64$$

8. $\beta = 180^\circ - 47.5^\circ = 132.5^\circ$

$$\theta = 180^\circ - (43.2^\circ + 132.5^\circ) = 180^\circ - 175.7^\circ = 4.3^\circ$$

$$\frac{\sin 4.3^\circ}{467} = \frac{\sin 43.2^\circ}{c} \rightarrow c = \frac{(\sin 43.2^\circ)467}{\sin 4.3^\circ} = \frac{319.68}{0.075} = 4262.4 \text{ ft}$$

$$\sin 47.5^\circ = \frac{h}{4262.4} \rightarrow h = 3142.57 \text{ ft}$$

9. $a^2 = b^2 + c^2 - 2bc\cos A$

$d^2 = (5)^2 + (5.3)^2 - 2(5)(5.3)\cos 65°$

$d = \sqrt{25 + 28.09 - 22.40} = 5.54$

$\frac{\sin 65°}{5.54} = \frac{\sin \theta°}{5.3} \rightarrow \sin \theta° = 5.3\left(\frac{\sin 65°}{5.54}\right) = 0.87$

$\theta = \sin^{-1}(0.87) = 60.5°$

$\phi = 180° - (65° + 60.5°) = 180° - 125.5° = 54.5°$

10. $a^2 = b^2 + c^2 - 2bc \cos A$

$d^2 = (72)^2 + (53)^2 - 2(72)(53)\cos 125°$

$d = \sqrt{5184 + 2809 + 4377.54} = 111.22$

$\frac{\sin 125°}{111.22} = \frac{\sin \theta°}{72} \rightarrow \sin \theta° = 72\left(\frac{\sin 125°}{111.22}\right) = 0.53$

$\theta = \sin^{-1}(0.53) = 32°$

$\phi = 180° - (125° + 32°) = 180° - 157° = 23°$

11. $a^2 = b^2 + c^2 - 2bc \cos A$

$d^2 = (12)^2 + (7)^2 - 2(12)(7) \cos 65°$

$d = \sqrt{144 + 49 - 70.10} = 11.09$ ft

12. $a^2 = b^2 + c^2 - 2bc \cos A$

$h^2 = (36)^2 + (29)^2 - 2(36)(29) \cos 150°$

$h = \sqrt{1296 + 841 + 1808.26} = 62.81$ kft

Review Exercises (Chapter 7—Answers)

1. $30° \times \frac{\pi \text{ rad}}{180°} = \frac{\pi}{6}$ rad

2. $72° \times \frac{\pi \text{ rad}}{180°} = \frac{6\pi}{15}$ rad

3. $12° \times \frac{\pi \text{ rad}}{180°} = \frac{\pi}{15}$ rad

4. $75° \times \frac{\pi \text{ rad}}{180°} = \frac{5\pi}{12}$ rad

5. $240° \times \frac{\pi \text{ rad}}{180°} = \frac{4\pi}{3}$ rad

6. $253^\circ \times \dfrac{\pi \text{ rad}}{180^\circ} = 4.416 \text{ rad}$

7. $135^\circ \times \dfrac{\pi \text{ rad}}{180^\circ} = \dfrac{3\pi}{4} \text{rad}$

8. $300^\circ \times \dfrac{\pi \text{ rad}}{180^\circ} = \dfrac{5\pi}{3} \text{ rad}$

9. $\dfrac{3\not{\pi}}{4} \times \dfrac{180^\circ}{\not{\pi}} = 135^\circ$

10. $\dfrac{7\not{\pi}}{9} \times \dfrac{180^\circ}{\not{\pi}} = 140^\circ$

11. $\dfrac{2\not{\pi}}{9} \times \dfrac{180^\circ}{\not{\pi}} = 40^\circ$

12. $\dfrac{3\not{\pi}}{2} \times \dfrac{180^\circ}{\not{\pi}} = 270^\circ$

13. $\dfrac{6\not{\pi}}{5} \times \dfrac{180^\circ}{\not{\pi}} = 216^\circ$

14. $\dfrac{4\not{\pi}}{5} \times \dfrac{180^\circ}{\not{\pi}} = 144^\circ$

15. $b = \sqrt{50^2 - 20^2} = \sqrt{2500 - 400} = \sqrt{2100} = 45.83 \text{ cm}$

16. $a = \sqrt{70^2 - 30^2} = \sqrt{4900 - 900} = \sqrt{4000} = 63.25 \text{ cm}$

17. $c = \sqrt{40^2 + 65^2} = \sqrt{1600 + 4225} = 76.32 \text{ m}$

18. $c = \sqrt{10^2 + 25^2} = \sqrt{100 + 625} = \sqrt{725} = 26.93 \text{ m}$

19. $b = \sqrt{(2.6)^2 - 2^2} = \sqrt{6.76 - 4} = \sqrt{2.76} = 1.66 \text{ mi}$

20. The distance $= \sqrt{6^2 + 12^2} = \sqrt{36 + 144} = \sqrt{180} = 13.42$

21. Coil current $= \sqrt{46^2 - 22^2} = \sqrt{2116 - 484} = \sqrt{1632} = 40.40 \text{ A}$

22. Total current $= \sqrt{(60.4)^2 + (75.6)^2} = \sqrt{3648.16 + 5715.36} = \sqrt{9363.52}$
$= 96.77 \text{ A}$

23. Reactance $= \sqrt{170^2 - 116^2} = \sqrt{28900 - 13456} = \sqrt{15444} = 124.27\ \Omega.$

24. Impedance $= \sqrt{(30.2)^2 + (56.5)^2} = \sqrt{912.04 + 3192.25} = \sqrt{4104.29}$
$= 64.07\ \Omega$

25. Impedance $= \sqrt{(40.3)^2 + (73.4)^2} = \sqrt{1624.09 + 5387.56} = \sqrt{7011.65}$
$= 83.74\ \Omega$

26. $d = \sqrt{(7)^2 + (11)^2} = \sqrt{49 + 121} = \sqrt{170} = 13.04$ in.

27. $a = \sqrt{24^2 - 21^2} = \sqrt{576 - 441} = \sqrt{135} = 11.62$

$$\sin\theta = \frac{11.62}{24} = 0.48$$

$$\csc\phi = \frac{24}{21} = 1.14$$

28. $b = \sqrt{6.8^2 - 4^2} = \sqrt{46.24 - 16} = \sqrt{30.42} = 5.5$

$$\tan\theta = \frac{4}{5.5} = 0.73$$

$$\sin\phi = \frac{5.5}{6.8} = 0.81$$

29. **(a)** $\cos 270° = 0$

(b) $\tan 270° = \frac{-1}{0} = \infty$

(c) $\csc 270° = \frac{1}{\sin 270°} = \frac{1}{-1} = -1$

(d) $\sec 270° = \frac{1}{\cos 270°} = \frac{1}{0} = \infty$

30. **(a)** $\cos 160° = -\cos(180 - 160) = -\cos 20° = -0.94$

(b) $\tan 126° = -\tan(180 - 126) = -\tan 54° = -1.38$

(c) $\csc 117° = \csc(180 - 117) = \csc 63° = \frac{1}{\sin 63°} = \frac{1}{0.89} = 1.12$

(d) $\sec 108° = -\sec(180 - 108) = -\sec 72° = -\frac{1}{\cos 72°} = -\frac{1}{0.31} = -3.23$

31. $r = \sqrt{x^2 + y^2} = \sqrt{(-\sqrt{3})^2 + (1)^2} = \sqrt{4} = 2$. So now we know, $x = -\sqrt{3}$, $y = 1$, $r = 2$, we can find the values of the trigonometric functions of the angle α

$$\sin\alpha = \frac{1}{2} = 0.5, \quad \cos\alpha = \frac{-\sqrt{3}}{2} = -0.87$$

$$\tan\alpha = -\frac{1}{\sqrt{3}} = -0.58, \quad \cot\alpha = \frac{-\sqrt{3}}{1} = -1.73$$

$$\sec\alpha = -\frac{2}{\sqrt{3}} = -1.15, \quad \csc\alpha = \frac{2}{1} = 2$$

32. $r = \sqrt{x^2+y^2} = \sqrt{(-2)^2+(-2)^2} = \sqrt{8} = 2.83$. So now we know, $x = -2$, $y = -2$, $r = 2.83$, we can find the values of the trigonometric functions of the angle α

$$\sin\alpha = \frac{-2}{2.83} = -0.71, \quad \cos\alpha = \frac{-2}{2.83} = -0.71,$$

$$\tan\alpha = \frac{-2}{-2} = 1, \quad \cot\alpha = \frac{-2}{-2} = 1,$$

$$\sec\alpha = -\frac{2.83}{2} = -1.42, \quad \csc\alpha = -\frac{2.83}{2} = -1.42$$

33. $\beta = 180° - 56° = 124°$
$\theta = 180° - (124° + 33°) = 180° - 157° = 23°$
$$\frac{\sin 124°}{L} = \frac{\sin 23°}{20} \to L = \frac{(\sin 124°)20}{\sin 23°} = \frac{16.58}{0.39} = 42.51 \text{ in.}$$

34. $\sin 35° = \frac{4}{L} \to L = \frac{4}{\sin 35°} \cong 7$

CHAPTER 8

Matrices, Determinants, and Vectors

Section 8.1 (Answers)

1. *Row* matrix
2. *Column* matrix
3. *Square* matrix
4. *Null* (zero) matrix
5. *Diagonal* matrix
6. *Unit* (identity) matrix
7. *Order* $= 2 \times 3$
8. *Order* $= 3 \times 3$
9. *Order* $= 2 \times 1$
10. *Order* $= 1 \times 2$
11. *Order* $= 3 \times 2$
12. *Order* $= 2 \times 2$
13. *Order* $= 3 \times 3$
14. *Order* $= 1 \times 2$

15. $Order = 2 \times 1$

16. $Order = 1 \times 1$

17. $Order = 1 \times 5$

18. $Order = 4 \times 1$

19. $x = 6, y = 3, t = 11$

20. $x = -1, y = 0, t = 11$

21. $a = 8, b = 9, c = 7$

22. $x = 21$

23. $a = 0, b = 0, f = 3, g = -5$

24. $t = 3, w = 4, y = 11, u = 13, v = 16$

25.
$$\begin{aligned} t &= 9, \\ x + 5 &= 24 \rightarrow x = 19 \\ y - 3 &= 7 \rightarrow y = 10 \end{aligned}$$

26.
$$\begin{aligned} k &= 11 \\ p - 4 &= 15 \rightarrow p = 19 \\ q + 8 &= 28 \rightarrow q = 20 \end{aligned}$$

27.
$$\begin{aligned} t - 3 &= 6 + 11 \rightarrow t = 20 \\ m + 12 - 4 &= 16 \rightarrow m = 8 \\ n + 3 + 8 &= 2n \rightarrow n = 11 \end{aligned}$$

28.
$$\begin{aligned} -6 + p &= 11 \rightarrow p = 17 \\ 4 + q &= 12 \rightarrow q = 8 \end{aligned}$$

29. $A + B = \begin{bmatrix} 3 & 4 \\ 5 & 6 \end{bmatrix} + \begin{bmatrix} -3 & -7 \\ 13 & -2 \end{bmatrix} = \begin{bmatrix} 0 & -3 \\ 18 & 4 \end{bmatrix}$

30. $B + C = \begin{bmatrix} -3 & -7 \\ 13 & -2 \end{bmatrix} + \begin{bmatrix} 2 & -1 \\ 1 & 4 \end{bmatrix} = \begin{bmatrix} -1 & -8 \\ 14 & 2 \end{bmatrix}$

31. $A - B = \begin{bmatrix} 3 & 4 \\ 5 & 6 \end{bmatrix} - \begin{bmatrix} -3 & -7 \\ 13 & -2 \end{bmatrix} = \begin{bmatrix} 6 & 11 \\ -8 & 8 \end{bmatrix}$

32. $B - C = \begin{bmatrix} -3 & -7 \\ 13 & -2 \end{bmatrix} - \begin{bmatrix} 2 & -1 \\ 1 & 4 \end{bmatrix} = \begin{bmatrix} -5 & -6 \\ 12 & -6 \end{bmatrix}$

33. $2A = 2\begin{bmatrix} 3 & 4 \\ 5 & 6 \end{bmatrix} = \begin{bmatrix} 6 & 8 \\ 10 & 12 \end{bmatrix}$

34. $-3B = -3\begin{bmatrix} -3 & -7 \\ 13 & -2 \end{bmatrix} = \begin{bmatrix} 9 & 21 \\ -39 & 6 \end{bmatrix}$

35. $2A + C = \begin{bmatrix} 6 & 8 \\ 10 & 12 \end{bmatrix} + \begin{bmatrix} 2 & -1 \\ 1 & 4 \end{bmatrix} = \begin{bmatrix} 8 & 7 \\ 11 & 16 \end{bmatrix}$

36. $-3B + A = \begin{bmatrix} 9 & 21 \\ -39 & 6 \end{bmatrix} + \begin{bmatrix} 3 & 4 \\ 5 & 6 \end{bmatrix} = \begin{bmatrix} 12 & 25 \\ -34 & 12 \end{bmatrix}$

37. $MN = (3 \times 3) \times (3 \times 3) = 3 \times 3$
$NM = (3 \times 3) \times (3 \times 3) = 3 \times 3$

38. $MN = (4 \times 4) \times (4 \times 4) = 4 \times 4$
$NM = (4 \times 4) \times (4 \times 4) = 4 \times 4$

39. $MN = (2 \times 4) \times (4 \times 2) = 2 \times 2$
$NM = (4 \times 2) \times (2 \times 4) = 4 \times 4$

40. $MN = (1 \times 3) \times (3 \times 1) = 1 \times 1$
$NM = (3 \times 1) \times (1 \times 3) = 3 \times 3$

41. $MN = (2 \times 5) \times (4 \times 3) =$ Not allowed
$NM = (4 \times 3) \times (2 \times 5) =$ Not allowed

42. $MN = (1 \times 5) \times (2 \times 4) =$ Not allowed
$NM = (2 \times 4) \times (1 \times 5) =$ Not allowed

43. $AB = \begin{bmatrix} -1 & 3 \\ 1 & 4 \end{bmatrix}\begin{bmatrix} 1 & 3 \\ 5 & -2 \end{bmatrix} = \begin{bmatrix} -1+15 & -3-6 \\ 1+20 & 3-8 \end{bmatrix} = \begin{bmatrix} 14 & -9 \\ 21 & -5 \end{bmatrix}$

44. $BA = \begin{bmatrix} 1 & 3 \\ 5 & -2 \end{bmatrix}\begin{bmatrix} -1 & 3 \\ 1 & 4 \end{bmatrix} = \begin{bmatrix} -1+3 & 3+12 \\ -5-2 & 15-8 \end{bmatrix} = \begin{bmatrix} 2 & 15 \\ -7 & 7 \end{bmatrix}$

45. $AC = \begin{bmatrix} -1 & 3 \\ 1 & 4 \end{bmatrix}\begin{bmatrix} 0 & 1 & -3 \\ 4 & 2 & 5 \end{bmatrix} = \begin{bmatrix} 0+12 & -1+6 & 3+15 \\ 0+16 & 1+8 & -3+20 \end{bmatrix} = \begin{bmatrix} 12 & 5 & 18 \\ 16 & 9 & 17 \end{bmatrix}$

46. $CA = \begin{bmatrix} 0 & 1 & -3 \\ 4 & 2 & 5 \end{bmatrix}\begin{bmatrix} -1 & 3 \\ 1 & 4 \end{bmatrix} =$ not allowed

47. Let $A^{-1} = \begin{bmatrix} a & b \\ c & d \end{bmatrix}$, then $AA^{-1} = \begin{bmatrix} 2 & 1 \\ 4 & 2 \end{bmatrix}\begin{bmatrix} a & b \\ c & d \end{bmatrix} = \begin{bmatrix} 1 & 0 \\ 0 & 1 \end{bmatrix}$

$A^{-1} = \begin{bmatrix} -0.5 & 0.25 \\ 0 & 0.5 \end{bmatrix}$

48. Let $A^{-1} = \begin{bmatrix} a & b \\ c & d \end{bmatrix}$, then $AA^{-1} = \begin{bmatrix} 1 & 2 \\ 3 & 4 \end{bmatrix}\begin{bmatrix} a & b \\ c & d \end{bmatrix} = \begin{bmatrix} 1 & 0 \\ 0 & 1 \end{bmatrix}$

$A^{-1} = \begin{bmatrix} -2 & 1 \\ 1.5 & -0.5 \end{bmatrix}$

49. $A^{-1} = \begin{bmatrix} 0.67 & -0.5 & 0.17 \\ -0.33 & 1 & -0.33 \\ -0.33 & 0.5 & 0.17 \end{bmatrix}$

50. $A^{-1} = \begin{bmatrix} -0.33 & 0.33 & 0.33 \\ 1 & 0 & -1 \\ 1.33 & -0.33 & -0.33 \end{bmatrix}$

51. $\begin{bmatrix} 260 & 230 & 215 \\ 242 & 219 & 235 \end{bmatrix}$

52.

	New York	Houston	Boston	Las Angelus
New York	0	1419	190	2451
Houston	1419	0	1605	1374
Boston	190	1605	0	2596
Los Angeles	2451	1374	2596	0

53. The total manufacturing and shipping costs for each product:

$$M_1 + M_2 = \begin{bmatrix} 43 & 51 \\ 60 & 90 \\ 80 & 30 \end{bmatrix} + \begin{bmatrix} 44 & 65 \\ 65 & 60 \\ 150 & 35 \end{bmatrix}$$

$$M_1 + M_2 = \begin{bmatrix} 87 & 116 \\ 125 & 150 \\ 230 & 65 \end{bmatrix}$$

The total manufacturing and shipping costs of model A product are \$87 and \$116, respectively.
The total manufacturing and shipping costs of model B product are \$125 and \$150, respectively.
The total manufacturing and shipping costs of model C product are \$230 and \$65, respectively.

54. For *January* the total dollar sales price of brand X items is

Retail price of iPads 815	×	Number of iPads sold 38	+	Retail price of laptops 610	+	Number of laptops sold 35t

and the total dollar cost is

Cost of iPads 690	×	Number of iPads sold 38	+	Cost of laptops 475	+	Number of laptops sold 35

For *February* the total dollar sales price of brand X items is

Retail price of iPads 815	×	Number of iPads sold 56	+	Retail price of laptops 610	+	Number of laptops sold 41

and the total dollar cost is

Cost of iPads 690	×	Number of iPads sold 56	+	Cost of laptops 475	+	Number of laptops sold 41

For *March* the total dollar sales price of brand X items is

Retail price of iPads 815	×	Number of iPads sold 47	+	Retail price of laptops 610	+	Number of laptops sold 25

and the total dollar cost is

Cost of iPads 690	$\times$	Number of iPads sold 47	$+$	Cost of laptops 475	$+$	Number of laptops sold 25

This yields for the following product:

$$\begin{bmatrix} 815 & 610 \\ 690 & 475 \end{bmatrix}\begin{bmatrix} 38 & 56 & 47 \\ 35 & 41 & 25 \end{bmatrix} = \begin{bmatrix} 52,320 & 70,650 & 53,555 \\ 42,845 & 58,115 & 44,305 \end{bmatrix}$$

Therefore, the requested matrix is

$$\begin{array}{c} \begin{array}{ccc} \text{Jan.} & \text{Feb.} & \text{Mar.} \end{array} \\ \begin{bmatrix} 52{,}320 & 70{,}650 & 53{,}555 \\ 42{,}845 & 58{,}115 & 44{,}305 \end{bmatrix} \end{array} \begin{array}{l} \text{Total dollar sales} \\ \text{Total dealer cost} \end{array}$$

55. $[125 \quad 134 \quad 112]\begin{bmatrix} 8 \\ 5 \\ 3 \end{bmatrix} = 125(8) + 134(5) + 112(3) = 1000 + 670 + 336 = 2006$

56. $[3210 \quad 870 \quad 89]\begin{bmatrix} 3 & 5 \\ 2 & 5 \\ 4 & 3 \end{bmatrix} = [3210(3) + 870(2) + 89(4) \quad 3210(5) + 870(5) + 89(3)]$

$= [11726 \quad 20667]$

Therefore, it is \$11,726 for the first day and \$20,667 for the next day.

Section 8.2 (Answers)

1. $\det\begin{bmatrix} 5 & 1 \\ 2 & 3 \end{bmatrix} = 5 \times 3 - 1 \times 2 = 15 - 2 = 13$

2. $\det\begin{bmatrix} 4 & -3 \\ 2 & 6 \end{bmatrix} = 4 \times 6 - (-3) \times 2 = 24 + 6 = 30$

3. $\det\begin{bmatrix} 6 & 1 \\ -3 & 4 \end{bmatrix} = 4 \times 6 - (-3) \times 1 = 24 + 3 = 27$

4. $\det\begin{bmatrix} 7 & -6 \\ -2 & -5 \end{bmatrix} = 7 \times (-5) - (-6) \times (-2) = -35 - 12 = -47$

5. $\det\begin{bmatrix} 1 & 0 \\ -6 & 4 \end{bmatrix} = 1 \times (4) - (0) \times (-6) = 4 - 0 = 4$

6. $\det\begin{bmatrix} 5 & 4 \\ 2 & 1 \end{bmatrix} = 5 \times (1) - (4) \times (2) = 5 - 8 = -3$

7. $\det\begin{bmatrix} -4 & 3 \\ 4 & -7 \end{bmatrix} = (-4) \times (-7) - (3) \times (4) = 28 - 12 = 16$

8. $\det\begin{bmatrix} 0 & -3 \\ -2 & 9 \end{bmatrix} = 0 \times (9) - (-3) \times (-2) = 0 - 6 = -6$

9. $\det\begin{bmatrix} 1 & 2 & 3 \\ 4 & -1 & 2 \\ 5 & -2 & 1 \end{bmatrix} = 6$

10. $\det\begin{bmatrix} 2 & 1 & 3 \\ -8 & -3 & -1 \\ 7 & 1 & 1 \end{bmatrix} = 36$

11. $\det\begin{bmatrix} 2 & 1 & -1 \\ 3 & 2 & 2 \\ 4 & 3 & 1 \end{bmatrix} = -4$

12. $\det\begin{bmatrix} 0 & 2 & -1 \\ -2 & 6 & 1 \\ 7 & 4 & 2 \end{bmatrix} = 72$

13. $\det\begin{bmatrix} 1 & 1 & 1 \\ 2 & 0 & 3 \\ 5 & -2 & 4 \end{bmatrix} = 9$

14. $\det\begin{bmatrix} 0 & 0 & 4 \\ 1 & 1 & 1 \\ 3 & 3 & 3 \end{bmatrix} = 0$

15. $(x-1)^2 - 6 = 4 \rightarrow (x-1)^2 = 10 \rightarrow x^2 - 2x - 9 = 0$
$x_1 = 4.16, x_2 = -2.16$

16. $4x^2 + 4x - 5x = 1$
$4x^2 - x - 1 = 0$
$x_1 = 0.64$
$x_2 = -0.39$

17. $x^2 + x - 2x - 2 - 0 = 5$
$x^2 - x - 7 = 0$
$x_1 = 3.19$
$x_2 = -2.19$

18. $0 - 2x = 12$
$x = -6$

19. $x^2 + x - x - 1 - 0 = 0$
$x = \pm 1$

20. $x^2 + 2x - 6 = 0$
$x_1 = 1.64$
$x_2 = -3.64$

21. $5x - 5 - x^2 - 2x = 0$

$$x^2 - 3x + 5 = 0$$

$$x = \frac{3 \pm \sqrt{9 - 4(1)(5)}}{2} = \frac{3 \pm i\sqrt{11}}{2}$$

22. $0 - x^2 - x = 0$

$x^2 + x = 0$

$x(x+1) = 0$

$x_1 = 0$

$x_2 = -1$

23. **Singular** because row 1 is all zeros.

24. **Singular** because column 2 is all zeros

25. **Singular** because row 3 is 3 times row 1.

26. **Singular** because row 2 is 2 times row 1.

27. $|MN| = |M||N| = -4 \times 3 = -12$

28. $|MM^t| = |MM| = |M||M| = -4 \times -4 = 16$

29. $|M^tN| = |M||N| = -4 \times 3 = -12$

30. $|2MN^{-1}| = 2^3|M||N^{-1}| = \frac{8 \times -4}{3} = \frac{-32}{3} = -10.67$

Section 8.3 (Answers)

1. vector

2. scalar

3. scalar

4. scalar

5. scalar

6. vector

7. $\vec{u} + \vec{v} = \langle -3, 5\rangle + \langle 2, -1\rangle = \langle -1, 4\rangle$

8. $\vec{u} - \vec{v} = \langle -3, 5\rangle - \langle 2, -1\rangle = \langle -5, 6\rangle$

9. $5\vec{u} - \vec{v} = 5\langle -3, 5\rangle - \langle 2, -1\rangle = \langle -17, 6\rangle$

10. $3\vec{u} + 2\vec{v} = 3\langle -3, 5\rangle + 2\langle 2, -1\rangle = \langle -5, 13\rangle$

11. $\|\vec{u}\| = \sqrt{(u_1)^2 + (u_2)^2} = \sqrt{(-3)^2 + (5)^2} = \sqrt{34}$

12. $9\|\vec{v}\| = 9\sqrt{(v_1)^2 + (v_2)^2} = 9\sqrt{(2)^2 + (1)^2} = 9\sqrt{5}$

13. $\|\vec{u}\| + \|\vec{v}\| = \sqrt{34} + \sqrt{5} = 5.83 + 2.24 = 8.07$

14. $\vec{u} + \vec{v} = \langle -3, 5\rangle + \langle 2, -1\rangle = \langle -1, 4\rangle$

$\|\vec{u} + \vec{v}\| = \sqrt{(-1)^2 + (4)^2} = \sqrt{17}$

15. $\vec{u} + \vec{v} = \langle -3, 4\rangle + \langle 2, -5\rangle = \langle -1, -1\rangle$

$\|\vec{u} + \vec{v}\| = \sqrt{(-1)^2 + (-1)^2} = \sqrt{2}$

16. $\vec{u} + \vec{v} = \langle 7, -1\rangle + \langle -3, 6\rangle = \langle 4, 5\rangle$

$\|\vec{u} + \vec{v}\| = \sqrt{(4)^2 + (5)^2} = \sqrt{41}$

17. $\vec{u} + \vec{v} = \langle -1, 9\rangle + \langle 8, 2\rangle = \langle 7, -7\rangle$

$\|\vec{u} + \vec{v}\| = \sqrt{(7)^2 + (-7)^2} = \sqrt{98}$

18. $\vec{u} + \vec{v} = \langle -1, 3\rangle + \langle -2, -8\rangle = \langle -3, -5\rangle$

$\|\vec{u} + \vec{v}\| = \sqrt{(-3)^2 + (-5)^2} = \sqrt{34}$

19. $\vec{u} = \frac{\vec{w}}{\|w\|} = \frac{\langle 4, 3\rangle}{\sqrt{16+9}} = \left\langle \frac{4}{5}, \frac{3}{5} \right\rangle$

20. $\vec{u} = \frac{\vec{w}}{\|w\|} = \frac{\langle 2, 5\rangle}{\sqrt{4+25}} = \left\langle \frac{2}{\sqrt{29}}, \frac{5}{\sqrt{29}} \right\rangle$

21. $\vec{u} = \frac{\vec{w}}{\|w\|} = \frac{\langle -2, -6\rangle}{\sqrt{4+36}} = \left\langle \frac{-2}{\sqrt{40}}, \frac{-6}{\sqrt{40}} \right\rangle$

22. $\vec{u} = \frac{\vec{w}}{\|w\|} = \frac{\langle 1, 2\rangle}{\sqrt{1+4}} = \left\langle \frac{1}{\sqrt{5}}, \frac{2}{\sqrt{5}} \right\rangle$

23. $\vec{w} = \langle 3, -1\rangle = 3\vec{i} - \vec{j}$

24. $\vec{w} = \langle -6, 4\rangle = -6\vec{i} + 4\vec{j}$

25. $\vec{w} = \langle 1, 9\rangle = \vec{i} + 9\vec{j}$

26. $\vec{w} = \langle -17, -25\rangle = -17\vec{i} - 25\vec{j}$

27. $\vec{u} \cdot \vec{v} = -3 \times 2 + 4 \times -5 = -26$

28. $\vec{u} \cdot \vec{v} = 7 \times -2 + 11 \times 1 = -3$

29. $\vec{u} \cdot \vec{v} = 4 \times 0 + -1 \times 9 = -9$

30. $\vec{u} \cdot \vec{v} = 1 \times 1 + -1 \times 0 = 1$

31. $\vec{u} \cdot \vec{v} = 1 \times 1 + -3 \times 1 = -2$

32. $\vec{u} \cdot \vec{v} = 1 \times -1 + 2 \times -1 = -3$

33. Since $\cos\theta = \dfrac{\vec{u}\cdot\vec{v}}{\|\vec{u}\|\|\vec{v}\|}$,

$$\cos\theta = \frac{\langle 1,3\rangle\cdot\langle 2,1\rangle}{\sqrt{1^2+3^2}\sqrt{2^2+(1)^2}} = \frac{2+3}{\sqrt{10}\sqrt{5}} = \frac{5}{\sqrt{50}}$$

Therefore, the angle between the vectors $\vec{u}$ and $\vec{v}$ is

$$\theta = \arccos\left(\frac{5}{\sqrt{50}}\right) = \cos^{-1}\theta\left(\frac{5}{\sqrt{50}}\right) = 45^\circ.$$

34. Since $\cos\theta = \dfrac{\vec{u}\cdot\vec{v}}{\|\vec{u}\|\|\vec{v}\|}$,

$$\cos\theta = \frac{\langle 4,1\rangle\cdot\langle 1,-2\rangle}{\sqrt{4^2+1^2}\sqrt{1^2+(-2)^2}} = \frac{4-2}{\sqrt{17}\sqrt{5}} = \frac{2}{\sqrt{85}}$$

Therefore, the angle between the vectors $\vec{u}$ and $\vec{v}$ is

$$\theta = \arccos\left(\frac{2}{\sqrt{85}}\right) = \cos^{-1}\theta\left(\frac{2}{\sqrt{85}}\right) \approx 77.47^\circ.$$

35. Since $\cos\theta = \dfrac{\vec{u}\cdot\vec{v}}{\|\vec{u}\|\|\vec{v}\|}$,

$$\cos\theta = \frac{\langle -2,5\rangle\cdot\langle 1,6\rangle}{\sqrt{(-2)^2+5^2}\sqrt{1^2+(6)^2}} = \frac{-2+30}{\sqrt{29}\sqrt{37}} = \frac{28}{\sqrt{1073}}$$

Therefore, the angle between the vectors $\vec{u}$ and $\vec{v}$ is

$$\theta = \arccos\left(\frac{28}{\sqrt{1073}}\right) = \cos^{-1}\theta\left(\frac{28}{\sqrt{1073}}\right) \approx 31.26^\circ.$$

36. Since $\cos\theta = \dfrac{\vec{u}\cdot\vec{v}}{\|\vec{u}\|\|\vec{v}\|}$,

$$\cos\theta = \frac{\langle -2,1\rangle\cdot\langle 1,3\rangle}{\sqrt{(-2)^2+1^2}\sqrt{1^2+(3)^2}} = \frac{-2+3}{\sqrt{5}\sqrt{10}} = \frac{1}{\sqrt{50}}$$

Therefore, the angle between the vectors $\vec{u}$ and $\vec{v}$ is

$$\theta = \arccos\left(\frac{1}{\sqrt{50}}\right) = \cos^{-1}\theta\left(\frac{1}{\sqrt{50}}\right) \approx 81.87^\circ.$$

37. Since $\cos\theta = \dfrac{\vec{u}\cdot\vec{v}}{\|\vec{u}\|\|\vec{v}\|}$,

$$\cos\theta = \frac{\langle -5,2\rangle\cdot\langle 0,-3\rangle}{\sqrt{(-5)^2+2^2}\sqrt{0^2+(-3)^2}} = \frac{0-6}{\sqrt{29}\sqrt{9}} = \frac{-6}{3\sqrt{29}} = \frac{-2}{\sqrt{29}}$$

Therefore, the angle between the vectors $\vec{u}$ and $\vec{v}$ is

$$\theta = \arccos\left(\frac{-2}{\sqrt{29}}\right) = \cos^{-1}\theta\left(\frac{-2}{\sqrt{29}}\right) \approx 111.8^\circ.$$

38. Since $\cos\theta = \dfrac{\vec{u}\cdot\vec{v}}{\|\vec{u}\|\|\vec{v}\|}$,

$$\cos\theta = \frac{\langle 1,2\rangle\cdot\langle -1,1\rangle}{\sqrt{1^2+(-2)^2}\sqrt{(-1)^2+(1)^2}} = \frac{-1-2}{\sqrt{5}\sqrt{2}} = \frac{-3}{\sqrt{10}}$$

Therefore, the angle between the vectors $\vec{u}$ and $\vec{v}$ is

$$\theta = \arccos\left(\frac{-3}{\sqrt{10}}\right) = \cos^{-1}\theta\left(\frac{-3}{\sqrt{10}}\right) \approx 161.57^\circ.$$

39. Since $\cos\theta = \dfrac{\vec{u}\cdot\vec{v}}{\|\vec{u}\|\|\vec{v}\|}$,

$$\cos\theta = \frac{\langle -1,-1\rangle\cdot\langle -2,2\rangle}{\sqrt{(-1)^2+(-1)^2}\sqrt{(-2)^2+(2)^2}} = \frac{2-2}{\sqrt{2}\sqrt{8}} = \frac{0}{\sqrt{16}} = 0$$

The angle θ between the vectors $\vec{u}$ and $\vec{v}$ is $\theta = \cos^{-1}(0) = \frac{\pi}{2}$.

Thus, the vectors $\vec{u}$ and $\vec{v}$ are orthogonal.

40. Since $\cos\theta = \dfrac{\vec{u}\cdot\vec{v}}{\|\vec{u}\|\|\vec{v}\|}$,

$$\cos\theta = \frac{\langle 0,1\rangle\cdot\langle 1,-2\rangle}{\sqrt{(0)^2+(1)^2}\sqrt{(1)^2+(-2)^2}} = \frac{0-2}{\sqrt{1}\sqrt{5}} = \frac{-2}{\sqrt{5}}$$

$$\theta = \cos^{-1}\left(\frac{-2}{\sqrt{5}}\right) = 153.43^\circ$$

neither

41. Since $\cos\theta = \dfrac{\vec{u}\cdot\vec{v}}{\|\vec{u}\|\|\vec{v}\|}$,

$$\cos\theta = \frac{\langle 5,0\rangle\cdot\langle 3,3\rangle}{\sqrt{(5)^2+(0)^2}\sqrt{(3)^2+(3)^2}} = \frac{15+0}{\sqrt{25}\sqrt{18}} = \frac{15}{5\sqrt{18}} = \frac{3}{\sqrt{18}} = \frac{1}{\sqrt{2}}$$

$$\theta = \cos^{-1}\left(\frac{1}{\sqrt{2}}\right) = 45^\circ$$

neither

42. Since $\cos\theta = \dfrac{\vec{u}\cdot\vec{v}}{\|\vec{u}\|\|\vec{v}\|}$,

$$\cos\theta = \frac{\langle 1,-2\rangle\cdot\langle 2,-1\rangle}{\sqrt{(1)^2+(-2)^2}\sqrt{(2)^2+(-1)^2}} = \frac{2+2}{\sqrt{5}\sqrt{5}} = \frac{4}{5}$$

$$\theta = \cos^{-1}\left(\frac{4}{5}\right) = 36.87^\circ$$

neither

43. Since $\cos\theta = \dfrac{\vec{u}\cdot\vec{v}}{\|\vec{u}\|\|\vec{v}\|}$,

$$\cos\theta = \frac{\langle 1/2,-2/3\rangle\cdot\langle 4,3\rangle}{\sqrt{(1/2)^2+(-2/3)^2}\sqrt{(4)^2+(3)^2}} = \frac{2-2}{\sqrt{(1/2)^2+(-2/3)^2}\sqrt{(4)^2+(3)^2}} = 0$$

The angle θ between the vectors $\vec{u}$ and $\vec{v}$ is $\theta = \cos^{-1}(0) = \frac{\pi}{2}$.

Thus, the vectors $\vec{u}$ and $\vec{v}$ are orthogonal.

44. Since $\cos\theta = \dfrac{\vec{u}\cdot\vec{v}}{\|\vec{u}\|\|\vec{v}\|}$,

$$\cos\theta = \frac{\langle 2,-4\rangle\cdot\langle -1,2\rangle}{\sqrt{(2)^2+(-4)^2}\sqrt{(-1)^2+(2)^2}} = \frac{-2-8}{\sqrt{20}\sqrt{5}} = \frac{-10}{10} = -1$$

The angle θ between the vectors $\vec{u}$ and $\vec{v}$ is $\theta = \cos^{-1}(-1) = \pi$.
Thus, the vectors $\vec{u}$ and $\vec{v}$ are parallel.

45. $W = \vec{F}\cdot\vec{d} = (70)(4) = 280$ ft·lb

46. $W = \vec{F}\cdot\vec{d} = (22\vec{i} - 5\vec{j})\cdot(40\vec{i} - 8\vec{j}) = (22)(40) + (-5)(-8) = 920$

47. $W = \vec{F}\cdot\vec{d} = \langle 20\cos 30° - 20\sin 30°\rangle\cdot\langle 60,0\rangle$
$W = 1200\cos 30° = 1039.23$ ft·lb

48. $W = \vec{F}\cdot\vec{d} = \langle 80\cos 20°, 80\sin 20°\rangle\cdot\langle 120,0\rangle$
$W = 9600\cos 20° = 9021.01$ ft·lb

49.

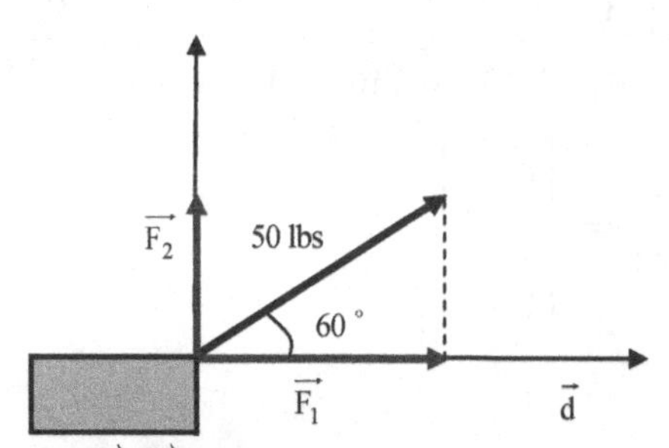

$W = \|\vec{F}\|\|\vec{d}\|\cos\theta = \vec{F}\cdot\vec{d}$
$W = (50)(90)\cos 60° = (50)(90)(0.5) = 2250$ ft·lb

50. The force magnitude is $\|\vec{F}\| = 8$ and the $\theta = \tan^{-1}\left(\frac{1}{1}\right) = 45°$ with $\vec{i}$.

$$\vec{F} = 8(\cos 45°\,\vec{i} + \sin 45°\,\vec{j}) = 8\left(\tfrac{\sqrt{2}}{2}\vec{i} + \tfrac{\sqrt{2}}{2}\vec{j}\right) = 4\sqrt{2}(\vec{i}+\vec{j})$$
$$= \sqrt{32}(\vec{i}+\vec{j})$$

The line of motion of the object from A = (0,0) to B = (1,0), so $\overrightarrow{AB} = \vec{i}$. The work W is therefore,

$W = \vec{F}\cdot\overrightarrow{AB} = \sqrt{32}(\vec{i}+\vec{j})\cdot\vec{i} = \sqrt{32}$ ft·lb.

Review Exercises (Chapter 8—Answers)

1. $AB = \begin{bmatrix} -3 & 2 \\ 4 & 0 \end{bmatrix}\begin{bmatrix} 5 & -2 \\ 1 & 6 \end{bmatrix} = \begin{bmatrix} -13 & 18 \\ 20 & -8 \end{bmatrix}$

2. $CD = \begin{bmatrix} -2 & 3 \end{bmatrix}\begin{bmatrix} 4 \\ -5 \end{bmatrix} = [7]$

3. $CB = \begin{bmatrix} -2 & 3 \end{bmatrix}\begin{bmatrix} 5 & -2 \\ 1 & 6 \end{bmatrix} = \begin{bmatrix} -7 & 22 \end{bmatrix}$

4. $AD = \begin{bmatrix} -3 & 2 \\ 4 & 0 \end{bmatrix}\begin{bmatrix} 4 \\ -5 \end{bmatrix} = \begin{bmatrix} -22 \\ 16 \end{bmatrix}$

5. $A + B = \begin{bmatrix} -3 & 2 \\ 4 & 0 \end{bmatrix} + \begin{bmatrix} 5 & -2 \\ 1 & 6 \end{bmatrix} = \begin{bmatrix} 2 & 0 \\ 5 & 6 \end{bmatrix}$

6. $C + D = \begin{bmatrix} -2 & 3 \end{bmatrix}\begin{bmatrix} 4 \\ -5 \end{bmatrix} \rightarrow$ not allowed

7. $A + C = \begin{bmatrix} -3 & 2 \\ 4 & 0 \end{bmatrix}\begin{bmatrix} -2 & 3 \end{bmatrix} =$ not allowed

8. $3A - 2B = 3\begin{bmatrix} -3 & 2 \\ 4 & 0 \end{bmatrix} - 2\begin{bmatrix} 5 & -2 \\ 1 & 6 \end{bmatrix} = \begin{bmatrix} -9 & 6 \\ 12 & 0 \end{bmatrix} - \begin{bmatrix} 10 & -4 \\ 2 & 12 \end{bmatrix} = \begin{bmatrix} -19 & 10 \\ 10 & -12 \end{bmatrix}$

9. $CA + C = \begin{bmatrix} -2 & 3 \end{bmatrix}\begin{bmatrix} -3 & 2 \\ 4 & 0 \end{bmatrix} + \begin{bmatrix} -2 & 3 \end{bmatrix}$

$CA + C = \begin{bmatrix} 18 & -4 \end{bmatrix} + \begin{bmatrix} -2 & 3 \end{bmatrix} = \begin{bmatrix} 16 & -1 \end{bmatrix}$

10. $A^{-1} = \frac{1}{-8}\begin{bmatrix} 0 & -2 \\ -4 & -3 \end{bmatrix} = \begin{bmatrix} 0 & \frac{1}{4} \\ \frac{1}{2} & \frac{3}{8} \end{bmatrix}$

11. $B^t = \begin{bmatrix} 5 & 1 \\ -2 & 6 \end{bmatrix}$

12. $I_B = \begin{bmatrix} 1 & 0 \\ 0 & 1 \end{bmatrix}$

13. $\det A = (-3)(0) - (2)(4) = -8$

14. $\det E = 2\begin{bmatrix} -2 & 3 \\ 5 & 7 \end{bmatrix} - 1\begin{bmatrix} -1 & 3 \\ 4 & 7 \end{bmatrix} + 0\begin{bmatrix} -1 & -2 \\ 4 & 5 \end{bmatrix}$

$\det E = \begin{bmatrix} -4 & 6 \\ 10 & 14 \end{bmatrix} - \begin{bmatrix} -1 & 3 \\ 4 & 7 \end{bmatrix} + 0 = \begin{bmatrix} -3 & 3 \\ 6 & 7 \end{bmatrix}$

15. Size of $A = 2 \times 2$, Size of $B = 2 \times 2$, Size of $C = 1 \times 2$, Size of $D = 2 \times 1$,
Size of $E = 3 \times 3$

16. Metric A is square matrix
Metric B is square matrix
Metric C is row matrix
Metric D is column matrix
Metric E is square matrix

17. Size $= 2 \times 2$

18. Size $= 1 \times 3$

19. Size $= 1 \times 2$

20. Size $= 2 \times 1$

21. Size $= 4 \times 1$

22. Size $= 1 \times 1$

23. Size $= 1 \times 1$

24. Size $= 3 \times 2$

25. Size of $A = 3 \times 2$
Size of $B = 2 \times 3$
Size of $C = 3 \times 2$

26. **(a)** $A + C = \begin{bmatrix} 0 & 2 \\ 3 & -1 \\ 5 & 1 \end{bmatrix} + \begin{bmatrix} -1 & 2 \\ 4 & 5 \\ 3 & -3 \end{bmatrix} = \begin{bmatrix} -1 & 4 \\ 7 & 4 \\ 8 & -2 \end{bmatrix},$

(b) $A - C = \begin{bmatrix} 0 & 2 \\ 3 & -1 \\ 5 & 1 \end{bmatrix} - \begin{bmatrix} -1 & 2 \\ 4 & 5 \\ 3 & -3 \end{bmatrix} = \begin{bmatrix} 1 & 0 \\ -1 & -6 \\ 2 & 4 \end{bmatrix}$

(c) $-3A = -3\begin{bmatrix} 0 & 2 \\ 3 & -1 \\ 5 & 1 \end{bmatrix} = \begin{bmatrix} 0 & -6 \\ -9 & 3 \\ -15 & -3 \end{bmatrix}$

(d) $BC = \begin{bmatrix} -3 & 4 & 1 \\ 1 & 2 & 0 \end{bmatrix}\begin{bmatrix} -1 & 2 \\ 4 & 5 \\ 3 & -3 \end{bmatrix} = \begin{bmatrix} 22 & 11 \\ 7 & 12 \end{bmatrix}$

(e) $B + 0 = \begin{bmatrix} -3 & 4 & 1 \\ 1 & 2 & 0 \end{bmatrix} + \begin{bmatrix} 0 & 0 & 0 \\ 0 & 0 & 0 \end{bmatrix} = \begin{bmatrix} -3 & 4 & 1 \\ 1 & 2 & 0 \end{bmatrix}$

(f) $A^t = \begin{bmatrix} 0 & 3 & 5 \\ 2 & -1 & 1 \end{bmatrix}$

27. $A^{-1} = \begin{bmatrix} 0.6 & 0.2 \\ -0.4 & 0.2 \end{bmatrix}$

28. $x+y = 6 \rightarrow x = 6-y$
$x+2y = 2 \rightarrow 6-y+2y = 2 \rightarrow y = -4$
$x = 10$

29. $zT = \begin{bmatrix} z_1 & z_2 \end{bmatrix}\begin{bmatrix} \frac{1}{3} & \frac{2}{3} \\ \frac{1}{5} & \frac{4}{5} \end{bmatrix} = \begin{bmatrix} z_1 & z_2 \end{bmatrix}$

$z_1 + z_2 = 1$

$z_2 = 1 - z_1$

$\frac{z_1}{3} + \frac{z_2}{5} = z_1$

$\frac{2z_1}{3} + \frac{4z_2}{5} = z_2$

$\frac{5z_1 + 3z_2}{15} = z_1 \rightarrow z_1 = \frac{3}{10}z_2$

$z_1 = \frac{3}{13}, z_2 = \frac{10}{13}$

30. $M+N = \begin{bmatrix} c_1 & c_2 \\ 0 & 1 \end{bmatrix} + \begin{bmatrix} -1 & 2 \\ 1 & -2 \end{bmatrix} = \begin{bmatrix} 5 & 4 \\ 1 & -1 \end{bmatrix}$

$c_1 - 1 = 5 \rightarrow c_1 = 6$

$c_2 + 2 = 4 \rightarrow c_2 = 2$

31. $x = 3; y = -4$

32. $t = 10, z = 12$

33. $0+x = 10 \rightarrow x = 10$
$-3+x+y = 6 \rightarrow y = -1$
$8+t = 2 \rightarrow t = -6$

34. $x = 12,$
$-12+x-y = 8 \rightarrow y = -12+x-8 \rightarrow y = -8$

35. $x = 1$

$x+y = 0 \rightarrow y = -1$

$z = -1$

$5k+1 = 4 \rightarrow k = \frac{3}{5}$

36. $AA^{-1} = \begin{bmatrix} 3 & 1 \\ 2 & 0 \end{bmatrix} \begin{bmatrix} 2x+1 & 3y \\ z & k-1 \end{bmatrix} = \begin{bmatrix} 1 & 0 \\ 0 & 1 \end{bmatrix}$

$$AA^{-1} = \begin{bmatrix} 6x+3+z & 9y+k-1 \\ 4x+2 & 6y \end{bmatrix} = \begin{bmatrix} 1 & 0 \\ 0 & 1 \end{bmatrix}$$

$$y = \frac{1}{6}$$

$$x = -\frac{1}{2}$$

$$9\left(\frac{1}{6}\right) + k - 1 = 0 \rightarrow k = 1 - \frac{3}{2} = -\frac{1}{2}$$

$$6\left(-\frac{1}{2}\right) + 3 + z = 1 \rightarrow z = 1$$

37. $W = \|\vec{F}\|\|\vec{d}\|\cos\theta = \vec{F} \cdot \vec{d}$

$W = (40)(10)\cos 60° = (40)(10)(0.5) = 200 \text{ ft} \cdot \text{lbs}$

38. $W = \|\vec{F}\|\|\vec{d}\|\cos\theta = \vec{F} \cdot \vec{d}$

$W = (30)(8)\cos 45° = (30)(8)(0.7) = 168 \text{ ft} \cdot \text{lbs}$

Glossary

Common Math Symbols

Math symbol	Definition
$=$	Equal to
$\neq$	Not equal to
$\approx$	Approximately equal to
$>$	Greater than
$<$	Less than
$\geq$	Greater than or equal to
$\leq$	Less than or equal to
$+$	Plus (addition)
$-$	Minus (subtraction)
$\times$, $\cdot$, *	Time or multiply by (multiplication)
$\div$, /	Divide by or over (quotient)
:	Ratio or proportion
$\rightarrow$	Implies
$\leftrightarrow$	If and only if
( )	Parenthesis
[ ]	Bracket
{ }	Braces, indicate set membership
$\varnothing$	Empty set
π	Pi, (3.14...)
e	Exponent, (2.72...)
log	Logarithm
| |	Absolute value
$\sqrt{}$	Square root
|	Such that

Set of natural number is $\mathbb{N} = \{1, 2, 3, ...\}$.

Set of integer numbers is $\mathbb{Z} = \{.., -3, -2, -1, 0, 1, 2, 3, ...\}$.

Set of rational numbers is $\mathbb{Q} = \frac{m}{n}; m, n \in \mathbb{Z}$ with $n \neq 0$.

Note: $\sqrt{2}, \sqrt{3}, \sqrt[3]{7}, and\ \pi \notin \mathbb{Q}$.

$\mathbb{R}$ is set of all numbers $\mathbb{Q}$ and not rational with their negative and zeros.

$\mathbb{C}$ is set of the complex Numbers written in form $a + bi$, where $a, b \in \mathbb{R}$.

The Greek Alphabet

Name	Lower case character	Upper case character	Name	Lower case character	Upper case character
Alpha	α	A	Nu	ν	N
Beta	β	B	Xi	ξ	Ξ
Gamma	γ	Γ	Omicron	o	O
Delta	δ	Δ	Pi	π	Π
Epsilon	ε	E	Rho	ρ	P
Zeta	ζ	Z	Sigma	σ	Σ
Eta	η	H	Tau	τ	T
Theta	θ	Θ	Upsilon	υ	Υ
Iota	ι	I	Phi	ϕ	Φ
Kappa	κ	K	Chi	χ	X
Lambda	λ	Λ	Psi	ψ	Ψ
Mu	μ	M	Omega	ω	Ω

Index

Note: Page numbers followed by "f", "t" and "b" indicates figures, tables and boxes respectively.

Printed in the United States
By Bookmasters